Multi-UAV Planning and Task Allocation

Artificial Intelligence and Robotics Series

Series Editor:
Roman V. Yampolskiy

Intelligent Autonomy of UAVs
Advanced Missions and Future Use
Yasmina Bestaoui Sebbane

Artificial Intelligence
With an Introduction to Machine Learning, Second Edition
Richard E. Neapolitan, Xia Jiang

Artificial Intelligence and the Two Singularities
Calum Chace

Behavior Trees in Robotics and AI
An Introduction
Michele Collendanchise, Petter Ögren

Artificial Intelligence Safety and Security
Roman V. Yampolskiy

Artificial Intelligence for Autonomous Networks
Mazin Gilbert

Virtual Humans
David Burden, Maggi Savin-Baden

Deep Neural Networks: WASD Neuronet Models, Algorithms, and Applications
Yunong Zhang, Dechao Chen, Chengxu Ye

Introduction to Self-Driving Vehicle Technology
Hanky Sjafrie

Digital Afterlife: Death Matters in a Digital Age
Maggi Savin-Baden, Victoria Mason-Robbie

Cunning Machines: Your Pocket Guide to the World of Artificial Intelligence
James Osinski

For more information about this series please visit: https://www.crcpress.com/Chapman--HallCRC-Artificial-Intelligence-and-Robotics-Series/book-series/ARTILRO

Multi-UAV Planning and Task Allocation

Yasmina Bestaoui Sebbane

CRC Press
Taylor & Francis Group
Boca Raton London New York

CRC Press is an imprint of the
Taylor & Francis Group, an **informa** business

A CHAPMAN & HALL BOOK

First edition published 2020
by CRC Press
6000 Broken Sound Parkway NW, Suite 300, Boca Raton, FL 33487-2742
and by CRC Press
2 Park Square, Milton Park, Abingdon, Oxon, OX14 4RN

CRC Press is an imprint of Taylor & Francis Group, LLC

© 2020 Taylor & Francis Group, LLC

ISBN: 978-0-367-45782-2 (hbk)
ISBN: 978-1-003-02668-6 (ebk)

Typeset in Times
by codeMantra

Contents

Author

Yasmina Bestaoui Sebbane (1960–2018) was a full professor in Automatic and Robotics at the University of Evry, Val d'Essonne (France). She was also the head of the pôle Drones of IBISC (Informatique, Biologie Intégrative et Systèmes Complexes), PEDR laboratory in the University of Evry, and also served as the department head (2006–2015). She was an Academic Palms recipient (2017) and led and contributed to many scientific committees and collaborations with other universities to create new teaching courses. She earned her PhD in Control and Computer Engineering from École Nationale Supérieure de Mécanique, Nantes, France, in 1989 (currently École Centrale de Nantes), and her Research Advisor Qualification (HDR: Habilitation to Direct Research) in Robotics from the University of Evry, France, in 2000. Her research interests included control, planning, and decision-making of unmanned systems, particularly unmanned aerial vehicles and robots.

1 Multi-Aerial-Robot Planning

1.1 INTRODUCTION

Multi-robot systems are a major research topic in robotics due to their intrinsic robustness and versatility [434]. Designing, testing and deploying in the real world a large number of aerial robots is a concrete possibility due to the recent technological advances. Tasks that have been of particular interest include synergetic mission planning, patrolling, fault-tolerant cooperation, swarm control, role assignment, multi-robot path planning, exploration and mapping, navigation, coverage, detection and localization, tracking and pursuit and perimeter surveillance [21,451]. Significant research effort has been invested during the last few years in design and simulation of multiagent robotics and intelligent swarm systems. Swarm-based robotics system can be generally defined as highly decentralized collectives, i.e., groups of extremely simple robotic agents, with limited communication computation and sensing abilities, designed to be deployed together in order to accomplish various tasks [419]. As regions with storms and hazardous weather are analogous as obstacles in the airspace, the problem of control of multiple aerial robots through hazardous weather is similar to the problem of control of multiple agents through obstacles [412,415]. Modularization is considered as the most naturally occurring structure able to exhibit complex behaviors from a network of individually simple components, which interact with each other in relatively simple ways. Collaborative elements are usually information processing software units, such as neurons or dynamically simulated entities such as ant colonies or bees with swarm intelligence, so that the coordinated behavior of the group emerges from the interaction of these simple agent behaviors [423]. The approach proposed is also based on simple collaborative information processing components, as close as possible to the hardware devices (sensors and actuators) called intelligent hardware units (IHUs), embodying a physical autonomous agent to design a new control architecture concept [425,443,450].

The following problem is a challenge.

Problem 1. *Given that the networked system is to accomplish a particular task, how should the interaction and control protocols be structured to ensure that the task is in fact achieved?*

The particular focus in this chapter is on graph theoretic methods for the analysis and synthesis of networked dynamic systems [442,455]. By abstracting away the complex interaction geometries associated with the sensing and communication footprints of the individual aerial robots, and instead identifying aerial robots with nodes in a graph and encoding the existence of an interaction between nodes as an edge, a set of tools for networked system are available [432]. Recently, there have been great efforts to develop cooperative systems of multiple aerial robots and investigate their benefits [444,447,449]. In terms of practical value, one of the greatest benefits of the cooperative system is that it can provide users with better information superiority [454]. However, developing such an autonomous cooperative system is quite challenging because of technical and operational issues such as decision-making and planning to be solved. Architectures and interaction protocols must be developed for multiagent systems [465]. The main differences between algorithms designed for static environments and those designed to work in dynamic environments is the fact that the aerial robot's knowledge base (either centralized or decentralized) becomes unreliable due to the changes that take place in the environment. Hence, the aerial robot's behavior must ensure that the aerial robots generate a desired effect, regardless of the changing environment. The majority of these works discuss challenges involving a multiagent system operating on static domains. Such models, however, are often too limited to capture real-world problems which in many cases involve external elements, which may influence their environment, activities and goals. Hence, after team approach introduction, deterministic decision-making and association with limited communications are presented.

Then, multiagent decision-making under uncertainty techniques are discussed. Finally, some case studies are discussed.

1.2   TEAM APPROACH

Multiagent systems are systems in which many autonomous intelligent agents interact with each other. Agents can be either cooperative, pursuing a common goal, or selfish, going after their own interests [147]. Their cooperation can be either without negotiation (**implicit communication**) or with negotiation. Their coordination can be at the lower level or **explicit communication** (partitioned or synchronized), or this complex system is organized into well-designed subsystems.

1. An **Indifferent agent** does not take into account any information about the other agents. It does not even take into consideration its own preferences.
2. An **Oracle agent** knows the roles and strategies of the other agents (i.e., it has the correct models about the others). Furthermore, it even knows the others' calendars so that an oracle agent is able to see in advance the other's moves, and then it just chooses to propose the slot that maximizes its utility in each round of the game.
3. A **Self-centered agent** does not consider information about the other agents, but it takes into account its own role.
4. A **Collaborative agent** also takes into account the agent's own role. However, it also takes into consideration information about the previous round trying to join in the biggest observed team.

This section treats the different aspects of cooperation in multiagent systems and the links between them in two phases: the cooperation model and the cooperative agent model. The first phase consists in defining the concepts permitting to qualify the activity in a cooperative multiagent system, while the second describes the internal aspects of a cooperative agent as well as its basic properties. To define the cooperation model, this presentation is structured according to three levels [422].

- **Cooperation** means that if there is a team of aerial robots that intend to reach a common objective, this team requires a set of roles organized under an organizational structure that determines the responsibility of each aerial robot. A team of aerial robots is composed of a finite set of aerial robots that collaborate to achieve a common objective; each one is defined as a simple aerial robot or a subset of aerial robots that inter-communicate and cooperate under an internal control. Properties of an individual aerial robot are classified in three subsets: the **capabilities**, the **knowledge** and the **beliefs**.
- A **common objective** is defined by a non-empty set of global goals and a set of constraints related to the running modes of this objective. Global goal is a set of local goals; each one is described by a set of elementary tasks. The global goals of a same common objective are related by a variable dependence ratio such as **a total dependence, a partial dependence or an independence**.
- A **role** is defined according to the aerial robot's capabilities, in terms of domain capabilities and control capabilities. A generic organizational structure is defined by a finite set of relations between the different roles required for the achievement of a common objective. The cooperation model verifies for each role that there exists at least one aerial robot endowed with competence allowing it to take it in charge.

A **cooperative activity** can take place if particular set of aerial robots converge toward a global goal where each one achieves its proper local goals. The local goals of an aerial robot express the tasks assigned to it in order to perform a global goal. The assignment is based on the distribution of roles described by the aerial robot goal. This later verifies the completeness property related to the attribution of the local goal of a same global goal. Two properties related to the quantification of the efficiency and the performance of a cooperative activity may be defined.

- **Efficiency:** In order to qualify the cooperative activity to be efficient, performing only local goals that constitute the common objective is not enough; the generic constraints to reach this objective must be satisfied. The ability of a team of aerial robots to achieve the objective according to the constraints is measured by a function. Thus, for each aerial robot, the existence of a sequence of tasks execution that respects the retained constraints is verified. The mission function assigns a mission to each aerial robot that defines its part of responsibility in the team.
- **Performance** of the organization is about the verification for each aerial robot of the presence of a subset of organizational links that allows it to acquire knowledge, through others, in order to achieve its task and avoid all deadlock situations.

In opposition to the cooperation model which treats collective aspect, the cooperating agent model, which treats the collective aspect, deals with the individual aspects of the aerial robot.

Definition 2. *Cooperating Aerial Robot A cooperating aerial robot is an entity moving in an environment (may include other aerial robots) which has capabilities and resources enabling it to perform individual tasks and to cooperate with other aerial robots in order to reach a common objective. This common objective is reached by achieving aerial robot's local goals.*

Two representations exist:

- the **acquaintances** of the aerial robot which represent its partial views on organizations to which it belongs.
- its **agenda**, which describes the set of the aerial robot's tasks to be performed on time. This agenda can be updated at any time by the aerial robot itself.

For the aerial robot to progress and be evaluated in a cooperative team, some properties must be considered to adequately perform the cooperative activity:

- autonomy
- perception
- auto-organization.

The distribution of the roles induces an instantiation of the generic organizational structure called **concrete organizational structure** [40]. This structure can establish additional links needed for the resolution of the exceptions which can occur during a task process [417]. Formation Control can have different shapes:

- leader–follower controllers
- virtual-structure
- group/shape control.

1.2.1 COOPERATION

The problem of how to cooperatively assign and schedule demands in a multi-aerial-robot system can be formulated in a static mathematical model that fits within the framework of the vehicle routing problem [347,411].

- A team of m aerial robots is required to service a set of n demands in a 2D space.
- Each demand requires a certain amount of on-site service (for example, surveillance of a neighborhood of this point).
- The goal is to compute a set of routes that optimizes the cost of servicing (according to some quality-of-service metric).

The **cooperation model** and the **cooperation aerial robot model** are linked. The cooperation model consists in defining the concepts permitting to qualify the activity in a cooperative multiagent system, while the second describes the internal aspects of a cooperative aerial robot as well as its basic properties. A **cooperative activity** can take place if a particular set of aerial robots converge toward a global goal where each one achieves its proper local goals. The local goals of an aerial robot express the task assigned to it in order to perform a global goal. The assignment is based on the distribution of roles described by the aerial robot goal. This later verifies the completeness property related to the attribution of the local goal of a same global goal. Cooperative control involves a collection of decision-making components with limited processing capabilities, sensed information and limited inter-component communications, all seeking to achieve a collective objective. An important differentiation in team autonomy is their degree of centralization.

- **Centralized** means that all the information about the aerial robots is sent to a single server where the controls are calculated for each aerial robot and sent back. Centralized teams have one or more of the following attributes:
  - single unit responsible for information fusion (similar) and integration (disparate)
  - single unit responsible for coordinating decision-making or control
  - common communication facility.
- **Decentralized** means that no central server exists and each aerial robot computes its own control based only on other aerial robots within a particular range of itself. The situation is ideal from the computational perspective because the number of aerial robots that can be within range at a given time is bounded and therefore so is the time the algorithm will take to run, regardless of the total number of aerial robots involved.
- **Distributed** means that no central server is used and each aerial robot computes its own control, but an aerial robot may need more than local information to do so. In this case, computations scale with (n), the number of aerial robots involved, but the scaling is at least $O(n)$ better than the centralized case because now n separate processors are computing in parallel rather than one server doing all the work. Consequently, these systems have the advantages of being robust, scalable and flexible, and they allow the development of control algorithms often based on the swarm intelligence approach that describes "any attempt to design algorithms or distributed problem solving devices inspired by the collective behavior of social insect colonies and other animal societies" [41].

Cooperative control refers to a group of dynamic entities exchanging information to accomplish a common objective. Cooperative control entails planning, coordination and execution of a mission by two or more aerial robots [440]. A typical formation includes a leader and a number of followers. Control schemes are usually designed to maintain the geometry of the formation. Followers try to maintain a constant relative distance from neighboring aerial robots, while the leader is responsible for trajectory tracking. When a number of aerial robots are flying in formation, their on-board systems establish their relative position, speeds and attitudes by exchanging the necessary information via the communication network [430]. Alternatively, they may use on-board proximity sensors. The on-board computers, control system then use this information to produce a cohesive flight. The **Reynolds boids model** illustrates the basic premise behind the multiagent problem, in which a collection of aerial robots are to collectively solve a global task using local interaction rules. Each aerial robot is designed to react to its neighbors following a protocol consisting of three rules operating at different spatial scales. These rules are

1. **Separation:** avoid colliding with neighbors.
2. **Alignment:** align velocity with neighbors' velocities.
3. **Cohesion:** avoid becoming isolated from neighbors.

Over the past years, many variants of these three rules have been developed; e.g., a Lyapunov-based approach is proposed for switching joint interconnection; the graph theory is used to analyze consensus and synchronization problems; a local attractive/repulsive function is proposed to deal with separation and cohesion problems; artificial potentials and gyroscopic forces are designed for collision avoidance [431,459,462].

The primary distinguishing feature of a cooperative control system is **distribution of information**. As opposed to 'centralized' solutions, no one decision-maker has access to the information gathered by all agents. Furthermore, there is typically a **communication cost** in distributing locally gathered information. A secondary distinguishing feature is complexity. Even if information was centrally available, the inherent complexity of the decision problem makes centralized solution computationally infeasible. Therefore, decomposition approaches are used as in cooperative control [447]. Both these features result in the **distributed decision architecture of cooperative control**. The potential benefits of such architectures include the opportunity for real-time adaptation (or self-organization) and robustness to dynamic uncertainties such as **individual component failures, non-stationary environments and adversarial elements**. These benefits come with significant challenges such as the complexity associated with a potentially large number of interacting agents and the analytical difficulties of dealing with overlapping and partial information [26]. The team of n aerial robots can be described by a triple (g, r, H):

- the position and orientation of the team: $g \in SE(n)$ – motion group.
- the shape of the formation shape variables is given by $r \in R^m$.
- H is a directed graph – control graph: the graph describing the control structure of the team.

Formation dynamics can be important even if aerial robots move slowly. Decisions made by aerial robots propagate through the system, and the communication delays can exacerbate the effects of dynamics.

1.2.2 CASCADE-TYPE GUIDANCE LAW

The technical issues that must be addressed for formation keeping range from single-aerial-robot-level issues to higher system-level issues, where the latter defines the system control hierarchy, advanced decision-making and performance monitoring. With respect to control and guidance problems in formation flying, each aerial robot must be able to accept flight commands from an outside source and carry out the assigned commands using its own flight control system. The level of complexity and robustness of the overall formation depend on the communication and computation architecture among the distributed aerial robots. **Formation keeping** means that the relative positions between aerial robots are tightly controlled so that the overall formation geometry remains unchanged [435]. A leaderless coordination scheme can be suggested to secure a formation robust to failures in one or more aerial robots. The leader/follower formulation has advantages in which a control problem can be simplified as a standard tracking problem if the leaders information is available to the followers. Treating formation keeping as a **dynamic 3D tracking problem**, the classical compensation-type control law decomposes the control law into three subtasks:

1. **forward** distance control
2. **lateral** distance control
3. **vertical** distance control.

Guidance-based approaches have also been applied to the formation keeping problem. Naturally the leader is treated as a target, and the follower is guided so that it maintains a required range from the leader instead of intercepting it. This method requires no information flow from the leader to the follower vehicles. More specifically, a method is proposed to generate the flight command inputs to aerial robot k which will maintain the formation geometry. The concept of **branch global leader, local leader and local follower** is introduced. The whole formation is divided into several branches. Each branch is simply a chain of multiple aerial robots that starts from the global leader aerial robot. Except for the global leader, other aerial robots will take the role of a local leader, a local follower or both depending on their position in the branch. A particular aerial robot need not necessarily belong to only one specific **branch**, but it may be a leader of several followers belonging to different branches. It is assumed that one aerial robot can receive its local leader's flight information and also is able to transmit its own flight status to the local follower. First-order system is adopted to represent three control channels including the aerial robot flight dynamics as follows:

$$\dot{V} = \frac{V_c - V}{\tau_v} \tag{1.1}$$

$$\dot{\chi} = \frac{\chi_c - \chi}{\tau_\chi} \tag{1.2}$$

$$\dot{\gamma} = \frac{\gamma_c - \gamma}{\tau_\gamma} \tag{1.3}$$

where (V_c, χ_c, γ_c) are the control command inputs to each control loop and τ_v, τ_χ and τ_γ are the time constants, respectively, for each variable. The first relation in equation (1.1) is speed control channel, the second relation in equation (1.2) the heading angle control channel and finally the third relation in equation (1.3) the flight path angle control channel. If d_i and V_i denote the current position of the aerial robot i and its velocity vectors and if two aerial robots j and k maintain their current flight directions and speeds, then their relative position vector at a later t_{go} can be written as

$$\Delta d = (d_j - d_k) + (V_j - V_k)t_{go} \tag{1.4}$$

If neither of these aerial robots takes corrective measures, the expected formation error at future time $t + t_{go}$ will be

$$\delta d = \Delta d - \Delta d^* \tag{1.5}$$

where Δd is the required relative position vector between aerial robots j and k for formation flying. Using (e_x, e_y, e_z) to denote the unit vectors of the fixed inertial frame, the absolute velocity and the acceleration vectors error of an aerial robot in terms of its speed and flight direction angles (heading and flight path), respectively, can be written as

$$\delta d = M_x e_x + M_y e_y + M_z e_z \tag{1.6}$$

$$\begin{aligned}
\mathbf{V} &= V_x e_x + V_y e_y + V_z e_z \\
&= (V \cos\gamma \cos\chi)e_x + (V \cos\gamma \sin\chi)e_y + (V \sin\gamma)e_z
\end{aligned} \tag{1.7}$$

$$\begin{aligned}
\dot{\mathbf{V}} &= (\dot{V} \cos\gamma \cos\chi - V\dot{\chi} \cos\gamma \sin\chi - V\dot{\gamma}\sin\gamma \cos\chi)e_x \\
&+ (\dot{V} \cos\gamma \sin\chi + V\dot{\chi} \cos\gamma \cos\chi - V\dot{\gamma}\sin\gamma \sin\chi)e_y \\
&+ (\dot{V} \sin\gamma + V\dot{\gamma}\cos\gamma)e_z
\end{aligned} \tag{1.8}$$

An aerial robot control frame (e_V, e_χ, e_γ) is introduced where e_V denotes a unit direction vector along the current velocity vector, e_χ is a unit direction vector perpendicular to e_V and positive in the sense of increasing heading angle, and e_γ completes the right handed system. Then, the coordination transformation matrix D from the fixed inertial frame to the aerial robot control frame can be defined as

$$D = \begin{pmatrix} \cos\chi \cos\gamma & \sin\chi \cos\gamma & -\sin\gamma \\ -\sin\chi & \cos\chi & 0 \\ \cos\chi \sin\gamma & \sin\chi \sin\gamma & \cos\gamma \end{pmatrix} \tag{1.9}$$

Rewriting relations in equation (1.4) in the aerial robot control frame using the transformation matrix D in equation (1.9) yields

$$\delta d = M_V e_V + M_\chi e_\chi + M_\gamma e_\gamma$$

$$\mathbf{V} = V e_V \tag{1.10}$$

$$\dot{\mathbf{V}} = \dot{V} e_V + V \dot{\chi} \cos\gamma e_\chi - V \dot{\gamma} e_\gamma$$

Assuming that the difference in the flight path and heading angles of both the leader aerial robot j and follower aerial robot k are small, another coordinated transformation matrix can be defined from the follower aerial robot k control frame to the leader aerial robot j frame as

$$\delta D = \begin{pmatrix} 1 & \delta\chi & -\delta\gamma \\ -\delta\chi & 1 & 0 \\ \delta\gamma & 0 & 1 \end{pmatrix} \tag{1.11}$$

where $\delta\gamma = \gamma_j - \gamma_k$ is the flight path angle difference and $\delta\chi = \chi_j - \chi_k$ is the heading angle difference. The velocity vector of the leader $\mathbf{V}_j$ in the follower control frame aerial robot k control frame $(e_{V_k}, e_{\chi_k}, e_{\gamma_k})$

$$\mathbf{V}_j = V_j e_{V_k} + V_j \delta\chi e_{\chi_k} - V_j \delta\gamma e_{\gamma_k} \tag{1.12}$$

Similarly, the acceleration vector of the leader aerial robot j in the follower aerial robot k control frame can be written as

$$\frac{d\mathbf{V}_j}{dt} = \left(\dot{V}_j - V_j \dot{\chi}_j \cos\gamma_j \delta\chi - V_j \dot{\gamma}_j \delta\gamma \right) e_{V_k} + \left(V_j \dot{\chi}_j \cos\gamma_j + \dot{V}_j \delta\chi \right) e_{\chi_k} - \left(V_j \dot{\gamma}_j + \dot{V}_j \delta\gamma \right) e_{\gamma_k} \tag{1.13}$$

Next, a positive definite function is defined:

$$L = \frac{1}{2} \delta d \cdot \delta d = \frac{1}{2} \left(\Delta d - \Delta d^* \right) \cdot \left(\Delta d - \Delta d^* \right) \tag{1.14}$$

If the follower aerial robot k is controlled so that the value of equation (1.14) remains small for a sufficiently short time t_{go}, then Δd will remain close to Δd^*, and this implies that their relative positions are maintained. Among many alternatives to make $\frac{dL}{dt}$ negative, the following choice can be made:

$$\dot{L} = -2N^k L \tag{1.15}$$

where N^k is a positive constant. The guidance commands yield

$$\dot{V}_k = \left(\dot{V}_j - V_j \dot{\chi}_j \cos\gamma_j \delta\chi - V_j \dot{\gamma}_j \delta\gamma \right) + \frac{N_k}{t_{go}} M_V^k \tag{1.16}$$

$$\dot{\chi}_k = \frac{V_j \dot{\chi}_j \cos\gamma_j + \dot{V}_j \delta\chi}{V_k \cos\gamma_k} + \frac{N_k}{V_k \cos\gamma_k t_{go}} M_\chi^k \tag{1.17}$$

$$\dot{\gamma}_k = \frac{V_j \dot{\gamma}_j + \dot{V}_j \delta\gamma}{V_k} + \frac{N_k}{V_k t_{go}} M_\gamma^k \tag{1.18}$$

which are the required rates of change in the speed and flight direction angles of aerial robot k to decrease the formation error.

1.2.3 CONSENSUS APPROACH

1.2.3.1 Consensus Opinion

Consider a group of aerial robots with dynamics approximated by

$$\dot{X}_i = U_i \quad i = 1, \ldots, m \tag{1.19}$$

where every node only interacts with its neighbors $N_i = \{j : (i,j) \in E\}$ on a graph $G = (V,E)$. In general, G is a directed graph with the set of nodes $V = \{1..n\}$, the set of edges $E \subset V \times V$ and the adjacency matrix $A = [a_{ij}], i = 1,\ldots,n, j = 1,\ldots,n$, with non-negative elements. In reaching a consensus, the objective of the aerial robots is to converge to the following space:

$$X_1 = X_2 = \ldots = X_n \tag{1.20}$$

where the **opinion configuration** of all aerial robots is the same. This can be achieved by applying a **linear consensus protocol** such as

$$\dot{X}_i = \sum_{j \in N_i} a_{ij}(X_j - X_i) \tag{1.21}$$

For **undirected graphs**, the linear system in equation (1.20) is identical to a gradient system

$$\dot{X} = \nabla \Phi_G(X) \tag{1.22}$$

with a potential function called the **disagreement function:**

$$\Phi_G(X) = \frac{1}{2} \sum_{i<j} a_{ij}(X_j - X_i)^2 \tag{1.23}$$

The disagreement function quantifies the total disagreement for a group of aerial robots with **opinion vector** X and **information flow** G. The consensus dynamics in equation (1.20) can be rewritten as

$$\dot{X} = -LX \tag{1.24}$$

where $L = D - A$ is the graph **Laplacian matrix**, $D = diag(A)$ is the degree matrix with diagonal elements $d_i = \sum_j a_{ij}$ and $\mathbf{1} = (1,\ldots,1)$ is the vector of ones. By definition, $L\mathbf{1}$ and thus graph Laplacian always have a zero eigenvalue $\lambda_1 = 0$. If G is a strongly connected graph or has a directed spanning tree, $rank(K) = n - 1$ and all other eigenvalues of L are non-zero [83]. An equilibrium of system equation (1.20) is in a state $X^* = (\zeta,\ldots,\zeta) = \zeta\mathbf{1} = (1,\ldots,1)$ where all nodes agree.

Theorem 3. *Let G be a strongly connected graph digraph. Then all aerial robots asymptotically reach a consensus opinion in ζ. Moreover, assuming that the digraph G is balanced, $\sum_j a_{ij} = \sum_j a_{ji}, \forall i$, a consensus to*

$$\zeta = \bar{X}(0) = \frac{1}{n} \sum_{i=1}^{n} X_i(0) \tag{1.25}$$

is reached exponentially fast with a rate greater than or equal to $\frac{\lambda_2}{2}(L + L^T)$, the second largest eigenvalue or algebraic connectivity of the symmetric part of the Laplacian of the digraph G

Presence of obstacles that cut communication links between aerial robots can be detrimental to stability and performance of the overall system [448]. This effect can be modeled as a change in the topology that results in multiagent systems with switching networks. A **dynamic graph G** can be represented as $G(t) : [0, +\infty] \to \mathbf{G}$ with $\mathbf{G}$ being the superset of all directed graphs with edges (i,j) in $V \times V$. Let $L(t)$ denote the graph Laplacian of $G(t)$. Then the linear switching system

$$\dot{X} = -L(t)X \tag{1.26}$$

is capable of capturing the model of consensus in networks with link/node failures. The **loss of a node in a network with topology G** can be modeled as the loss of a subgraph G_i of G that is induced by node i and its neighbors V_i [9].

1.2.3.2 Reachability and Observability

If n aerial robots communicate according to a time-invariant undirected communication graph $G = (I, E)$ where $I = \{1, \ldots, n\}$, $E = \{(i, j) \in I \times I | i$ and j communicate$\}$ and $N_i = \{j \in I | (i, j) \in E\}$ is the set of neighbors of node I. A **path graph** is a graph in which there are only nodes of degree 2 except for two nodes of degree 1. The nodes of degree 1 are called **external nodes** and denoted 1 and n, while the others are called **internal nodes** [438]. A **cycle graph** is a graph in which all the nodes have degree 2. The aerial robots are modeled as single integrators running on a consensus algorithm based on a Laplacian control law [76]. The continuous time dynamics of the agents are given by

$$\dot{x}_i(t) = -\sum_{j \in N_i} x_i(t) - x_j(t) \quad i \in \{1, \ldots, n\} \tag{1.27}$$

or equivalently

$$\dot{x}(t) = -Lx(t) \quad t \leq 0 \tag{1.28}$$

where $x = [x_1, \ldots, x_n]^T$ is the vector of robot's states and L is the graph Laplacian.

Regarding the **reachability problem**, a subset of nodes $I_c = \{i_1, \ldots, i_m\} \subset I$, called leaders or control nodes can be controlled by an external input.

$$\dot{x}(t) = -Lx(t) + Bu(t) \tag{1.29}$$

where $u(t) = [u_{i_1}, \ldots, u_{i_m}]$ is the input and $B = [e_{i_1} |, \ldots, | e_{i_m}]$.

Regarding the **observability problem**, an external processor is assumed to collect information from a subset of nodes $I_o = \{i_1, \ldots, i_m\} \subset I$ in the network called **observation nodes**. The external processor has to reconstruct the entire network state from the states of the observation nodes. The system dynamics is given by

$$\dot{x}(t) = -Lx(t) + Bu(t) \quad y(t) = Cx(t) \tag{1.30}$$

where the output matrix is $C = [e_{i_1} |, \ldots, | e_{i_m}]$. The **reachable subspace** is the set X_r of states that are reachable from the origin. The **unobservable subspace** is the set X_{no} of initial states that produce an identically zero output.

Lemma 4. *Popov, Belevitch, Hautus lemma for symmetric matrices Let $L \in \mathbb{R}^{n \times n}$, $B \in \mathbb{R}^{n \times m}$ and $C \in \mathbb{R}^{m \times n}$ be the state, input and output matrices of a linear time-invariant system where L is symmetric. Then the orthogonal complements to the reachable subspace X_r (associated to the pair (L, B)) and the unobservable subspace X_{no} (associated to the pair (L, C)) are spanned by vectors satisfying, for some $\lambda \in \mathbb{R}$,*

$$B^T v_l = 0 \quad L v_l = \lambda v_l \tag{1.31}$$

and

$$C v_l = 0 \quad L v_l = \lambda v_l \tag{1.32}$$

The eigenvalues and eigenvectors for which equation (1.31) holds are called **unreachable eigenvalues and eigenvectors**. The eigenvalues and eigenvectors for which equation (1.32) holds are called **unobservable eigenvalues and eigenvectors**.

1.2.4 FLOCKING BEHAVIOR

The flocking behavior relies on the aerial robot's cohesion ability to maintain the **group's connectivity**, its **alignment ability** to synchronize velocity and, if necessary, its **separation ability** to avoid collision. Networks of many aerial robots constitute swarms. Due to their mobility, the resulting swarm has a dynamic topology that varies as a function of the configuration of the aerial robots. This spatially induced network is called a **proximity graph**.

Definition 5. *Proximity graph The proximity graph is the union of the proximity subgraph of all aerial robots: let $q_i \in \mathbb{R}^m$ denote the configuration of aerial robot i in an m-dimensional space. Then the set of neighbors of i are $N_i(q) = j : |q_j - q_i| < r, j \neq i$ and the set of vertices and edges of the proximity graph $G(q) = (V, E(q))$ are defined as*

$$V = \{1, 2, ..., n\}$$
$$E(q) = \left\{ (i, j) \in V \times V : \|q_j - q_i\| < r, j \neq i \right\} \tag{1.33}$$

where $\|;\|$ denotes the 2-norm in $\mathbb{R}^m$ and $q = col(q_1, q_2, ..., q_n)^T$.

Inspired by cohesive collective behavior of flocks of birds and schools of fish in nature, a distributed algorithm for area coverage relies on creation of flocking behavior for swarms. The fundamental challenge in this approach is to provide a mathematical model of flocks and a formal representation of key features in flocking behavior. In addition, one needs to develop a meaningful notion of stability of flocking motion that can be verified for a proposed flocking algorithm that is rather similar to defining it when a nonlinear control system $\dot{X} = f(X, U)$ in the sense of Lyapunov. If the group of aerial robots have the following dynamics

$$\dot{q}_i = p_i$$
$$\dot{p}_i = U_i \tag{1.34}$$

where $q_i, p_i, U_i \in \mathbb{R}^m$ are position, velocity and control input of robot i, respectively, flocks with local interactions can be created such that their emergent behaviors satisfy the following conditions in steady state:

- **Alignment:** the velocity of the aerial robots is aligned.
- **Spatial self-organization:** the robots form lattice-shaped meshes, structures in space.
- **Collision and obstacle avoidance:** nearby aerial robots have a tendency to avoid collision with each other and obstacles (both in transient state and steady state).
- **Cohesion:** there exists a ball of radius $R > 0$ that contains all the aerial robots.
- **Connectivity:** the aerial robots self-assemble connected networks of aerial robots.

Collective potential of flocks is designed, and virtual agents are introduced that are embedded in the computing infrastructure of physical aerial robots.

- **Physical agents:** Aerial robots are referred to as α-agents.
- **Intermediate Agents** that move on the boundary of obstacles and are obtained from the projection of α-agents on their neighboring obstacles are β-agents.
- **Virtual agents** that are embedded in α-agents as an internal dynamics are γ-agents. They contain the information regarding a desired trajectory for the center of mass of the entire flock for migration/tracking.

1.2.4.1   Collective Potential of Flocks

The following set of algebraic constraints based on pairwise distance defines a class of lattice-shaped configurations q called an α-**lattice**, i.e., $-\delta < |q_j - q_i| - d < \delta$:

$$\|q_j - q_i\| = d \quad \forall j \in N_i(q) \tag{1.35}$$

Parameters d and $k = \frac{r}{d}$ are called **scale** and **ratio** for an α-lattice. A **map** $\|.\|_\sigma : \mathbb{R}^m \to \mathbb{R}^+$ called σ-norm for an m-vector z is defined by

$$\|z\|_\sigma = \frac{-1}{\varepsilon}\left(1 - \sqrt{1 + \varepsilon|z|^2}\right) \tag{1.36}$$

where $\varepsilon > 0$ is a parameter of the σ-norm and the gradient of this map is given by

$$\sigma_\varepsilon = \frac{z}{\sqrt{1 + \varepsilon\|z\|^2}} = \frac{z}{1 + \varepsilon\|z\|_\sigma} \tag{1.37}$$

The benefits of defining σ-norm is that $\|z\|_\sigma$ is differentiable everywhere but the 2-norm $\|z\|$ is not differentiable at $z = 0$. Later, this feature becomes useful in introducing flocking algorithm based on a smooth gradient. Using a smooth bump function $\rho_h : \mathbb{R} \to [0,1]$, smooth spatial adjacency (or interaction) coefficient $a_{ij}(q)$ can be defined for a swarm of particles as

$$a_{ij}(q) = \rho_h\frac{\|q_j - q_i\|_\sigma}{r_\alpha} \in [0,1], j \neq i \tag{1.38}$$

where $r_\alpha = \|r\|_\sigma, h \in [0,1]$ is a parameter and ρ_h is defined by

$$\rho_h(z) = \left\{ \begin{array}{ll} 1 & z \in [0,h] \\ \frac{1}{2}\left(1 + \cos\pi\frac{z-h}{1-h}\right) & z \in [h,1] \\ 0 & z > 1 \end{array} \right\} \tag{1.39}$$

$\rho_h(z)$ is a C^1 smooth function with the property that $\rho_h'(z) = 0$ over the interval $[1,\infty]$ and $|\rho_h'(z)|$ is uniformly bounded in z.

Now, the smooth collective potential function $V(q)$ of a flock can be expressed as

$$V(q) = \frac{1}{2}\sum_i\sum_{j\neq i}\psi_\alpha\left(\|q_j - q_i\|_\sigma\right) \tag{1.40}$$

where ψ_α is a smooth pairwise attractive/repulsive potential with a finite cut-off at $r_\alpha = \|r\|$ and a global minimum at $z = d_\alpha = \|d\|$. This potential takes its minimum at a configuration q that is an α-lattice, i.e., the set of q satisfying $\|q_j - q_i\|_\sigma = d_\alpha, \forall j \in N_i(q)$. To construct a smooth pairwise potential $\psi_\alpha(z)$ with a finite cut-off, an action function $\phi_{\alpha(z)}$ is integrated, and it vanishes for all $z \geq r_\alpha$:

$$\phi_\alpha(z) = \rho_h\frac{z}{r_\alpha}\phi\left(z - d_\alpha\right) \tag{1.41}$$

$$\phi(z) = \frac{1}{2}\left((a+b)\sigma_1(z+c) + (a-b)\right) \tag{1.42}$$

where

$$\sigma_1 = \frac{z}{\sqrt{1+z^2}} \tag{1.43}$$

and $\phi(z)$ is an uneven sigmoidal function with parameters that satisfy $0 < a \leq b, c = \frac{|a-b|}{\sqrt{4ab}}$ to guarantee $\phi(0) = 0$. The pairwise attractive/repulsive potential ϕ_α in equation 1.40 is defined as

$$\psi_\alpha(z) = \int_{d_\alpha}^z\phi_\alpha(s)ds \tag{1.44}$$

1.2.4.2 Distributed Flocking Algorithms

The following distributed algorithm is proposed for flocking:

$$u_i = u_i^\alpha + u_i^\beta + u_i^\gamma \tag{1.45}$$

where interactions are between neighboring α- and β-agents and also their interval γ-agents.

$$u_i^\alpha = \sum_{j \in N_i} \phi_\alpha \left\| q_j - q_i \right\|_\sigma n_{ij} + \sum_{j \in N_i} a_{ij}(q) \left\| p_j - p_i \right\|_\sigma n_{ij} \tag{1.46}$$

The first term is a gradient-based term, while the second one is the consensus term, where

$$n_{ij} = \sigma_\varepsilon(q_j - q_i) = \frac{q_j - q_i}{\sqrt{1 + \varepsilon \| q_j - q_i \|^2}} \tag{1.47}$$

is a vector along the line connecting q_i to q_j and $\varepsilon \in (0,1)$ is a fixed parameter of the σ norm. The gradient-based term is equal to $\left| -\nabla_{q_i} V(q) \right|$. The term u_i^α represents all the interactions among α-agents; u_i^β is a repulsive force that keeps an α-agent away from its projection on obstacles (or β-agents); and

$$u_i = -C_1(q_i - q_r) - C_2(p_i - p_r); C_1, C_2 > 0 \tag{1.48}$$

is the tracking control that allows α-agents to track their internal γ-agents with dynamics

$$\dot{q}_r = p_r$$
$$\dot{p}_r = f_r(q_r, p_r) \tag{1.49}$$

that are initialized to $(q_r(0), p_r(0)) = (q_d, p_d)$, for all γ-agents.

1.2.5 CONNECTIVITY AND CONVERGENCE OF FORMATIONS

The interdependence of formation stability and the underlying communication structure are considered in this section.

1.2.5.1 Problem Formulation

A moving formation of N homogeneous aerial robots has the following discrete time dynamics:

$$x_i(k+1) = A_v x_i(k) + B_v u_i(k), \ i = 1, \ldots, N \tag{1.50}$$

The n configuration variables for aerial robot i, referred to as position-like variables and their derivatives, are denoted by x_i. The control inputs for the aerial robots are denoted by u_i. For each aerial robot, the error signal $z_i(k)$ for coordination is given by the following relation:

$$z_i(k) = \sum_{j \in J_i(k)} (x_i(k) - x_j(k)) - (h_i - h_j) \tag{1.51}$$

where $J_i(k)$ is the set of neighbors of aerial robot i at time k and h is a parameter defined below.

Definition 6. *Formation* *A formation of N aerial robots is given by an offset vector* $h = h_p \otimes \begin{pmatrix} 1 \\ 0 \end{pmatrix} \in \mathbb{R}^{2nN}$.

The N aerial robots are said to be in formation at time k if there exists $\mathbb{R}^n$ valued vectors q and w such that $(x_p)_i(k) - (h_p)_i = q$ and $(x_v)_i(k) = w$ for $i = 1, \ldots, N$, where the subscript p refers to position and the subscript v to the corresponding velocities. The aerial robots converge to formation h if there exist real-valued functions $q(.)$ and $w(.)$ such that $(x_p)_i(k) - (h_p)_i - q(k) \to 0$ and $(x_v)_i(k) - w(k) \to 0$ as $k \to \infty$ for $i = 1, \ldots, N$.

This definition covers both moving $|w| > 0$ and hovering ($w = 0$) aerial robot formations. At any time k, the information exchange between aerial robots is captured by a **communication graph** $G(k)$, where the N vertices represent the aerial robots and the edges represent the **communication links**. Aerial robot j is considered to be a neighbor of aerial robot i at a time k if $G(k)$ has the edge (i, j). However, if an aerial robot is a neighbor at time k, it is not guaranteed that the same aerial robot will be a neighbor of i at time $k + 1$. In fact, depending on the formation requirements, aerial robots may move in such a fashion so that they change neighbors often while converging to formation. For a formation of homogeneous aerial robots, a stabilizing matrix F_v for each aerial robot can be sought. The system of N homogeneous aerial robots is given by

$$
\begin{aligned}
x(k+1) &= Ax(k) + Bu(k) \\
z(k) &= L(k)(x(k) - h)
\end{aligned}
\tag{1.52}
$$

where x is the augmented state vector, $A = I_N \otimes A_v, B = I_N \otimes B_v$ and $L(k) = L_G(k) \otimes I_{2n}$ where the matrix $L_G(k)$ belongs to a finite set of constant matrices L. Assuming that each aerial robot has a stable control law already designed, the control law at the formation level should be defined with the intervehicle communication requirements so that formation convergence and stability can be guaranteed. The coordinated control is implemented using a decentralized feedback control matrix

$$
F = \operatorname{diag}(F_v)
\tag{1.53}
$$

such that if $u(k) = Fz(k)$, then the aerial robots converge to formation. This feedback matrix F should be the same regardless of the communication graph between the aerial robots. The closed-loop system becomes

$$
x(k+1) = Ax(k) + BFL(k)(x(k) - h)
\tag{1.54}
$$

In the following analysis, algebraic graph theory is used to analyze the stability of a dynamical system that is of the form

$$
x(k+1) = (A + BFL(k))x(k)
\tag{1.55}
$$

Let $\sigma : N \to L$ be a piecewise constant function that determines the L matrix at each time k, $\sigma(k) = L_s, L_s \in L$. Then the system defined above can be considered as a switched dynamical system. Its stability depends in part on the stability of the individual system

$$
x(k+1) = (A + BFL_s)x(k)
\tag{1.56}
$$

But it is not guaranteed by that. The properties of the eigenvalues of matrices L_s are used to prove the stability of a related system matrix. A matrix is stable in the discrete Hurwitz sense if all its eigenvalues are located in the interior of the circle or radius 1 in the complex plane.

1.2.5.2 Stability of Formations in Time-Invariant Communication

First, the communication graph $G(k)$ is considered to be the same at every instance of time. This is the case where the matrix L_s is time invariant; it is a constant function. A necessary and sufficient condition for this system to converge is that the communication graph has a rooted directed spanning tree. This ensures that the algebraic multiplicity of the zero eigenvalue of L_s is one, which in turn guarantees that the system above is **formation stable**. The concept of formations can be extended to hierarchical formations, and formations with a special case of hierarchical communication that is time invariant can be considered. Given several stable formations that communicate with each other, the results from formation convergence and stability can be extended to stability of formations [31].

Definition 7. *Hierarchical Formation* A group of N aerial robots is said to be in a hierarchical formation if, at all times, they are the vertices of a pre-specified geometrical shape and the communication graph is a sum of hierarchical products of graphs.*

At each layer in the hierarchy, the vertices represent a subformation. Every subformation is constructed of individual or subformation of aerial robots. The subformation (or aerial robot) labeled 1 is assumed to be the leader of its subformations and is the only one that can receive or transmit information to other subformations. The leader of each subformation is assumed to communicate with aerial robots at the next layer, and the rest of the vertices in its corresponding graph have positive indegrees. Practically, this is interpreted as follows: the non-leader subformation receives information from at least one other subformation in their group. In the subformation graph, this is captured by having the first diagonal entry of the Laplacian equal to zero and all other diagonal entries positive. It is necessary to work with **Laplacian matrices** from the set

$$S = \{I_n \otimes L_1 + L_2 \otimes One\} \tag{1.57}$$

where $I_n \in \mathbb{R}^{n \times n}$ is an identity matrix, $L_1 \in \mathbb{R}^{m \times m}$ is a Laplacian matrix, $L_2 \in \mathbb{R}^{n \times n}$ is a Laplacian matrix, and $One \in \mathbb{R}^{m \times m}$ is a matrix with the first entry $= 1$ and all other entries are zero. For a matrix $L \in S$, its eigenvalues depend on the Laplacian L_1, L_2. In the case of undirected graphs, some bounds for the eigenvalues of the matrix L are derived as follows:

1. The graph G corresponding to the Laplacian L has maximum vertex indegree

$$d = \max\{d_1, \text{indegree}(1) + d_2\} \tag{1.58}$$

   where d_1 is the maximum indegree of the graph G_1, corresponding to Laplacian L_1, indegree(1) is the indegree of the vertex labeled 1 in G_1 and d_2 is the maximum vertex indegree of the graph G_2 corresponding to the Laplacian L_2.
2. So, for all eigenvalues λ of L,

$$|\lambda| \leq 2\max\{d_1, \text{indegree}(1) + d_2\} \tag{1.59}$$

In the case of directed graphs, a bound for the eigenvalues can be derived as follows.

Theorem 8. *Let $L_1 \in \mathbb{R}^{m \times m}$ and $L_2 \in \mathbb{R}^{n \times n}$ be Laplacian matrices of two directed graphs and suppose L_1 has its diagonal entry equal to zero and all other diagonal entries positive. If $L = I_n \otimes L_1 + L_2 \otimes One \in S$, then the eigenvalues of L are those of L_2 and the non-zero eigenvalues of L_1 repeated n times.*

Theorem 9. *Suppose a two-layer formation hierarchy, with n aerial robots at the subformation level and m subformations. Let each subformation be represented by the following controllable system:*

$$x_{k+1} = A_s x_k + B_s u_k \tag{1.60}$$

$$z_k = L_s(x_k - h_s) \tag{1.61}$$

$$u_k = F_s z_k \tag{1.62}$$

where A_s and B_s are the individual subformation dynamics, L_s is the information exchange graph Laplacian, F_s is the subformation control feedback matrix and h the corresponding formation offset vector. With F_s such that each subformation is stable, if G is a graph on m vertices whose Laplacian eigenvalues λ_f have the property that $A + \lambda_f BF$ is stable, then G can be used as the communication infrastructure for the second layer of the formation so that the entire system converges to a new overall stable formation.

When converging to formation, the aerial robots achieve consensus on the location of the center of the final formation. This consensus is achieved at both the second layer level, i.e., the subformation leaders, as well as the first layer levels. This is done independently of the relative position of the aerial robots in its group. In the case when the subformations are not identical, i.e., the communication graph is a sum of hierarchical graphs, the choice of L_f should be such that each $A + \lambda BF$ should be stable. An example of such is the case when, at all levels, the information exchange graphs can be modeled using directed tree graphs.

1.3 DETERMINISTIC DECISION-MAKING

Plans may be required for a team of robots to deal with the specifics of their sensors (**planning for sensing**) or to assist in the development of appropriate communication paths between elements of the collection (**planning for communication**) or planning so that the elements of the collection cause some motion of objects in the environment (**planning for action**). Team Model takes into account the **state** of the aerial robots and the environment, **decision variables** (controls), **decision rules** (controllers), **process model**, **information structure** and **utility structure** [452].

Definition 10. *Path planning for multiple robots: Let $G = (V, E)$ be an undirected graph, and let $R = \{\bar{r}_1, \bar{r}_2, \ldots, \bar{r}_v\}$ be a set of aerial robots where $v < |V|$. The initial arrangement of aerial robots is defined by a simple function $S_R^0 : R \to V$; that is, $S_R^0(r) \neq S_R^0(s)$ for every $r, s \in R$ such that $r \neq s$; the goal arrangement of robots is defined by another simple function $S_R^+(r) \neq S_R^0(s)$. A problem of multi-robot path planning is the task to find a number ζ called a makespan and a sequence $S_R = \left[S_R^0, S_R^1, \ldots, S_R^\zeta \right]$ where $S_R^k : R \to V$ is a simple function for every $k = 1, 2, \ldots, \zeta$ while S_R must satisfy the following constraints:*

1. *$S_R^\zeta = S_R^+$; that is, aerial robots finally reach their destinations.*
2. *Either $S_R^k(r) = S_R^{k+1}(r)$ or $\{S_R^k(r), S_R^{k+1}(r)\} \in E$ for every $r \in R$ and $k = 1, 2, \ldots, \zeta - 1$.*
3. *If $S_R^k(r) \neq S_R^{k+1}(r)$ and $S_R^k(s) \neq S_R^{k+1}(r), \forall s \in R$ such that $s \neq r$, then the move of r at the time step k is allowed (a move into an occupied vertex) $S_R^k(r) \neq S_R^{k+1}(r)$, and if $s \in R$ such that $s \neq r$ and $S_R^k(s) = S_R^{k+1}(r)$ and $S_R^k(s) \neq S_R^{k+1}(s)$ and the move of s at the time step k is allowed, then the move of r at the time step k is also allowed. All the moves of aerial robots must be allowed.*

The problem described above is formally a quadruple $\Sigma = \left\langle G = (V, E), R, S_R^0, S_R^+ \right\rangle$.

An **allowed move** can be performed either into an unoccupied vertex or into a vertex that has just been vacated using an allowed move. Of course, in aerial robotics applications, some solutions are preferred to others. Typically, solutions with the small makespan are required. This immediately raises the question whether it is possible to compute a solution of the smallest possible makespan. Teams of aerial robots can be engaged in elaborate missions. In the case of aerial robots, **weight** is a key measure from a hardware perspective, while **source lines of codes** (SLOCs) are used as the primary size measure for software. For instance, a multi-vehicle search mission consists of many resources interacting to concurrently execute many tasks in a dynamic fashion. In another mission, it is assumed that the details of the terrain such as location of targets and physical obstacles are known. There are multiple objectives and multiple aerial robots launched from different locations. Teams of aerial robot are assembled by mission control, and each aerial robot is assigned to a particular objective. In a typical scenario, to successfully achieve mission objectives, all members of a team have to reach their assigned objective simultaneously. The key challenge in this kind of mission is to deal with the timing constraints stemming from the variable speed range of aerial robots teammates [410]. In another example, aerial robots might leave and rejoin the team, and missions start and terminate. A mission can be modeled as a directed application graph, where nodes represent service providers and edges represent services. The search mission algorithm then builds such application graph dynamically, and it reconfigures them as service providers depart or arrive.

The **decision table** obtained off-line, prior to a mission, is essential for the online decision-making of the cooperative control system. Depending on the state of an agent and its teammates, different decisions may be obtained. Information exchanged over the network is therefore essential to guarantee optimality of the path planning process and forms the feedback needed for the cooperation control system. A decision table depends on agents location and the number of vehicles per formation which are operational. One could design a multiagent path planning system by solving a minimization problem with dynamic programming carried out offline prior to the mission. The following are needed for the planning system:

1. **perfect knowledge** of the environment
2. **networked information** on the state of all agents available when needed.

The desired **attributes** are numerous: modularity, sensing, actuation and computation must be done locally. **Robustness** should consider failures of nodes and failures of communication between nodes. **Scalability** should consider bottlenecks due to computation and bottlenecks due to communication. The level of complexity and robustness of the overall formation depend on the communication and computation architecture among the distributed aerial robots.

1.3.1 DISTRIBUTED RECEDING HORIZON CONTROL

One control approach that accommodates a general cooperative objective is Receiving Horizon Control (RHC) also known as Model Predictive Control. In RHC, the current control action is determined by solving a finite horizon optimal control problem at each sampling instant. In continuous time formulations, each optimization yields an open loop control trajectory, and the initial position of the trajectory is applied to the system until the next sampling instant.

In this section, a distributed implementation of RHC is presented in which each an aerial robot is assigned its own optimal control problem, optimized only for its own control at each update, and exchanges information with neighboring agents. The approach here uses a **move suppression term** in each local cost function to penalize the deviation of the computed state trajectory from the assumed state trajectory. **Closed-loop stability** follows the weight when the move suppression term is larger than a parameter that bounds the amount of coupling in the cost function between neighbors, while move suppression terms follow traditionally the rate of change of the control inputs, specifically in discrete time applications of model predictive process control, the move suppression term here involves the state trajectory.

Cooperation between aerial robots can be incorporated in the optimal control problem by including coupling terms in the cost function. Aerial robots that are coupled in the cost function are referred to as neighbors. When the aerial robots are operating in real-time distributed environments, as is typically the case with multi-vehicle systems, a centralized RHC implementation is generally not viable due to the computational requirements of solving the centralized problem at each update. In the control problem, subsystems will be coupled in an integrated cost function of an optimal control problem. For example, aerial robots i and j might be coupled in the integrated cost by the term $\|q_i - q_j + d_{ij}\|$ where q_ℓ is the position of the vehicle ℓ and d_{ij} is a constant relative position vector that points from i to j. The purpose of the distributed RHC approach is to decompose the overall cost so that in this example i and j would each take a fraction of the term $\|q_i - q_j + d_{ij}\|^2$ among other terms in defining their local cost functions. The aerial robots i and j update their Receiving Horizon controllers, in parallel, exchanging information about each other's anticipated position trajectory so that each local cost can be calculated. More generally, the coupling cost terms are non-separable, so that aerial robots i and j must exchange trajectory information if they are coupled via the cost function.

Definition 11. *Non-separable Function* *A non-negative function* $g : \mathbb{R}^n \times \mathbb{R}^n \to \mathbb{R}^+$ *is said to be non-separable in* $x \in \mathbb{R}^n$ *and* $y \in \mathbb{R}^n$ *if g is not additively separable* $\forall x, y \in \mathbb{R}^n$, *g cannot be written as the sum of two non-negative functions* $g_1 : \mathbb{R}^n \to \mathbb{R}^+$, $g_2 : \mathbb{R}^n \to \mathbb{R}^+$ *such that* $g(x,y) = g_1(x) + g_2(y), \forall x, y \in \mathbb{R}^n$.

The objective is to stabilize this team of aerial robots to an equilibrium point using RHC. In addition, each aerial robot is required to cooperate with a set of other aerial robots where cooperation refers to the fact that every aerial robot has incentive to optimize the collective cost function that couples their state to the states of other aerial robots. For each aerial robot $i \in v = \{1, \ldots, N_a\}$, the state and control vectors are denoted $z_i(t) \in \mathbb{R}^n$, respectively, at any time $t \geq t_0 \in \mathbb{R}^n$. The decoupled, time-invariant nonlinear system dynamics are given by

$$\dot{z}_i(t) = f_i(z_i(t), u_i(t)) \quad t \geq t_0 \tag{1.63}$$

While the system dynamics can be different for each aerial robot, the dimension of every aerial robot's state and control is assumed to be the same. For aerial robot i, the decoupled state and input constraints and $z_i(t) \in Z, u_i(t) \in U, \forall t \geq t_0$, where Z, U are assumed to be common. The Cartesian product is denoted $Z^{N_a} = Z \times \cdots \times Z$. The concatenated vectors are denoted $z = (z_1, \ldots, z_{N_a})$. In concatenated vector form, the system dynamics are

$$\dot{z}(t) = f(z(t), u(t)) \quad t \geq t_0 \tag{1.64}$$

where $f(z, u) = (f_1(z_1, u_1), \ldots, f_{N_a}(z_{N_a}, u_{N_a}))$. The desired equilibrium point is the origin.

Remark 12. *Assumption: The following hold $\forall i \in v$:*

1. $f : \mathbb{R}^n \to \mathbb{R}^n$ *is continuous,* $0 = f_i(0, 0)$*, and f_i is locally Lipschitz in z_i.*
2. Z *is a closed, connected subset of $\mathbb{R}^n$ containing the origin in its interior.*
3. U *is a compact, convex subset of $\mathbb{R}^m$ containing the origin in its interior.*

It is assumed that a single collective cost function $L : \mathbb{R}^{nN_a} \to \mathbb{R}^+$ is provided that couples the states of the aerial robots and that each aerial robot has incentive to minimize this function with respect to its own state. For each $i \in v$, let $N_- \subseteq v/\{i\}$ be the set of other aerial robots whose states are coupled non-separably to z_i in $L(z)$. By definition, $j \in N_i$ if and only if $i \in N_j, \forall i, j \in v$. Denote $N_i = |N_i|$, and let $z_i = \left(z_{j_1}, \ldots, z_{j_{N_i}}\right)$ be the collective vector of states coupled non-separably to z_i in $L(z)$.

Remark 13. *Assumption: The function $L : \mathbb{R}^{nN_a} \to \mathbb{R}^+$ is continuous positive definite and can be decomposed as follows: $\forall i \in v$, there exists an integer $N_i \in (1, \ldots, N_a - 1)$ and a continuous and non-negative function $L_i : \mathbb{R}^n \times \mathbb{R}^{nN_a} \to \mathbb{R}^+$ not identically zero such that*

1. $L_i(z_i, z_{i-1})$ *is non-separable in $z_i \in \mathbb{R}^n$ and $z_{i-1} \in \mathbb{R}^{nN_i}$*
2. $\sum_{i \in v} L_i(z_i, z_{i-1}) = L(z)$
3. *there exists a positive constant $c_i \in]0, \infty[$ such that*

$$L_i(x, y) \leq L_i(x, w) + c_i \|y - w\| \, ; \forall x \in \mathbb{R}^n, \forall y, w \in \mathbb{R}^{nN_i} \tag{1.65}$$

The constant c_i is referred to as the strength of coupling parameter, and the term $c_i \|.\|$ is referred to as the cost coupling bound.

1.3.2 CONFLICT RESOLUTION

A wide range of conflict resolutions have addressed the collision avoidance problem that can be split into three types:

- **Prescribed:** All aerial robots follow a set of protocols that tend to yield a discrete event controller, which when combined with the aerial robot's continuous dynamics forms a hybrid system.
- **Optimized:** These methods attempt to find the best route for all the aerial robots to take to avoid each other while minimizing a cost function. Generally, they use a look ahead or time horizon so that the solution does not have to be recalculated often. The collision cone concept is a first order look ahead for detecting conflict. The collision cone (or velocity obstacle) is a set of velocities for one aerial robot that will cause it to collide with another, assuming each of their velocities are constant.

- **Force field:** These methods use continuous feedback mechanism to compute the control. The force field between two aerial robots is similar to the repulsion between two like charged particles. However, many possible alternatives are available for feedback schemes. These methods are generally reactive in that the control reacts to the current state of the system, rather than planning a trajectory ahead of time.

The aerial robot model used in each formulation is also important for the algorithm applicability to real systems. First, the dynamics must be at least second order and the inputs must be bounded to model the difficulties of overcoming inertia. Aerial robots have nonholonomic constraints and a maximum speed, but some aerial robots such as fixed wing aircraft must have a positive minimum speed. Another important aspect of a collision avoidance algorithm is whether it works for a heterogeneous group of aerial robots. Heterogeneity can be due to differences in aerial robot size dynamics, speed range, control authority. Aerial robot size, maximum speed and control authority are often important parameters for the other aerial robots in the system in order to properly avoid each other. In a homogeneous system, these parameters are known implicitly because they are the same for every aerial robot, but in a heterogeneous system, those parameters must be exchanged. **Communication** is the most obvious tool, though sensors could be employed to identify the aerial robot type and compare it to some known list. A final term important to collision avoidance is liveness which denotes the ability of the aerial robots to attain their goals. **Liveness** is important to consider because one simple way to avoid collisions is to have everyone stop moving: a **deadlock**. Another problem scenario is a **livelock** where the aerial robots continue to move but in such a way that they cannot reach their goals.

1.3.2.1   Distributed Reactive Collision Avoidance

This algorithm is a two-step process consisting of an optimization-based **deconfliction maneuver**, followed by the longer-term **deconfliction maintenance** phase, which is a reactive, force-field-type approach. Both of the steps are based upon the collision cone concept. The aerial robots considered here are modeled as point masses with 3D double integrator dynamics; however, physical aerial robots have finite size, finite velocity and finite acceleration. Therefore to account for physical constraints in the theoretical model, the condition for collision is to come within a minimum allowed distance at some point in time d_{sep}. **Deconfliction difficulty parameter** η is defined as

$$\eta = \frac{V_{\text{max}}^2}{U_{\text{max}} d_{\text{sep}}} \tag{1.66}$$

where V_{max} is the maximum velocity and U_{max} the maximum control (acceleration). This factor is the ratio of the worst-case turning radius to the required separation distance d_{sep} or stopping distance. Let $\tilde{r}_{ij}$ be the distance between aerial robots i and j and $\tilde{v}_{ij}$ the relative velocity between aerial robot i and j. A collision occurs between aerial robots i and j when $||\tilde{r}_{ij}|| < d_{\text{sep}}$, and a conflict occurs between aerial robots i and j if they are not currently in collision, but with zero control input (i.e., with constant velocity) at some future point in time, they will enter into a collision:

$$\min_{t>0} ||\tilde{r}_{ij} - t\tilde{v}_{ij}|| \, d_{sep_{ij}} \tag{1.67}$$

A useful way to check for conflicts is to verify if $|\beta| \geq \alpha$ where

$$\alpha = \arcsin\left(\frac{d_{\text{sep}}}{||\tilde{r}_{ij}||}\right) \qquad \beta = t\tilde{v} - \tilde{r} \tag{1.68}$$

The angle α represents the half width of the collision cone.

The basic algorithm is to first check the separation criterion to see if other aerial robots are close enough to worry about. If so, conflict is checked. If a conflict is found, a maneuver is performed until a conflict-free criterion is reached. Once conflict free, the **deconfliction maintenance controller** is used to keep them that way. The deconfliction maintenance phase also takes the desired control into account, though there is no guarantee that it is followed at all times.

1.3.2.2   Deconfliction Maintenance

This controller allows each aerial robot to use its desired control input unless that input would cause the aerial robot to come into conflict with another. In order to smoothly transition between the desired control and the avoidance control, each aerial robot needs a way to measure how close its velocity vector is to causing a conflict. The first step is to construct a unit vector $\hat{c}$ representing the side of the collision cone nearest $\tilde{v}$. The vector $\hat{c}$ is found by rotating $\tilde{r}$ by α around a vector $q = \tilde{r} \times \tilde{v}$ and normalizing

$$\hat{c} = \frac{\tilde{r}}{\|\tilde{r}\|} \cos \alpha + \left(\frac{q \times \tilde{r}}{\|q\| \|\tilde{r}\|} \right) \sin \alpha \tag{1.69}$$

Next a normal vector e is constructed from the collision cone to the relative vector $\tilde{v}$:

$$e = \left\{ \begin{array}{ll} \tilde{v} & \text{if } \hat{c}^T \tilde{v} \leq 0 \\ (I - \hat{c}\hat{c}^T)\tilde{v} & \text{if } \hat{c}^T \tilde{v} > 0 \end{array} \right\} \tag{1.70}$$

In order to combine the effects of multiple collision cones, the system is decomposed into three component directions, and those directions will be analyzed separately.

The next step is to determine how much control (change in velocity) can be applied in each of these directions T, N, B of the Frenet frame before a conflict forms.

1.3.3   ARTIFICIAL POTENTIAL

The artificial potential field method is an **architecture based on fusion behavior** that combines several behaviors together. This algorithm is used to generate a first- or second-order dynamical system and can often be used to mathematically prove the stability of the emergent behaviors, thereby replacing traditional algorithm validation. By using a steering and repulsive artificial potential field, a swarm of aerial robots can be successfully controlled such that desired patterns are formed, with the approach of **bifurcating potential fields** allowing for a transition between different patterns through a simple parameter switch. Using a first-order dynamical system, the desired swarm velocity field is transformed in guidance commands for forward control speed and heading angle. This section extends the analysis by developing a **bounded bifurcating potential field**. To prevent the guidance velocity commands from becoming saturated, a bounded bifurcating hyperbolic potential is introduced. In addition, the artificial potential field is generalized so that the swarm can get attracted to a variety of different states.

1.3.3.1   Velocity Field

A swarm of N homogeneous aerial robots are considered where x_i, v_i and x_j, v_j define the position and velocity vectors of the ith and j^{th} aerial robots, respectively, and x_{ij} defines the separation distance between the ith and j^{th} robots. Each aerial robot is treated as a particle acted upon by a velocity field as described in

$$v_i = -\nabla_i U^s(x_i) - \nabla_i U^r(x_{ij}) \tag{1.71}$$

where U^s and U^r are defined as the steering and repulsive artificial potential fields $x_i = (x_i, y_i, z_i)^T$ and $x_{ij} = x_i - x_j$. The gradients of the steering and repulsive potentials define a velocity field acting on each aerial robot, where the steering potential is used to control the formation and the repulsive potential is used for collision avoidance and an equally spaced final formation [108].

1.3.3.2 Artificial Potential Field

By using a guidance algorithm, based on classical bifurcation theory, a formation of aerial robots can create autonomous desired patterns, switching between patterns through a simple parameter change. Using **Lyapunov** stability methods, the desired autonomous patterns can be analytically proven as opposed to traditional algorithm validation methods. To ensure the stability of real safety or mission critical systems, it is important to consider the issue of **saturation**. A new **bounded bifurcating potential field** is therefore developed. Considering equation (1.71), through the triangle inequality, the maximum control velocity for the jth aerial robot is

$$|v_i| \le |\nabla_i U^s(x_i)| + |\nabla_i U^r(x_{ij})| \tag{1.72}$$

Therefore, the maximum control velocity each aerial robot will experience is a combination of the maximum gradient of the steering and repulsive potentials.

1.3.3.2.1 Bifurcating Steering Potential

Bifurcating steering potential fields allow the manipulation of the shape of the potential through a simple parameter change. This change alters the stability properties of the potential and thus the emergent patterns that the swarm relaxes into. An example of bifurcating steering potential is shown in equation (1.73) based on the **pitchfork bifurcation equation:**

$$U_i^s(x_i, \mu) = -\frac{1}{2}\mu(|x_i| - r)^2 + \frac{1}{4}\mu(|x_i| - r)^4 \tag{1.73}$$

where μ is the bifurcation parameter and r is a scalar.

As stated in equation (1.71), the gradient of the potential defines a velocity field acting on each aerial robot. Therefore, equation (1.74) shows the velocity field v_i^s due to the pitchfork potential:

$$v_i^s(x_i, \mu) = -\nabla_i U^s(x_i) = \left[\mu(|x_i| - r) + (|x_i| - r)^3\right]\widehat{x}_i \tag{1.74}$$

where $\widehat{x}_i$ denotes a unit vector.

The steering control velocity is unbounded as distance from the equilibrium position increases; therefore the issue of velocity saturation may occur in a real system. To overcome this problem, a hyperbolic potential field can be used. This function has a bound, constant gradient as the distance from the equilibrium position increases and has a smooth shape at the goal. Equation (1.75) shows the hyperbolic potential field $U_i^{s,h}(x_i)$ that can be used to steer the aerial robot:

$$U_i^{s,h}(x_i) = C_h[(|x_i| - r)^2 + 1]^{0.5} \tag{1.75}$$

where C_h controls the amplitude of this function.

To achieve a bifurcating potential field, an additional **exponential steering potential** term is added as shown in equation (1.76):

$$U_i^{s,h}(x_i) = \mu C_e \exp^{-(|x_i|-r)^2/L_e} \tag{1.76}$$

where C_e and L_e control the amplitude and range of the potential, respectively, and μ is the bifurcation parameter.

Combining equations (1.75) and (1.76) results in a bound steering potential U_i^{s,h_e} given in equation (1.77). Again, if $\mu < 0$, the potential will bifurcate into two stable equilibrium positions:

$$U_i^{s,h}(x_i) = C_h[(|x_i| - r)^2 + 1]^{0.5} + \mu C_e \exp^{-(|x_i|-r)^2/L_e} \tag{1.77}$$

The maximum value of the new bound velocity field can be found analytically. First, consider the hyperbolic potential function given in equation (1.75); the velocity $v_i^{s,h}$ due to this term is given in equation (1.78):

$$v_i^{s,h}(x_i) = -\nabla_i U^{s,h}(x_i) = \frac{C_h(|x_i| - r)}{[(|x_i| - r)^2 + 1]^{0.5}} \widehat{x}_i \tag{1.78}$$

Therefore, the maximum velocity due to this term is

$$|v_i^{s,h}|_{max} = C_h \tag{1.79}$$

Next, consider the exponential steering potential given in equation (1.76). The maximum velocity due to the exponential term $|v_i^{s,e}|_{max}$ occurs when $|x_i| = r \pm \sqrt{L_e/2}$ giving

$$|v_i^{s,e}|_{max} = -\sqrt{2}\mu \exp^{0.5} \frac{C_e}{\sqrt{L_e}} \tag{1.80}$$

Depending upon the constants chosen in equation (1.77), the maximum bound velocity will be controlled through either the hyperbolic term or the exponential term. If the hyperbolic term dominates, then $|v_i^{s,e}|_{max} = C_h$. If however, the exponential term dominates, then $|v_i^{s,e}|_{max}$ can be found numerically.

1.3.3.2.2   Repulsive Potential

The repulsive potential is a simple pairwise exponential function that is based on a **generalized Morse potential** as follows:

$$U_i^r = \sum_{j,j\neq i} C_r \exp^{\frac{-|x_{ij}|}{L_r}} \tag{1.81}$$

where $|x_{ij}| = |x_i - x_j|$ and C_r, L_r are constants controlling the amplitude and length scale of the potential, respectively. The repulsive potential is a bound velocity that has a maximum value equal to $\frac{C_r}{L_r}$ that occurs when $x_{ij} = 0$. This would, however, occur when two aerial robots are in the same position and therefore would collide. The realistic maximum control velocity can therefore be expressed as

$$|v_i^r|_{max} = \frac{C_r}{\sqrt{L_r}} \exp^{-|x_{ij}|_{min}/L_r} \tag{1.82}$$

where $|x_{ij}|_{min} = |x_i - x_j|_{min}$ is the minimum separation distance between two aerial robots without colliding.

1.3.3.3   Pattern Formation and Reconfigurability

To allow 3D formation patterns, the following steering potential can be used based on the bound bifurcating potential:

$$U^s = C_h \left[\sqrt{(|x_i| - r)^2 + 1} + \sqrt{\sigma_i^2 + 1} + \mu C_e \exp^{\frac{-(|x_i| - r)^2}{L_e}} \right] \tag{1.83}$$

where

$$|x_i| = \sqrt{x_i^2 + y_i^2 + z_i^2} \tag{1.84}$$

$$\sigma_i = k.x_i \tag{1.85}$$

$$k = (a, b, c)^T \tag{1.86}$$

where (a, b, c) are constants.

The purpose of this steering potential is to drive the aerial robot to a distance r from the origin, with the manipulation of the free parameters μ, a, b, c allowing for different formations. For example, if $a = 1, b = 0$ and $c = 0$, then a ring formation that is parallel to the $y - z$ plane is obtained. In the case of $a = 0, b = 0$ and $c = 0$, each aerial robot will be driven to distance r in the x–y–z plane, thus creating a sphere pattern.

For multiagent scenarios in dynamic environments, an algorithm should be able to replan for a new path to perform updated tasks without any collision with obstacles or other agents during the mission. In [433], the generation of an **intersection-based algorithm** and a **negotiation-based algorithm** for task assignment are proposed. A hierarchical framework for task assignment and path planning in a dynamic environment is presented. The path planning algorithm is augmented with a trajectory replanner based on potential field which solves for a detouring trajectory around other agents or pop-up obstacles. Task assignment is based on the negotiation algorithm; it should generate a feasible solution within a reasonable time for real-time applications. In addition, this algorithm is an **process based on event trigger**, which means that it runs only when an agent sends a message to invoke this algorithm. This is a desirable attribute for decentralized systems running in real time [132].

1.3.4 SYMBOLIC PLANNING

Local communication and control strategies inspired from natural systems such as flocks of birds or schools of fish can lead to useful and predictable global behaviors. Alternatively, such communication and control protocols can be achieved through the use of embedded graph grammars [456,457,460]. In this formalism, aerial robots are represented by vertices in a graph. The edges in the graph represent coordination or communication. Labels on the vertices or edges represent internal states, which can be used, for example, by the communication protocol. A grammatical rule is a rewritten schema of the form $L \longrightarrow R$ where L and R are small graphs. If a subgraph of the global state graph matches L, that subgraph is updated to match R thereby arriving at a (non-deterministic) sequence of graphs representing a trajectory $G_0 \to G_1 \to G_3 \ldots$. To associate motion controllers with each robot requires more machinery:

1. A **workspace position** that denotes the position of the corresponding aerial robot is associated to each vertex in the graph (i.e., the graph is embedded into the workspace).
2. A **continuous graph** is associated to each grammatical rule that states what condition on, for example, the locations of the robots must hold for the rule to be applicable.
3. A **motion controller** is associated with each symbol.

This results in an **embedded graph grammar**. An embedded graph grammar rule essentially allows for statements like "If there are robots i, j and k in the embedded graph such that $\|x_i - x_k\| \approx r$ and the communication subgraph for i, j and k matches the left-hand side of the rule, then change the subgraph according to the right-hand side." Graph grammars have been implemented directly with a variety of distributed robot systems. The advantage of the approach is that each aerial robot in the network can simply be a graph grammar interpreter. Low-level communication protocols can be separated from the task definition (via a graph grammar) by a layer of abstraction. This has also been accomplished with a stochastic version of graph grammars where rules can be thought of as programmable chemical reactions. In both these cases, the level of abstraction afforded by the graph grammar view has enabled straightforward specification and implementation of complex multi-vehicle tasks.

Recently there has been an increased interest in using temporal logics to specify mission plans [122]. Temporal logics are appealing because they provide a formal high-level language in which to describe a complex mission [414]. In addition, tools from model checking can be used to generate a robot path satisfying the specification if such a path exists [418]. However, frequently there are multiple robot paths that satisfy a given specification. In this case, one would like to choose the optimal path according to a cost function. The current tools from model checking do not provide a method for doing this. The linear temporal logic (LTL) specifications and a particular form of cost function provide a method for computing optimal robot paths. Each aerial robot moves along the vertices of an environment modeled as a graph. The main difficulty in moving from a single robot to multiple robots is in **synchronizing** the motion of the aerial robots or in allowing the aerial robots to move **asynchronously**. This section focuses on minimizing a cost function that captures the maximum time between satisfying instances or an optimizing proposition. The cost is motivated by problems in persistent monitoring and in pickup and delivery problems. The proposed solution relies on describing the motion of the group of robots in the environment as a **timed automaton**. This description allows us to represent the relative position between robots. Such information is necessary for optimizing the aerial robot motion.

Problem 14. *Determine an algorithm that takes as input a weighted transition system* **T**, *an LTL formula* ϕ *over its set of atomic propositions and an optimizing proposition* π *and outputs a run* $r_\mathbf{T}$ *minimizing the cost* $C(r_\mathbf{T})$.

The cost function can evaluate the steady state time between satisfying instances of Π. This form of the cost is motivated by persistent monitoring tasks, where the long-term behavior is optimized. A **timed automaton** can be considered. The semantics of the timed automaton can be understood as follows: starting from the initial state q_A^0, the values of all clocks increase at rate one, and the system remains at this state until a clock constraint corresponding to an outgoing transition is satisfied. When this happens, the transition is immediately taken, and the clocks in the clock regions are reset [456]. A timed automaton has a finite set of clock regions R_A, which is the set of equivalence classes of clock valuations induced by its clock constraints G_A. A clock region $r \in R_A$ is a subset of the infinite set of all clock valuations of C_A in which all clock valuations are equivalent in the sense that the future behavior of the system is the same. A clock region is either a corner point $(0, 1)$, an open line segment $x \in [0, 1]$ or an open region $0 \leq x_1 \leq x_2 \leq 1$. The clock region R_A of a timed automaton **A** induces an equivalence relation $\tilde{A}$ over its state space.

For example, to decide when replanning is required and the amount of time needed to calculate and broadcast the new flight paths to the various aerial robots, there are two non-trivial elements to consider. A possible solution is to use two separate controllers on a global level and a local level, respectively (plan globally, react locally). Even in this case, a good level of knowledge about the environment is still required. A distributed control is generally preferable since its non-critical reliance on any specific element can in turn guarantee increased reliability, safety and speed of response to the entire system. More information about this implementation can be found in [456].

The definition of motion description language (MDLn) is extended to a networked system in [429]. This construction MDLn supports **inter-agent specification rules** as well as desired topologies, enabling the specification of **high-level control** for group interactions. In particular, MDLn strings specify **multi-modal executions** of the system through a concatenation of modes. Each mode in the MDLn string is a triple specifying a control law, interrupting conditions and desired network dependencies. In addition to proposing MDLn as a specification language for a networked system, an architecture is also given in which MDLn strings can be effectively parsed and executed in multi-robot applications.

Mission specifications can be expressed in a high-level specification language [416]. In the generalized problem setup, these mission specifications are specified using the LTL language. Then the LTL specification is converted to a set of constraints suitable to a mixed-integer linear programming (MILP) which in turn can be incorporated into two MILP formulations of the standard vehicle routing problem (VRP). Solving the resulting MILP provides an optimal plan that satisfies the given mission specification.

1.4 ASSOCIATION WITH LIMITED COMMUNICATION

1.4.1 INTRODUCTION

The data association problem is addressed in this section. It consists in establishing correspondence between different measurements or estimates of a common element. It is very useful in localization, mapping, exploration and tracking applications. The **nearest neighbor** (NN) and the **maximum likelihood** (ML) are widely used methods which associate each observation with its closest feature in terms of the Euclidean distance. Another popular method is the **joint compatibility branch and bound** which considers the compatibility of many associations simultaneously [21]. The combined constraints data association builds a graph where the nodes are individually compatible associations and the edges relate binary compatible assignments. Over this graph, a maximal common subgraph problem is solved for finding the maximum clique in the graph. **Scan matching** and **iterative closest point** are popular methods for comparison. Other methods like the **multiple hypothesis tracking** and the **joint probabilistic data association** maintain many association hypotheses instead of selecting one of them. In solutions based on submaps, one of them is usually transformed into an observation of another. The local submaps are merged with the global map following a sequence or in a hierarchical binary tree fashion. All the mentioned data association approaches operate on elements from two sets. One set usually contains the current observations, and the other one consists of the feature estimates. Lately, many localization, mapping and exploration algorithms for multi-robot systems have been presented. However, they have not fully addressed the problem of **multi-robot data association**. Many approaches rely on broadcasting all controls and observations measured by the robots. Then, the data association is solved as in a single-robot scenario. In these methods, the problem of inconsistent data association is avoided by forcing a cycle-free merging order. Another work simultaneously considers the association of all local maps. It uses expectation maximization methods for both computing the data association and the final global map. The main limitation of this work is that the data from all sensors needs to be processed together, which implies a centralized scheme or a broadcast method [21].

Some methods rely on centralized schemes, full communication between the aerial robots or broadcasting methods. However, in multi-robot systems, distributed approaches are more interesting. They present a natural robustness to individual failures since there are no centralized nodes. Besides, they do not rely on any particular communication scheme, and they are robust to changes in the topology. On the other hand, distributed algorithms introduce an additional level of complexity in the algorithmic design. Although the aerial robots make decisions based on their local data, the system must exhibit a global behavior.

1.4.2 PROBLEM FORMULATION

A team is composed of $n \in N$ aerial robots. The n aerial robots have communication capabilities to exchange information about the other robots. However, this communication is limited. Let $G_{com} = \{R_{com}, \varepsilon_{com}\}$ be the undirected communication graph. The nodes are the aerial robots, $R_{com} = \{1, \ldots, n\}$. If two aerial robots i, j exchange information, then there is an edge between them, $(i, j) \in \varepsilon_{com}$. Let N_i be the set of neighbors of robot i.

$$N_i = \{j | (i, j) \in \varepsilon_{com}\} \tag{1.87}$$

Each aerial robot i has observed a set S_i of m_i features:

$$S_i = \{f_1^i, \ldots, f_{m_i}^i\} \tag{1.88}$$

It can compute the data association between its own set S_i and the set of its neighbors S_j with $j \in N_i$. However, these data associations are not perfect. There may appear inconsistent data associations relating different features from the same set S_i. If the aerial robots merge their data as soon as they solve the local data association, inconsistent associations cannot be managed since the merging cannot be undone. The goal of this algorithm is to detect and resolve these inconsistent associations before executing the merging. Along this section, the indices i, j, k are used to refer to aerial robots and indices r, r', s, s' to refer to features. The Nth feature observed by the ith aerial robot will be denoted as f_r^i. Given a matrix A, the notation $[A]_{r,s}$ will correspond to the component (r, s) of the matrix, whereas A_{ij} denotes the block (i, j) when the matrix is defined by blocks. Let F be a function that computes the data between any two sets of features S_i, S_j and returns an association matrix $A_{ij} \in \aleph_j^{m_i \times m_j}$ where

$$[A]_{r,s} = \left\{ \begin{array}{cc} 1 & \text{if } f_r^i \text{ and } f_s^j \text{ are associated} \\ 0 & \text{otherwise} \end{array} \right\} \tag{1.89}$$

for $r = 1, \ldots, m_i$ and $s = 1, \ldots, m_j$.

Remark 15. *It is assumed that*

1. *when F is applied to the same set S_i, it returns the identity $F(S_i, S_i) = A_{ii} = I$*
2. *the returned association A_{ij} has the property that the features are associated in a one-to-one way:*

$$\sum_{r=1}^{m_i} [A_{ij}]_{r,s} \leq 1 \qquad \sum_{s=1}^{m_j} [A_{ij}]_{r,s} \leq 1 \tag{1.90}$$

for all $r = 1, \ldots, m_i$ and $s = 1, \ldots, m_j$
3. *aerial robots i and j associate their features in the same way. Given two sets S_i and S_j, it holds that*

$$F(s_i, S_j) = A_{ij} = A_{ji}^T = (F(s_i, S_j))^T \tag{1.91}$$

Most of the existing data association methods satisfy these assumptions.

If F is applied to all pairs of sets of features S_i, S_j for $i, j \in \{1, \ldots, n\}$, its result can be represented by an undirected graph $G_{\text{map}} = \{V_{\text{map}}, E_{\text{map}}\}$. Each node in V_{map} is a feature f_r^i for $i = 1, \ldots, n_j, r = 1, \ldots, m_i$. There is an edge between two features f_r^i, f_s^j if and only if $[A_{ij}]_{r,s} = 1$. If F were providing the true data association, then G_{map}^{gt} would exclusively contain disjoint cliques, identifying features observed by multiple robots. Since F is not perfect, G_{map} is a perturbed version of G_{map}^{gt} that includes additional spurious edges, while it misses others. In addition, due to communication restrictions, the data associations available to this aerial robot are just a subset of G_{map}; the available association graph is termed $G = (V, E)$. It has the same set of nodes than $G_{\text{map}}, V = V_{\text{map}}$. It has an edge between two features f_r^i, f_s^j only if the edge exists in G_{map} and the robots i and j are neighbors in the communication graph. The goal of the presented algorithm is to detect and resolve inconsistencies in this graph $G \subseteq G_{\text{map}}$ in a decentralized fashion.

Definition 16. *Association set An association set is a set of features that form a connected component in G. This set is an inconsistent association or a conflictive set if there exists a path in G between two or more features from the same robot. A feature is inconsistent or conflictive if it belongs to an inconsistent association.*

The number of features in G is $m_{\text{sum}} = \sum_{i=1}^m m_i$; $d \leq m_{\text{sum}}$ is the diameter of G, the length of the longest path between any two nodes in G; the adjacency matrix of G is given by

$$W = \begin{bmatrix} W_{11} & ... & W_{1n} \\ ... & ... & ... \\ W_{n1} & ... & W_{nn} \end{bmatrix} \tag{1.92}$$

where W_{ij} is the block within W for the associations between robots i and j

$$W_{ij} = \left\{ \begin{array}{ll} A_{ij} & \text{for } j \in \{N_i \bigcup U_i\} \\ 0 & \text{otherwise} \end{array} \right\} \tag{1.93}$$

1.4.2.1 Decentralized Resolution of Inconsistent Association

The resolution of inconsistent associations consists of deleting edges from G so that the resulting graph is conflict free.

Definition 17. *Conflictive set* *Let c denote the number of conflictive sets in G. A conflictive set is detectable by an aerial robot i if there exists a $r \in \{1,...,m_i\}$ such that $f_r^i \in C$. The subteam that detects a conflictive set is $R \subseteq R_{\text{com}}$. The number of features from each aerial robot $i \in R$ involved in C is $\tilde{m}_i$. $\tilde{G}$ is conflict free if $c = 0$*

All the edges whose deletion transforms G into a conflict-free graph belong to any of the C conflictive sets of G. Since the conflictive sets are disjoint, they can be considered separately. The resolution of the problem consists of partitioning C into a set of disjoint conflict-free components C_q such that

$$\bigcup_q C_q = C \qquad C_q \bigcap C_{q'} = \emptyset \tag{1.94}$$

for all $q, q' = 1, 2, ...$

Theorem 18. *Let R be the set of aerial robots that detect C. Let i_* be the robot with the most features involved in C:*

$$i_* = argmax \tilde{m}_{i_*} \tag{1.95}$$

The number of conflict-free components into which C can be decomposed is lower bounded by $\tilde{m}_{i_*}$.

The Resolution algorithm constructs $\tilde{m}_{i_*}$ conflict-free components using a strategy close to a **breadth-first search** (BFS) tree construction. Initially, each aerial robot i detects the conflictive sets for which it is the root, using its local information $X_{i1}(t_i),...,X_{in}(t_i)$. The root aerial robot for a conflictive set is the one with the most inconsistent features involved. In case two aerial robots have the same number of inconsistent features, the one with the lowest robot identity is selected. Then, each aerial robot executes the resolution pseudocode given in Algorithm 1. The root creates $\tilde{m}_{i_*}$ components and initializes each component C_q with one of its features $f^{i_*} \in C$. Then, it tries to add to each component C_q the features directly associated to $f^{i_*} \in C_q$. If f_s^j is assigned to C_q, for all f_r^i such that $[W_{ij}]_{r,s} = 0$, robot j sends a component request message to robot i. When aerial robot i receives it, it may happen that

- **Case a:** f_r^i is already assigned to C_q
- **Case b:** f_r^i is assigned to a different component
- **Case c:** Another feature f_r^i is already assigned to C_q
- **Case d:** f_r^i is unassigned and no feature in i is assigned to C_q.

In **case a**, f_r^i already belongs to the component C_q and robot i does nothing. In **cases (b) and (c)**, f_r^i cannot be added to C_q, and aerial robot i deletes the edge $[W_{ij}]_{r,s}$ and replies with a reject message to robot j; when j receives the reject message, it deletes the equivalent edge $[W_{ij}]_{s,r}$. In **case (d)**, the aerial robot i assigns its feature f_r^i to the component C_q, and the process is repeated.

Algorithm 1 Resolution Algorithm

---
1. **Initialization**
2. For each collective set C for which i is root ($i = i_*$), do
3. Create $\tilde{m}_{i_*}$ components
4. Assign each inconsistent feature $f_r^{i_*} \in C$ to a different component C_q
5. Send component request to all its neighboring features.
6. endfor
7. **Algorithm**
8. For each component request from f_s^j to f_r^i, do
9. If case (b) or case (c), then
10. $[W_{ij}]_{r,s} = 0$
11. Send reject message to j
12. If case (d), then
13. Assign f_r^i to the component
14. Send component request to all its neighboring features.
15. endif
16. endfor
17. For each component rejected from f_s^j to f_r^i, do
18. $[W_{ij}]_{r,s} = 0$.
19. endfor

Theorem 19. *Let us consider that each aerial robot $i \in R_{\text{com}}$ executes the decentralized solution algorithm (Algorithm 1) on G obtaining G':*

1. *After $t = n$ iterations, n new features are added to any component C_q, and the algorithm finishes.*
2. *Each obtained C_q is a connected component in G'*
3. *C_q is conflict free.*
4. *C_q contains at least two features.*

for all $q \in \{1, \ldots, \tilde{m}_{i_*}\}$ and all conflictive sets.

This section has presented a technique to match several sets of features observed by a team of robots in a consistent way under limited communication.

1.4.3 GENETIC ALGORITHMS

One of the emerging approaches to manage cooperative task assignments and path planning for aerial robot is the use of genetic algorithms [114]. In the aerial robot example, a set of assigned target's identification (ID) combined with the ID of the selected path that can be used to reach the corresponding target would represent a chromosome. During the execution of a genetic algorithm, the objective is to achieve an appropriate configuration of the mentioned target/path data set, and this objective might be achieved with a series of genetic crossover operations [149].

Genetic algorithms are fundamentally combinatorial optimization tasks, and as a part of the solution, the **traveling salesman problem** (TSP) can be used. When it is used as a subsolution in the aerial robot context, they become the equivalents of targets. While it is a simple problem, it is well known that no fast solution exists and the time required to solve the problem using any currently known solution increases very quickly as the size of the problem (number of aerial robots and number of targets) grows. However, possible techniques such as **approximation**, **randomization** or **heuristics** might be used. All of these methods have inherently different constraints, running times and success characteristics. The VRP is static in the sense that routes of vehicles are computed assuming that no new demand arrives. Despite its frequent occurrence in practical problems, the VRP is often a static approximation to problems that are in reality dynamic, and therefore sometimes it fails to be an appropriate model. In these applications, demands often arrive sequentially in time, and planning algorithms should provide policies (in contrast to preplanned routes) that prescribe how the routes should evolve as a function of those inputs that evolve in real time. Dynamic demands varying over time, add **queuing phenomena** to the combinatorial nature of the VRP. For example, in a surveillance mission, where a team of aerial robots must ensure continued coverage of a certain area, as events occur, i.e., as new targets are detected by on-board sensors or other assets, these aerial robots must proceed to the location of the new event and provide close-range information about the target. Each request for close-range observation might also require an on-site service time (example a time interval to identify the target) which might be known a priori only through prior statistics. The aerial robot mission control is then a continuous process of detecting targets, planning routes and sending aerial robots. In such a dynamic and stochastic setting, the routing process might rely on a priori statistics about the future stream of targets, and stability of the overall system is an additional issue to be considered [14,436].

The assignment and scheduling problem can be addressed in a more general framework where new demands for service are generated over time by a stochastic process. The most general model for VRPs that have both a dynamic and a stochastic component is the *M*-**vehicle dynamic traveling repairman** (MDTRP). In this framework, *M* vehicles operating in a bounded environment and traveling with bounded velocity must service demands whose time of arrival, location and on-site service are stochastic: for example, the inter-arrival times might be exponential random variables; the demand's locations might be uniformly distributed within the workspace; and the time intervals required to identify the targets might be exponential random variables as well. The objective is to find a policy to service demands over an infinite horizon that minimizes the expected system time (wait plus service) of the demands. The best previously known control policies for the MDTRP rely on centralized task assignment and are not robust against changes in the environment. Therefore, they are of limited applicability in scenarios involving networks of autonomous vehicles operating in time-varying environments [441,461].

Interval analysis is used to incorporate uncertainties in decision models and in group decision-making to aggregate the preferences of all group members in a single model. The state variables and the parameters are allowed to vary within given ranges. One-on-one importance intervals are calculated with nonlinear programming by maximizing and minimizing the value function subject to the feasible ranges; then the prioritization approach gives an interval for the overall importance index of each aerial robot by aggregating the importance intervals. This allows the ranking of the state information of the aerial robot in an order of importance. The intervals can also be used in sensitivity analysis in which the impacts of the different factors affecting the prioritization can be found out.

1.4.4   GAMES THEORY REASONING

The aircraft is a huge system; many factors such as classical mechanics, aerodynamics, structural mechanics, control theory, etc. need to be considered during its design. So the configuration inference and decision-making process based on experts or expert system is very complicated, and the efficiency is often degraded [348]. When the decision is made in several fields, the game is called multi-dimensional game theory [445]. The gaming model including two players is $G = (S_1, S_2, u_1, u_2)$ where $S_i, i = 1, 2$ is the strategy space of each player and $u_i, i = 1, 2$ is the payoff function of each player. According to the factors and characters of game, the gaming process can be depicted in several types. Based on different players, the cooperative, non-cooperative and leader/follower protocols can be introduced [427].

1.4.4.1   Cooperative Protocol

A game is called a **cooperative game** when there is a binding agreement or it is a non-cooperative game. A cooperative game emphasizes the cooperation and the problem of how to assign the payoff from cooperation. A cooperative game consists of two elements: a set of players N and a characteristic function v specifying the value created by different subsets of the players in the game. The characteristic function v is a function that associates with every subset S of N a number denoted by $v(S)$. The $v(S)$ is interpreted as the value created when the members of S come together and interact. In this cooperative protocol, both players have information about each other, and they work together to find a **Pareto** solution. A pair (x_{1p}, x_{2p}) is Pareto optimal if no other pair (x_1, x_2) exists such that

$$u_i(x_1, x_2) \leq u_i(x_{1p}, x_{2p}) \quad i = 1, 2 \tag{1.96}$$

$$u_j(x_1, x_2) < u_j(x_{1p}, x_{2p}) \quad i = 1, 2 \tag{1.97}$$

for at least one $j = 1, 2$.

1.4.4.2   Non-Cooperative Protocol

This protocol occurs when the full coalition of players is not possible being emerged due to organization, information or process barriers. Players must make decisions by assuming that the choices of other decision-makers are made. In an iterative approach, the final **Nash** equilibrium solution will be obtained. A strategy pair (x_{1N}, x_{2N}) is a Nash solution if

$$u_1(x_{1N}, x_{2N}) = \max_{x_1} u_1(x_{1N}, x_{2N}) \tag{1.98}$$

$$u_2(x_{1N}, x_{2N}) = \max_{x_2} u_2(x_1) \tag{1.99}$$

The Nash equilibrium solution has the property of being the fixed points of two subsets of the feasible space $(x_{1N}, x_{2N}) \in S_{1N}(x_{2N}), S_{2N}(x_{1N})$ where

$$S_{1N} = \{x_{1N} | u_1(x_{1N}, x_{2N}) = \max_{x_1} u_1(x_1, x_2)\} \tag{1.100}$$

$$S_{2N} = \{x_{2N} | u_2(x_{1N}, x_{2N}) = \max_{x_2} u_2(x_1, x_2)\} \tag{1.101}$$

are called the **rational reaction sets** (RRSs) of the two players. The RRS of a player is a function that embodies his reactions to the decisions made by the other players.

1.4.4.3   Leader/Follower Protocol

When one player dominates another, they have a **Leader/follower relationship**. This is a common occurrence in a design process when one discipline dominates the design or in a design process that involves a sequential execution of interrelated disciplinary process. P1 is said to be the leader if she declares her strategy first by assuming that P2 behaves rationally. Thus, the model of P1 as a leader is as follows:

Max $u_1(x_1, x_2)$
subject to $x_2 \in S_{2N}(x_1)$
where $S_{2N}(x_1)$ is the rational reaction set of P2.

For exactly describing game theory, the above-mentioned protocols are represented by functions. The normal game is usually represented by a matrix which shows the players, strategies and payoffs.

1.5   MULTIAGENT DECISION-MAKING UNDER UNCERTAINTY

Dealing with uncertainties in complex dynamic environments is a basic challenge to the operation of real-world robotic systems [409]. Such a system must be able to monitor the state of its components and environment in order to form informed plans of intelligent action. Each aerial robot needs to make **inferences** about the other aerial robots as well, possibly under limited communication, over a course of repeated interactions [145].

Decision theoretic models for planning under uncertainty have been studied extensively in artificial intelligence and operations research since the 1950s. The Markov decision process (MDP) and partially observable Markov decision process (POMDP) in particular have emerged as useful frameworks for centralized decision-making in fully observable stochastic environments. An even more general problem results when two or more agents have to coordinate their actions [420]. If each agent has its own separate observation function, but the agents must work together to optimize a joint reward function, the problem that arises is called **decentralized control of a partially observable system**. The problem is particularly hard because each individual agent may have different partial information about the other agents and about the state of the world.

1.5.1   DECENTRALIZED TEAM DECISION PROBLEM

Decentralized control of multiple decision-makers under uncertainty arises in multi-robot coordination [77]. Such problems must be treated as decentralized decision problems because each aerial robot may have different partial information about the other aerial robots and about the state of the world. It has been shown that these problems are significantly harder than their centralized counterparts, requiring new formal models and algorithms to be developed. A key focus on modeling other aerial robots is concerned with the prediction of other aerial robots (exploiting the internal models about the other's preferences, strategies, intentions and so on). Then, the modeler aerial robot can use this prediction in order to behave in the best way according to its preferences. Research on modeling other aerial robots has been approached from different perspectives:

- An approach for tracking recursive aerial robot models based on a **plan recognition task**.
- **Recursive modeling method** uses nested models of other agents combining game-theoretic and decision-theoretic mechanisms.
- **Use of influence diagrams** for learning models about other agents.
- **Bayesian updating** of aerial robot models with the formalism of recursive modeling method.

1.5.1.1   Bayesian Strategy

An aerial robot model is a vector which records a probability distribution of the actual character of the modeled aerial robot [421]. Each aerial robot has two basic models about every other aerial robot a. The first one is the **role model** $r_a = (r_1, \ldots, r_n)$ r_i is the probability that aerial robot a has the particular role i and n is the amount of different predefined roles. The second model is the **strategy model** $s_a = (s_1, \ldots, s_m)$, s_i is the probability that aerial robot a has strategy i and m is the amount of different predefined strategies. Since independence between roles and strategies is assumed, a new combined model for every other aerial robot can be constructed: the **personality model**. This model is a 2D matrix rs_a such that $rs_a(i, j) = r_a(i)s_a(j)$. A decision theoretic strategy that takes explicit advantage of knowing the other's model can be constructed.

1.5.1.2   Semi-Modeler Strategy

This strategy tells the aerial robot to choose the slot which maximizes its expected utility based on predefined fixed models about the other aerial robots. A semi-modeler aerial robot is assumed to already have models about the others, and its strategy just uses these probabilities models to choose the action that maximizes its expected utility. The models are given to the semi-modeler aerial robot at the beginning of the game, and they never change during the game. It is also important to note that the given models are not necessarily correct models about the others. In order to build a modeler agent, model construction is required. Let's define a modeler strategy that uses a Bayesian updating mechanism in order to build the other's models in an incremental and iterative way [148].

1.5.1.2.1   Bayesian Modeler Strategy

An aerial robot using this strategy incrementally builds models about the others using a **Bayesian belief updating approach** and chooses the action which maximizes its expected utility. A Bayesian modeler aerial robot does not have any information about the others. However, the set of predefined roles and strategies are public knowledge. At the beginning, the modeler aerial robot can behave as a semi-modeler aerial robot with equi-probable models about the others. With no other knowledge about the others, it is reasonable to start with equi-probable probability distributions of the possible traits about the others. Then the modeler aerial robot can start updating those models based on the others' behavior. This aerial robot builds models about the other agents in an incremental and iterative way, updating those models after each round during the whole game. All the probabilities of each model are incrementally updated, trying to reach the actual character of the aerial robot being modeled. The probability $\mathrm{Prob}(\mathrm{per}_a(i,j))$ that an aerial robot a has the personality resulting from combining role i and strategy j is precisely the value $rs_a(i,j)$ in matrix rs_a, and the equation used to update each personality model can be written as

$$rs_a(i,j) = \frac{\mathrm{Prob}\left(prob_a(s_a)|\mathrm{per}_a(i,j)\right)\mathrm{Prob}\left(\mathrm{per}_a(i,j)\right)}{\mathrm{sum}_x \sum_y \mathrm{Prob}\left(\mathrm{prob}_a(s_a)|per_a(x,y)\right)\mathrm{Prob}\left(\mathrm{per}_a(x,y)\right)} \tag{1.102}$$

The prior probabilities $\mathrm{Prob}(\mathrm{per}_a(i,j))$ are taken from the last recorded value $rs_a(i,j)$. The conditional probabilities $\mathrm{Prob}(\mathrm{pro}_a(s_a)|\mathrm{per}_a(i,j))$ can be calculated from the known calendar density and the known aerial robot behavior due to the personality $per_a(i,j)$. Thus, the Bayesian modeler is able to get all the posterior probabilities from the calculated conditional probabilities and the known prior probabilities. Then this rs matrix is updated with these new probabilities in order to be used as prior probabilities in the following round.

In [133], the problem of information consensus in a team of networked agents is considered by presenting a generic **consensus method** that allows agreement to a Bayesian fusion of uncertain local parameter estimates. In particular, the method utilizes the concept of conjugacy of probability distributions to achieve a steady state estimate consistent with a Bayesian combination of each agent's local knowledge without requiring complex channel filters or being limited to normally distributed uncertainties. This algorithm termed **hyperparameter consensus** is adaptable to some local uncertainty distributions within the exponential family and should converge to a Bayesian fusion of local estimates with some standard assumptions on the network topology.

1.5.1.2.2   Reinforcement Modeler Strategy

Reinforcement modeler strategy is similar to the Bayesian modeler strategy. It learns the others' models and exploits them with a greedy decision theoretic approach. The idea is to keep vectors of values instead of probabilistic vectors. Three models exist for each aerial robot a. The reinforcement modeler aerial robot constructs the state signal:

- **Discrete time steps:** these correspond to the rounds of each game.
- **States:** at each time step t, the aerial robot receives a representation of the environment $s_t \in S$, where S is the set of possible states.
- **Actions:** based on the state, the learner will choose an action $a_t \in A(s_t)$ from a set of possible actions $A(s_t)$ available in state s_t.
- **Rewards:** as a consequence of its action, in this case, the reinforcement modeler aerial robot will have $n \cdot m$ rewards $r_{i,j,t+1}$ for every other aerial robot i of the n aerial robots with m different personalities. These rewards are not directly given by the referee aerial robot, but they are computed by the reinforcement modeler using the information of each state.

After updating all the personalities values, the reinforcement modeler aerial robot decomposes these models into two separate role and strategy models for each aerial robot. This decomposition is the reverse process of the personality model composition.

1.5.1.3 Communication Models

To formalize the problem of decentralized control of multiple agents, a number of different formal models have recently been introduced. In all these models, at each step, each aerial robot takes an action, a state transition occurs, and each aerial robot receives an individual observation, but the reward generated by the environment is the same for all aerial robots. Thus, the focus of this section is on cooperative systems in which each aerial robot wants to maximize a joint global reward function. In contrast, non-cooperative multi-aerial-robot systems, such as **partially observable stochastic games** (POSGs) allow each aerial robot to have its own private reward function. Solution techniques for cooperative multi-aerial-robot systems are quite different from the algorithms described in this section and often rely on techniques from game theory [446].

MDPs describe a mathematical formalism, decentralized Markov decision process (**DEC-MDP**), to model multi-robot systems acting under uncertainty and solve distributed decision-making problems. Their use for planning tasks in real applications, however, presents some difficulties.

Definition 20. *DEC-POMDP: A DEC-POMDP is a tuple $\langle I, S, \{A_i\}, P, \{\Omega_i\}, O, R, T \rangle$ where*

1. *I is a finite set of agents indexed $1, ..., n$*
2. *S is a finite set of states, with distinguished initial state s_0*
3. *A_i is a finite set of actions available to aerial robot i and $\vec{A}_i = \otimes_{i \in I} A_i$ is the set of joint actions, where $\vec{a} = (a_1, ..., a_n)$ denotes a joint action*
4. *$P : S \times \vec{A} \rightarrow \Delta S$ is a Markovian transition function and $P(s'|s, \vec{a})$ denotes the probability that after taking joint action $\vec{a}$ in state s, a transition to state s' occurs*
5. *Ω_i is a finite set of observations available to aerial robot i and $\vec{\Omega} = \otimes_{i \in I} \Omega_i$ is the set of joint observation where $\vec{o} = (o_1, ..., o_n)$ denotes a joint observation*
6. *$O : \vec{A} \times S \rightarrow \Delta \vec{\Omega}$ is an observation function and $O(\vec{o}|\vec{a}, s')$ denotes the probability of observing joint observation $\vec{o}$ given that joint action $\vec{a}$ was taken and led to state s', $s' \in S, \vec{a} \in \vec{A}, \vec{o} \in \vec{\Omega}$*
7. *$R : \vec{A} \times S \rightarrow \mathbb{R}$ is a reward function and $R(\vec{a}, s')$ denotes the reward obtained after joint action $\vec{a}$ was taken and a state transition to s' occurred*
8. *if the DEC-POMDP has a finite horizon, that horizon is represented by a positive integer T.*

The usual model of DEC-MDP does not take into account constraints on the execution of tasks. In addition, the complexity of their resolution is such that it is difficult to determine an optimal solution except for small problems. Some framework explicitly models the communication actions of the agents, and others subsume them under the general action sets. Each approach has different advantages and disadvantages depending on the focus of the analysis. A second aspect that differentiates the models is whether they use an implicit or explicit representation of agent's belief. The next four pages are mainly based on [446].

1.5.1.3.1 Models without Explicit Communication

Some useful definitions are introduced in the sequel.

Definition 21. *Local Policy for DEC-POMDP: A local policy δ_i for aerial robot i is a mapping from local histories of observations $\bar{o}_i = (o_{i_1}, ..., o_{i_\ell})$ over Ω_i to actions in A_i, $\delta_i : \Omega_i* \rightarrow A_i$.*

Definition 22. *Joint Policy for a DEC-POMDP: A joint policy $\delta = \langle \delta_1, ..., \delta_\ell \rangle$ is a tuple of local policies, one for each aerial robot.*

Solving a DEC-POMDP means finding a joint policy that maximizes the expected total reward.

Definition 23. *Solution Value for a Finite Horizon DEC-POMDP: For a finite horizon problem, the aerial robots act for a fixed number of steps, which is called the horizon and is denoted by T.*

The value of a joint policy δ for a finite DEC-POMDP with initial state s_0 is

$$V^{\delta}(s_0) = E\left[\sum_{t=0}^{T-1} R(\vec{a}_t, s_t)|s_0, \delta\right] \tag{1.103}$$

When the aerial robots operate over an unbounded number of time steps or the time horizon is so large that it can best be modeled as being infinite, the infinite horizon discounted performance criterion is used. A discount factor γ^t is used to weigh the reward collected in t time steps into the future.

Definition 24. *Solution Value for an Infinite Horizon DEC-POMDP: The value of a joint policy δ for an infinite horizon DEC-POMDP with initial state s_0 and discount factor $\gamma \in [0,1)$ is*

$$V^{\delta}(s_0) = E\left[\sum_{t=0}^{\infty} \gamma^t R(\vec{a}_t, s_t)|s_0, \delta\right] \tag{1.104}$$

Definition 25. *Perfect Recall: An aerial robot has perfect recall if it has access to all of its received information. (This includes all local observations as well as messages from other agents.)*

Definition 26. *Multiagent Team Decision Problem: A multiagent team decision problem (MTDP) is a tuple $\left\langle \alpha, S, \vec{A}_\alpha, P, \vec{\Omega}_\alpha, \vec{O}_\alpha, \vec{B}_\alpha, R, T \right\rangle$ where*

1. *α is a finite set of agents indexed $1,...,n$.*
2. *$S = \Xi_1 \times \cdots \times \Xi_m$ is a set of world states, expressed as a factored representation (a cross product of separate features) with distinguished initial state s_0.*
3. *$\{A_i\}_{i\in\alpha}$ is a set of actions for each aerial robot i, implicitly defining a set of combined actions $\vec{A}_\alpha = \prod_{i\in\alpha} A_i$.*
4. *$P : S \times \vec{A}_\alpha \times S \to [0,1]$ is a probabilistic distribution over successor states, given the initial state s_0 and a joint action:*

$$P(s,a,s') = \text{Prob}\left(s^{t+1} = s'|s^t = s, A_\alpha^t = a\right) \tag{1.105}$$

5. *$\{\Omega\}_{i\in\alpha}$ is a set of observations that each aerial robot i can experience implicitly defining a set of combined observations $\vec{\Omega}_\alpha = \prod_{i\in\alpha} \Omega_i$.*
6. *$\vec{O}_\alpha$ is a joint observation function, modeling the probability of receiving a joint observation ω after joint action $\vec{a}$ was taken and a state transition to s' occurred i.e., $(\vec{O}_\alpha(s',\vec{a},\omega)) = \text{Prob}\left(\vec{\Omega}_\alpha^{t+1} = \omega|S^{t+1} = s', \vec{A}_\alpha^t = \vec{a}\right)$.*
7. *$\vec{B}_\alpha$ is the set of possible combined belief states. Each aerial robot $i \in \alpha$ forms a belief state $b_i^t \in B_i$ based on its observations seen through time t, where B_i circumscribes the set of possible belief states of aerial robot i. This mapping of observations to belief states is performed by a state estimator function under the assumption of perfect recall. The resulting combined belief state is denoted $\vec{B}_\alpha = \prod_{i\in\alpha} B_i$. The corresponding random variable b_α^t represents the agent's combined belief state at time t.*
8. *$R : S \times \vec{A}_\alpha \to \mathbb{R}$ is a reward function representing a team's joint preferences.*
9. *If the MTDP has a finite horizon, that horizon is represented by a positive integer T.*

Definition 27. *Domain-Level Policy for an MTDP: The set of possible domain-level policies in an MTDP is defined as the set of all possible mappings from belief states to actions, $\pi_{iA} : B_i \to A_i$.*

Definition 28. *Joint Domain-level Policy for an MTDP: A joint domain-level policy for an MTDP $\pi_{\alpha A} = \langle \pi_{1A},...,\pi_{nA} \rangle$ is a tuple of domain-level policies, one for each agent.*

Solving an MTDP means finding a joint policy that maximizes the expected global reward.

1.5.1.3.2   Models with Explicit Communication

Both models presented above have been extended with explicit communication actions. In the resulting two models, the interaction between the aerial robots is a process in which aerial robots perform an action, then observe their environment and send a message that is instantaneously received by the other aerial robots (no delays in the system). Both models allow for a general syntax and semantics for communication messages. The aerial robots need to have conventions about how to interpret these messages and how to combine this information with their own local information. One example of a possible communication language is $\Sigma_i = \Omega_i$, where the aerial robots simply communicate their observations.

Definition 29. DEC-POMDP-COM: *A DEC-POMDP with communication (DEC-POMDP-COM) is a tuple*
$$\langle I, S, \{A_i\}, P, \{\Omega_i\}, O, \Sigma, C_\Sigma, R_A, R, T \rangle \text{ where}$$

1. $I, S, \{A_i\}, P, O, T$ *are defined as in the DEC-POMDP.*
2. Σ *is the alphabet of communication messages,* $\sigma_i \in \Sigma$ *is an atomic message sent by aerial robot i and* $\vec{\sigma} = (\sigma_1, \ldots, \sigma_n)$ *is a joint message, i.e., a tuple of all messages sent by the agents in one time step. A special message belonging to* Σ *is the null message* ε_σ *which is sent by an aerial robot that does not want to transmit anything to the others.*
3. $C_\Sigma : \Sigma \to \mathbb{R}$ *is the message cost function.* $C_\Sigma(\sigma_i)$ *denotes the cost for transmitting atomic message* σ_i *When agents incur no cost for sending a null message,* $C_\Sigma(\varepsilon_\sigma) = 0$.
4. $R_A : \vec{A} \times S \to \mathbb{R}$ *is the action reward function identical to the reward function in a DEC-POMDP, i.e.,* $R_A(\vec{a}, s')$ *denotes the reward obtained after joint action* $\vec{a}$ *was taken and a state transition to* s' *occurred.*
5. $R : \vec{A} \times S \times \vec{\Sigma} \to \mathbb{R}$ *denotes the total reward function defined via* R_a *and* C_Σ: $R(\vec{a}, s', \vec{\sigma}) = R_A(\vec{a}, s') - \sum_{i \in I} C_\Sigma(\sigma_i)$.

Solving a DEC-POMDP-COM means finding a joint policy that maximizes the expected total reward over either infinite or finite horizon.

Definition 30. Local Policy for Action for a DEC-POMDP-COM: *A local policy for action* δ_i^A *for aerial robot i is a mapping from local histories of observations* $\bar{o}_i$ *over* Ω_i *and histories of messages* $\bar{\sigma}_j$ *received* $j \neq i$ *to control actions in* A_i, $\delta_i^A : \Omega_i^* \times \vec{\Sigma} \to A_i$.

Definition 31. Joint Policy for a DEC-POMDP-COM: *A joint policy* $\delta = (\delta_1, \ldots, \delta_\ell)$ *is a tuple of local policies, one for each aerial robot, where each* δ_i *is composed of the communication and action policies for aerial robot i.*

Definition 32. Collective Observability: *A DEC-POMDP is jointly fully observable if the n-tuple of observations made by all the aerial robots uniquely determines the current global state:* $O(\vec{o}|\vec{a}, s') > 0 \Rightarrow Prob(s'|\vec{o}) = 1$. *Joint–full observability = collective observability.*

In a DEC-MDP, each aerial robot alone still only has partial observability and does not have full information about the global state.

1.5.1.3.3   I-POMDP Model

An interactive partially observable Markov decision process (I-POMDP) extends the POMDP model to the multi-aerial-robot case. Now, in addition to a belief over the underlying system state, a belief over the other agents is also maintained. To model this richer belief, an interactive state space is used. A belief over an interactive state subsumes the belief over the underlying state of the environment as well as the belief over the other aerial robots.

Remark 33. *Even if just two aerial robots are considered, expressing a belief over another aerial robot might include a belief over the other agent's belief. As the other aerial robot's belief might also include a belief over the first aerial robot's belief, this technique leads to a nesting of beliefs which makes finding optimal solutions within this model problematic.*

Definition 34. *Frame: A frame of an aerial robot is*
$\hat{\theta}_i = \{\{A_i\}, \{\Omega_i\}, T_i, O_i, R_i, OC_i\}$ *where*

1. A_i *is a set of actions aerial robot i can execute.*
2. Ω_i *is a set of observations the aerial robot i can make.*
3. T_i *is a transition function defined as* $T_i : S \times A_i \times S \to [0,1]$.
4. O_i *is the aerial robot observation function defined as* $O_i : S \times A_i \times \Omega_i \to [0,1]$.
5. R_i *is a reward function representing aerial robot i's preferences defined as* $R_i : S \times A_i \to \mathbb{R}$.
6. OC_i *is the aerial robot's optimality criterion. This specifies how rewards acquired over time are handled. For a finite horizon, the expected value of the sum is commonly used. For an infinite horizon, the expected value of the discounted sum of rewards is commonly used.*

Definition 35. *Type: The type of an aerial robot is* $\theta_i = (b_i, \hat{\theta}_i)$ *where*

- b_i *is agent's state of belief, an element of* $\Delta(S)$, *where S is the state space.*
- $\hat{\theta}_i$ *is agent's i frame.*

A type is an intentional model

Assuming that the aerial robot is Bayesian rational, given its type θ_i, the set of optimal actions denoted by $\mathrm{OTP}(\theta_i)$ can be computed.

Definition 36. *Models of an Aerial Robot: The set of possible models M_j of aerial robot j consists of the sub-intentional models SM_j and the intentional models IM_j. Thus, $M_j = SM_j \cup IM_j$. Each model $m_j \in M_j$ corresponds to a possible belief about the aerial robot, i.e., how aerial robot j maps possible histories of observations to a distribution of actions.*

- **Sub-intentional models** *SM_j are relatively simple as they do not imply any assumptions about the agent's belief. Common examples are no-information models and fictitious play models, both of which are history independent. A more powerful example of a sub-intentional model is a finite state controller.*
- **Intentional models** *IM_j are more advanced because they take into account the agent's beliefs, preferences and rationality in action selection. Intentional models are equivalent to types.*

I-POMDPs generalize POMDPs to handle the presence of, and interaction with, other aerial robots. This is done by including the types of the other aerial robots into the state space and then expressing a belief about the other agent's types.

Definition 37. *I-POMDP: An I-POMDP of aerial robot i is a tuple* $\langle IS_i, A, T_i, \Omega_i, O_i, R_i \rangle$ *where*

1. *IS_i is a set of interactive states defined as $IS_i = S \times M_j$ where S is the set of states of the physical environment and M_j the set of possible models of aerial robot j. Thus, aerial robot i's belief is now a probability distribution over states of the environment and the models of the other aerial robot $b_i \in \Delta(IS_i) = b_i \in (S \times M_j)$.*
2. *$A = A_i \times A_j$ is the set of joint actions of all aerial robots.*
3. *$T_i : S \times A \times S \to [0,1]$ is the transition function. Actions can only change the physical state and thereby may indirectly change the other robot's beliefs via received observations.*

4. Ω_i is the set of observations the aerial robot i can make.

5. $O_i : S \times A \times \Omega_i \to [0,1]$ is an observation function. Aerial robots cannot observe others' models directly.

6. $R_i : IS_i \times A \to \mathbb{R}$ is the reward function; it allows the aerial robots to have preferences depending on the physical states as well as on the other agents' models, although usually only the physical state matters [413].

1.5.2 ALGORITHMS FOR OPTIMAL PLANNING

1.5.2.1 Multiagent A* (MAA*): A Heuristic Search Algorithm for DEC-POMDP

This algorithm is based on the widely used A* and performs best first search of possible joint policies. It uses the following representation: q_i^t is a depth-t policy tree for aerial robot i and $\delta^t = (q_1^t, ..., q_n^t)$ is a policy vector of trees. Let $V(s_0, \delta)$ denote the expected value of executing policy vector δ from state s_0. Finding the optimal joint policy is thus identical to finding

$$\delta^{*T} = \text{argmax}_{\delta^t} V(s_0, \delta) \tag{1.106}$$

The algorithm searches in the space of policy vectors, where nodes at level t of the search tree correspond to partial solutions of the problem, namely policy vectors at horizon t. But, not all nodes at every level are fully expanded. Instead, a heuristic function is used to evaluate the leaf nodes of the search tree. The node with the highest heuristic estimate is expanded in each step. To compute the heuristic estimate of a search node, the evaluation function is decomposed into two parts: an exact evaluation of the partial solution (the policy vector up to the current level) and a heuristic estimate of the remaining part: the **completion**. The value of any depth-t policy vector also decouples into two parts:

$$V\left(s_0, \{\delta^t, \Delta^{T-1}\}\right) = V\left(s_0, \delta^t\right) + V\left(\Delta T - t | s_0, \delta^t\right) \tag{1.107}$$

The value of the completion depends on the state distribution that is reached after executing policy vector δ^t in state s_0. The value estimate of state s_0 and policy δ^t is

$$F(s_0, \delta^t) = V(s_0, \delta^t) + H^T(s_0, \delta^t) \tag{1.108}$$

For the heuristic search to be optimal and complete, the function H must be admissible, i.e., an overestimate of the exact value of the optimal completion of policy vector δ^t

$$\forall \Delta^{T-t} : H^{T-t}(s_0, \delta^t) \geq V\left(\Delta^{T-t} | s_0, \delta^t\right) \tag{1.109}$$

The algorithm must have an admissible heuristic function that is efficiently computable and as close as possible to the true value (to maximize the amount of pruning).

$$H^{T-t}(s_0, \delta^t) = \sum_{s \in S} \text{Prob}\left(s | s_0, \delta^t\right) h^{T-t}(s) \tag{1.110}$$

where $\text{Prob}(s|s_0, \delta^t)$ is the probability of being in state s after executing the policy tree vector δ from s_0. $h^t(s)$ is an optimistic value function heuristic for the expected sum of rewards when executing the best vector of depth-t policy trees from state $h^t(s) \geq V^t * (s)$. Among the heuristics that simulate revealing the underlying system state, the tightest heuristic possible is given by the optimal value itself $h^t(s) = V^t * (s)$. This value can be efficiently computed by applying MAA* recursively:

$$h^{T-t}(s) = \text{MAA}^{*(T-t)}(s) \tag{1.111}$$

1.5.2.2 Policy Iteration for Infinite Horizon

As there are infinitely many possible observation sequences for the infinite horizon case, any finite set of states of a correlated joint controller may be insufficient to produce an optimal solution. To guarantee ε-convergence, it is necessary to increase the number of controller states successively. This can be accomplished using an exhaustive backup. Local controllers are iteratively grown, for all aerial robots at once, leaving the correlation device unchanged. Let R_{max} denote the largest absolute value of a one-step reward in the DEC-POMDP. Pseudocode of this approach is presented in Algorithm 2.

Algorithm 2 Policy Iteration for Infinite Horizon DEC-POMDP

1. **Input:** DEC-POMDP problem, random correlated joint controller, convergence parameter ε
2. **Output:** A correlated joint controller that is ε optimal for all states
3. Begin
4. $t \leftarrow 0$
5. While $\gamma^{t+1}.|R_{max}|/(1-\gamma) > \varepsilon$, do
6. $t \leftarrow t+1$
7. Evaluate the correlated joint controller by solving a system of linear equations
8. Perform an exhaustive backup to add nodes to the local controllers
9. Perform value-preserving transformations on the controller
10. Return correlated joint controller
11. End

It is necessary to use the discount factor γ and the maximal number of iterations to define a simple ε-convergence test. Then the algorithm terminates after iteration t if $\gamma^{t+1}.|R_{max}|/(1-\gamma) \leq \varepsilon$, where R_{max} denotes the largest absolute value of a one-step reward in the DEC-POMDP. Intuitively, the algorithm exploits the fact that due to discounting, at some point, the future rewards collected are negligible. When looking for approximate solutions that are computationally tractable, the global value has to be sacrificed to lower the complexity and to handle larger problems.

1.5.2.3 Linear-Quadratic Approach

The early literature on team decision theory focuses on linear-quadratic models. In these models, a linear difference or differential equation represents the evolution of the decision environment in time. A quadratic objective function describes the goal of a team, and the probability theory is used for modeling uncertainties. Team decision models in which the objective function is more complex have proven to be computationally intractable.

The dynamic team decision problem can also be replaced with a set of static team problems. In the **prioritization approach**, the main building block of the objective function is a value function. It is a widely utilized method from decision analysis for analyzing multi-criteria decision problems. An **interval approach** instead of the probability theory can be used for modeling uncertainty [458]. A process by which individual entities take decisions autonomously can be developed so as to obtain a **rational collective behavior**. Each aerial robot will make decisions autonomously to maximize the performance of the team. The method of DEC-MDP, an adaptation of the MDP, is used in a multiagent framework. Currently, many researchers are designing effective multi-vehicle control concepts and algorithms. One of the scientific and technological challenges of multi-vehicle control is ensuring efficiency and safety in a context in which the conditions of the aerial robot, network and dynamic environment are changing and are potentially abnormal.

1.5.2.4   Decentralized Chance-Constrained Finite Horizon Optimal Control

This paragraph considers finite horizon optimal control for multiagent systems subject to additive Gaussian distributed stochastic disturbance and a chance constraint. The problem is particularly difficult when aerial robots are coupled through a joint chance constraint which limits the probability of constraint violation by any of the agents in the system. An algorithm based on dual decomposition, **Market-Based Iteration Risk Allocation** is proposed in [437] to solve the multiagent problem in a decentralized manner. The approach addresses the issue of **scalability** by letting each aerial robot optimize its own control input given a fixed value of a dual variable which is shared among agents. A central module optimizes the dual variable by solving a root finding problem iteratively. This method gives the same optimal solution as the centralized optimization approach since it reproduces the **Karush Kuhn Tucker** conditions of the centralized approach. This approach is analogous to a price adjustment process in a competitive market called **tatonnement** or **Walrasian auction**. Each agent optimizes its demand for risk at a given price which corresponds to the dual variable [426].

Multiagent systems under unbounded **stochastic uncertainty** with state and control constraints are an important application for aerial robot applications. Stochastic uncertainty with a probability distribution such a Gaussian is a natural model for exogenous disturbances such as wind gusts and turbulences. An effective framework to address robustness with stochastic unbounded uncertainty is optimization with a chance constraint. A chance constraint requires that the probability of failure is below a user-specified bound known as the risk bound. Users of aerial robots typically would like to bound the probability of system failure rather than the probabilities of individual aerial robot failure. A joint chance constraint must be imposed which limits the probability of having at least one aerial robot failing to satisfy any of its state and input constraints. In such cases, aerial robots are coupled through state constraints. It is important to find globally optimal control inputs in a decentralized manner while guaranteeing the satisfaction of the joint chance constraint.

1.5.3   TASK ALLOCATION: OPTIMAL ASSIGNMENT

Task allocation mechanisms are specific means for coordinating the actions of multiple aerial robots. These involve treating the work to be performed as a set of tasks and the aerial robot themselves as workers to be assigned to particular tasks. By estimating the **expected utility** of a particular robot's performance of a particular task, algorithms can optimize the allocation of robots to tasks or vice-versa in order to maximize expected collective performance. The complexity of the allocation problem depends on the particular capabilities required to achieve the task, the capabilities of the robots and whether the allocation must consider temporal scheduling aspects [424]. Generally, the appropriateness of the task allocation approach and the difficulty of the allocation problem both depend on the degree to which each of the robots and each of the tasks can be considered independent.

The archetype multi-robot task allocation problem, involving the performance of an instantaneous assignment of single aerial robots to single tasks, is considered. The problem reduces to an instance of the **optimal assignment problem** (OAP) for which the **Kuhn–Munkres Hungarian algorithm** is a solution. For multi-robot task allocation, however, outstanding issues remain. The Hungarian algorithm maximizes the utility for the team because it is provided with an estimate of each aerial robot's expected utility. Calculation of this estimate is costly because every robot must provide estimates for each task, and the optimality of the resultant allocation is only meaningful when these estimates are accurate. Furthermore, each aerial robot has to deal with uncertainty about the world in constructing these estimates [453]. Even if they maintain a representation of this uncertainty (e.g., a distribution over potential states), the expected utility is only the first moment of the utility distribution given a particular robot task assignment pair. Important questions are as follows:

1. How much effort should the aerial robots invest in constructing the utility estimates? For n robots and n tasks, n^2 estimates are provided, but only n elements make up the optimal assignment. Not all utility estimates need to be known with equal fidelity.
2. Once an allocation is computed, how stable is that allocation with respect to changes in the matrix of utility estimates?
3. If these utility estimates arise from an underlying probability distribution, what is the likelihood that the assignment is sub-optimal?

Centralized and distributed algorithms for performing allocations have been developed using greedy allocations and optimization techniques. The present study uses little information about the domain-specific aspects that lead to the structure of the coordination problem or even the source of the uncertainty. Schemes that use a richer model of the aerial model and task in order to construct a probabilistic model, for example, stochastic games/decentralized MDP, factored MDP, POMDP, allow to address explicitly the question of when to perform particular actions (movement, sensing, communication) in order to reduce uncertainty if doing so is beneficial for the task performance. However, these problems do not admit polynomial-time solutions, and often factorization or independence assumptions are introduced in order to make the problem tractable.

1.5.3.1   Hungarian Algorithm

The Hungarian algorithm treats the optimal assignment problem as a combinatorial problem in order to efficiently solve an $n \times n$ task assignment problem in $O(n^3)$ time. The utility estimates become edge weights in a **complete bipartite graph** in which each robot and task becomes a vertex. The Hungarian algorithm pseudocode is shown in Algorithm 3; it searches for a perfect matching in a sub-graph of the complete bipartite graph, where the perfect matching is exactly the optimal assignment problem. In step 7, the search process either increases the matching size or the so-called equality graph in which the matching resides.

Algorithm 3 Hungarian Algorithm

---

1. **Input:** A valid $n \times n$ assignment matrix represented as the equivalent complete weighted bipartite graph $G = (X, Y, E)$ where $|X| = |Y| = n$
2. **Output:** A perfect matching M
3. Generate an initial labeling l and matching M in G_e
4. If M is perfect, terminate algorithm. Otherwise, randomly pick an exposed vertex $u \in X$, set $S = \{u\}, T = \emptyset$
5. If $N(s) = T$, update labels,

$$\delta = \underbrace{\min_{x \in S, y \in Y-T}}\ l(x) + l(y) - w(x,y) \tag{1.112}$$

$$l'(v) = \begin{cases} l(v) - \delta & \text{if } v \in S \\ l(v) + \delta & \text{if } v \in T \\ l(v) & \text{otherwise} \end{cases} \tag{1.113}$$

6. If $N(S) \neq T$, pick $y \in N(S) - T$.
7. If y exposed, $u \to y$ is augmenting path. Augment M and go to step 2.
8. If y matched z, extend Hungarian tree: $S = S \cup \{z\}, T = T \cup \{y\}$, and go to step 3.

1.5.3.2 Interval Hungarian Algorithm

For each utility value, the interval in which the utility may be independently perturbed before the optimality of the computed assignment is violated is computed. Thus, given an input matrix of utilities, the algorithm characterizes a set of inputs which yield the same output. The intervals are computed based on the three categories of edges described below [424].

1.5.3.2.1 Interval Analysis for Matched Edges

The allowable interval for matched edge weights is analyzed as follows: for any such edge $e_m(r_\alpha, t_\beta)$, the interval can be described as $[\omega_{m\alpha\beta} - \varepsilon_m, +\infty)$ where $\omega_{m\alpha\beta}$ is the edge weight of $e_m(r_\alpha, t_\beta)$ and ε_m is the tolerance margin that the weight can decrease without violating the optimality of the current matching solution. It is safe to increase the weight as this is a maximization problem. A matched edge is hidden if its weight has decreased so as to no longer form part of a matching solution. Pseudocode of this approach is presented in Algorithm 4.

Algorithm 4 Interval of Matched Edges

---

1. **Input:** A matched edge $e_m(r_\alpha, t_\beta)$ and the corresponding resultant bipartite graph
2. **Output:** An Interval, lower bound ε_m for $e_m(r_\alpha, t_\beta)$
3. Hide $e_m(r_\alpha, t_\beta)$ by assigning it an unknown weight w_x. Set $S = \{r_\alpha\}, T = \emptyset$.
4. If $N(S) = T$, update labels.

$$\delta = \min_{x \in S, y \in Y - T, e(x,y) \neq e_m(r_\alpha, t_\beta)} l(x) + l(y) - w(x,y) \tag{1.114}$$

$$l'(v) = \begin{cases} l(v) - \delta & \text{if } v \in S \\ l(v) + \delta & \text{if } v \in T \\ l(v) & \text{otherwise} \end{cases} \tag{1.115}$$

$$l'(r_\alpha) + l'(t_\beta) - w_x > \delta \Rightarrow w_x < l'(r_\alpha) + l'(t_\beta) - \delta \tag{1.116}$$

Update $\varepsilon_m = l'(r_\alpha) + l'(t_\beta) - \delta$
5. If $N(S) \neq T$, pick $y \in N(S) - T$.
6. If $y = t_\beta$, there must be an augmenting path, $r_\alpha \to t_\beta$. Augment matching and terminate algorithm.
7. If y matched z, extend Hungarian tree: $S = S \cup \{z\}, T = T \cup \{y\}$ and go to step 2.

Theorems 38 and 39 allow computation of the interval of a matched edge $e_m(r_\alpha, t_\beta)$ in the following way. First, hide $e_m(r_\alpha, t_\beta)$ from the bipartite graph, and assign it an undecided weight w_x that satisfies the constraint: $w_x < l'(r_\alpha) + l'(t_\beta)$. Next, let the exposed vertex r_α be the root of a Hungarian tree, and construct an augmenting path excluding $e_m(r_\alpha, t_\beta)$. The algorithm terminates when such a path is found that generates a perfect matching. Because $l(t_\beta)$ stays unchanged but $l(r_\alpha)$ is decreased, w_x is decreased per iteration, and the lower bound $\omega_{m\alpha\beta} - \varepsilon_m$ occurs at the moment when the new perfect matching exists.

Theorem 38. *With the resultant matching solution M_0 and bipartite graph of Hungarian algorithm (Algorithm 4), if a matched edge $e_m(r_\alpha, t_\beta)$ is hidden, then the Hungarian algorithm can be completed with one iteration rooted at exposed node r_α. When a new perfect matching solution M' exists, the labeling reduction of the root satisfies $l(r_\alpha) - l'(r_\alpha) = m_0 - m'$.*

Theorem 39. *Matched Edge Interval: Hiding a matched edge from the Hungarian solution leads to a new solution and the labeling reduction ε_m at the root of ε_m. The Hungarian tree is the tolerance margin for this element, i.e., the safe interval for matched edge $e_m(r_\alpha, t_\beta)$, is $[\omega_{m\alpha\beta} - \varepsilon_m, +\infty)$*

1.5.3.2.2 Interval Analysis for Unmatched Edges

An unmatched edge $e_u(r_\alpha, t_\beta)$ has an interval $(-\infty, \varepsilon_u.]$ where the upper bound ε_u reflects the maximum value of the utility for it to remain an unmatched robot and task pair. Pseudocode of this approach can be found in Algorithm 5.

Algorithm 5 Interval of Unmatched Edges Algorithm

---

1. **Input:** An unmatched edge $e_u(r_\alpha, t_\beta)$ and the corresponding resultant bipartite graph
2. **Output:** An Interval, upper bound ε_u for $e_u(r_\alpha, t_\beta)$
3. Assume $e(r_\alpha$ and $\mathrm{mate}(r_\alpha))$, $e(t_\beta, \mathrm{mate}(t_\beta))$ are matched edges; then set $S = \{\mathrm{mate}(t_\beta)\}, T = \emptyset$.
4. Hide $e_u(r_\alpha, t_\beta)$ and all other edges incident to vertices r_α, t_β, and obtain the auxiliary bipartite graph G_a.
5. In G_a, if $N(S) = T$, update labels,

$$\delta = \underbrace{\min}_{x \in S, y \in Y - T} l(x) + l(y) - w(x, y) \tag{1.117}$$

and

$$l'(v) = \left\{ \begin{array}{ll} l(v) - \delta & \text{if } v \in S \\ l(v) + \delta & \text{if } v \in T \\ l(v) & \text{otherwise} \end{array} \right\} \tag{1.118}$$

6. In G_a, if $N(S) \neq T$, pick $y \in N(S) - T$.
7. If $y = \mathrm{mate}(r_\alpha)$, there must be an augmenting path, $\mathrm{mate}(t_\beta) \rightarrow \mathrm{mate}(r_\alpha)$. Augment matching and go to step 5.
8. If y is matched to z, extend Hungarian tree: set $S = S \cup \{z\}$ and $T = T \cup \{y\}$, and go to step 3.
9. $\varepsilon_u = m_0 - m_a$.

Theorem 40. *In the resultant bipartite graph of the Hungarian algorithm, the weight of any unmatched edge $e_u(r_\alpha, t_\beta)$ can be increased to the sum of two associated labeling values $l(rx) + l(ty)$ without affecting the assignment optimum.*

$\mathrm{mate}(v)$ is the other ending vertex with respect to vertex v, while m_0 is the optimum of the original solution, and m_a is the optimum of G_a.

To obtain ε_u for unmatched edge $e_u(r_\alpha, t_\beta)$, hide $e_u(r_\alpha, t_\beta)$ and all other edges incident to vertices r_α and t_β from the resultant bipartite graph. This yields a bipartite graph with $(n-1)$ vertices in each partition. This new bipartite graph is termed the auxiliary bipartite graph G_a. It is associated with a particular edge, and it has only $(n-2)$ matched edges. It therefore requires the addition of one edge for a matching solution.

Theorem 41. *Unmatched Edge Interval: Any unmatched edge $e_u(r_\alpha, t_\beta)$ in the Hungarian resultant bipartite graph has interval tolerance margin $\varepsilon_u = m_0 - (m_a + l(r_\alpha + l(r_\beta))$ where m_0 is the optimum of the original solution and m_a is the optimum of the auxiliary bipartite graph associated with $e_u(r_\alpha, t_\beta)$. The allowable interval edge $e_u(r_\alpha, t_\beta)$ is $(-\infty, m_0 - m_a]$.*

1.5.3.2.3 Interval Hungarian Algorithm

Combining the interval analysis of matched and unmatched edges, the interval Hungarian algorithm is obtained. Its pseudocode is given in Algorithm 6.

Algorithm 6 Interval Hungarian Algorithm

1. **Input:** A resultant bipartite graph from 4
2. **Output:** An Interval matrix $mx_{itv}(n,n)$ storing all intervals
3. $mx_{itv}(n,n) = NULL$.
4. For all edges e(i,j) in bipartite graph
5. If $e(i,j)$ is matched, then compute interval I(i,j) with Algorithm 4. Else compute interval I(i,j) with Algorithm 5.
6. $mx_{itv}(n,n) = I(i,j)$.
7. Return $mx_{itv}(n,n)$.

1.5.3.3   Quantifying the Effect of Uncertainty

When the Hungarian algorithm is applied to a matrix of expected utilities calculated from uncertain data, for example, using the mean of a utility distribution, the impact of the uncertainty on the resultant assignment is still unknown. The output from the interval Hungarian algorithm can be used to analyze the changes in optimal allocation as changes are made to particular utility values. This can be used to evaluate the likelihood that the calculated assignment will be suboptimal.

1.5.3.4   Uncertainty Measurement for a Single Utility

Theorem 42. *Uncertainty of a single interval: With regard to any specific single utility value, assuming other utilities are certain, the perfect matching solutions are identical if and only if any specific utility is within its allowable interval.*

To analyze the effect of uncertainty on a specific utility in the assignment matrix, the other values are assumed to be certain. Given a probability density function $f(x)$ for this specific expected utility, and associated interval I as output from the algorithm, the probability of a sub-optimal assignment is

$$P_I = \left\{ \begin{array}{ll} \int_{\varepsilon_m}^{+\infty} f(x) & \text{if} \quad I = [\varepsilon_m, +\infty) \\ \int_{-\infty}^{\varepsilon_m} f(x) & \text{if} \quad I = (-\infty, \varepsilon_m] \end{array} \right\} \tag{1.119}$$

For applications in which aerial robots are actively estimating quantities involved in producing $f(x)$, one may set some threshold T such that aerial robots only commit to an assignment if $P_I \geq T$ and instead resorts to reducing the uncertainty in the estimate if it is likely to have a major bearing on the particular assignment. High values of T will ensure the aerial robots only commit to allocations that are robust to errors in estimates of the expected utility.

1.5.3.4.1   *Uncertainty Measurement for Interval Utilities*

The previous subsection gives an approach for quantifying the effect of uncertainty on the robot-to-task assignment when only one utility is uncertain. Most often, however, multiple utilities are uncertain, and they may all be related if they involve inference over the same underlying state variables. The term **interrelated edges** is used to represent all directly related utilities in a single row or column. For the same assignment to be preserved, despite n interrelated edges, there must be one and only one edge that remains matched, and all the others should be unmatched.

Theorem 43. *Uncertainty of Interrelated Intervals: Given a set of n interrelated edges, assume e_m is the matched edge with interval $[\omega_m - \varepsilon_m, +\infty)$ and e_{u_i} are unmatched edges with intervals $(-\infty, \omega_{u_i} + \varepsilon_{u_i}], i = 1, \ldots, n-1$; for any $\varepsilon' \leq \varepsilon_m$, the weight of e_m can be safely substituted with $\omega_m - \varepsilon'$, and the interval for e_{u_i} becomes $(-\infty, \omega_{u_i} + \varepsilon_{u_i} - \varepsilon']$.*

The method for measuring the uncertainties of interrelated edges is designed in the following pseudocode given in Algorithm 7.

This approach uses a parameter k to balance the shrinking intervals. (Generally $k \to 0$ when the number of aerial robots is larger, which controls the loss of compromised ranges.) In practical applications, a conservative approach is to raise the threshold T so that the inaccuracy arising from multiple interrelated rows or columns can be compensated for.

Algorithm 7 Uncertainties of Interrelated Edges Algorithm

1. Determine $\varepsilon_{\min}$ for all interrelated edges:

$$\varepsilon_{\min} = \min\left(\varepsilon_m, \varepsilon_{u_i}\right), \quad i = 1, 2, \ldots, n-1 \qquad (1.120)$$

2. Determine each interrelated interval I_i:

$$I_i = \left\{ \begin{array}{l} [w_m - k\varepsilon_{\min}, +\infty) \\ (-\infty, w_{u_i} + \varepsilon_{u_i} - k\varepsilon_{\min}], \quad i = 1, 2, \ldots, n-1 \end{array} \right\} \qquad (1.121)$$

I_0 represents interval for the matched edge, k is an empirical coefficient and $k \in [0, 1]$ which effects the degree to which the matched and unmatched intervals are scaled.

3. Determine probability:

$$P_i = \left\{ \begin{array}{ll} \int_{\varepsilon'_m}^{+\infty} f(x) & \varepsilon'_m = w_m - k\varepsilon_{\min} \\ \int_{-\infty}^{\varepsilon'_{u_i}} f(x) & \varepsilon'_{u_i} = w_{u_i} + \varepsilon_{u_i} - k\varepsilon_{\min} \end{array} \right\} \qquad (1.122)$$

4. Determine reliability level: The assignment is reliable when $P_i \geq T$ and unreliable otherwise.

1.5.4   DISTRIBUTED CHANCE-CONSTRAINED TASK ALLOCATION

Given a list of N_a aerial robots and N_{tasks}, the goal of task allocation algorithm is to find a conflict-free matching of tasks to agents that maximize the global reward.

Definition 44. *Conflict-Free Assignment* *An assignment is conflict free if each task is assigned to no more than one agent.*

The objective function for the mission is given by a sum over local objective functions for each aerial robot, which are in turn functions of the tasks assigned to that agent, the times at which those tasks will be executed and the set of planning parameters. This section is mainly based on [439].

Problem 45. *This task assignment problem can be written as the following mixed-integer nonlinear program:*

$$\max_{x, \tau} \sum_{i=1}^{N_a} \sum_{j=1}^{N_t} c_{ij}(x, \tau, \theta) x_{ij} \qquad (1.123)$$

subject to

$$G(x, \tau, \theta) \leq b \quad x \in \{0, 1\}^{N_a \times N_t}, \tau \in \left\{ \mathbb{R}^+ \cup \emptyset \right\}^{N_a \times N_t} \qquad (1.124)$$

where x is a set of binary decision variables x_{ij} which are used to indicate whether or not task j is assigned to aerial robot i, τ is the set of real-positive decision variables τ_{ij} indicating when aerial robot i will execute its assigned task j, θ is the set of planning parameters that affect the *score calculation, c_{ij} is the reward aerial robot i receives for task j given the overall assignment and parameters and $G = [g_1, \ldots, g_{N_c}]^T$ with $b = [b_1, \ldots, b_{N_c}]^T$ is a set of N_c possibly nonlinear constraints of the form $g_k(x, \tau, \theta) \leq b_k$ that captures resource limitations of transition dynamics.*

This generalized problem formulation can accommodate different design objectives and constraints commonly used in multi-robot decision-making. In equations (1.123) and (1.124), the scoring and constraints functions are explicitly dependent upon the decision variables x, τ which makes this general mixed-integer programming problem very difficult to solve due to the inherent system inter-dependencies. In realistic mission scenarios, the planning parameters are typically a combination of deterministic and stochastic variables, and the above optimization must account for the uncertainty in θ increasing the dimensionality of the problem and further exacerbating computational intractability.

1.5.4.1 Chance-Constrained Task Allocation

Problem 46. *This chance-constrained task allocation problem can be written as the following mixed-integer nonlinear program:*

$$\underbrace{\max}_{x, \tau} y \tag{1.125}$$

subject to

$$\mathbb{P}_\theta\left(\left(\sum_{i=1}^{N_a}\sum_{j=1}^{N_t} c_{ij}(x, \tau, \theta)x_{ij}\right) > y\right) 1 - \varepsilon \tag{1.126}$$

$$\sum x_{ij} \leq 1, \forall j \quad x \in \{0,1\}^{N_a \times N_t}, \tau \in \{\mathbb{R}^+ \cup \emptyset\}^{N_a \times N_t} \tag{1.127}$$

The only hard constraint assumed in this formulation is that no task can be assigned to more than one aerial robot. If aerial robots must perform time-critical target search and track tasks, but task service times and aerial robot velocities are uncertain. The score functions are specified by

$$c_{ij}(\tau_{ij}) = \left\{ \begin{array}{ll} R_j exp^{-\lambda_j \Delta \tau_{ij}}\left(t_{j_{\text{start}}} + \bar{t}_{j_{\text{duration}}}\right) & \text{for } t \in [t_{j_{\text{start}}}, t_{j_{\text{end}}}] \\ 0 & \text{otherwise} \end{array} \right\} \tag{1.128}$$

where the task time window $[t_{j_{\text{start}}}, t_{j_{\text{end}}}]$ represents the period of time in which the task must be completed, τ_{ij} is the time at which aerial robot i finishes executing task j and

$$\Delta \tau_{ij} = \tau_{ij} - \left(t_{j_{\text{start}}} + \bar{t}_{j_{\text{duration}}}\right) \tag{1.129}$$

represents the time in excess of the expected task completion time. The exponential decay represents the time-critical nature of the task where the discount factor λ_j is used to reduce the nominal reward R_j. The arrival time at task j is a function of the aerial robot's assignment, its other task arrival times and the uncertain task durations $\tau_{ij}(x, \tau, \theta)$.

There are several issues that make this problem particularly hard to solve. The tasks are temporally coupled. Stochastic task durations and velocities for early tasks affect arrival times and thus scores for later tasks. When planning for teams of aerial robots, distributed planning strategies can offer advantages. The challenge with this problem involves developing expressions that relate each aerial robot's local risk to the global risk within a theoretically sound framework. Expressions for choosing each aerial robot's risk are analytically intractable and problem specific, so the challenge lies in developing good approximations to relating the global and local risk thresholds [111].

1.5.4.2   Distributed Approximation to the Chance-Constrained Task Allocation Problem

Problem 47. *The individual aerial robot problem is the following:*

$$\underbrace{max}_{x,\tau} y_i \tag{1.130}$$

subject to

$$\mathbb{P}_\theta \left(\left(\sum_{i=1}^{N_a} \sum_{j=1}^{N_t} c_{ij}(x,\tau,\theta) x_{ij} \right) > y_i \right) 1 - \varepsilon_i, \forall i \tag{1.131}$$

$$\sum x_{ij} \leq 1, \forall j \quad x \in \{0,1\}^{N_a \times N_t}, \tau \in \left\{ \mathbb{R}^+ \cup \emptyset \right\}^{N_a \times N_t} \tag{1.132}$$

where each aerial robot i solves its own chance-constrained optimization to maximize y_i subject to its individual risk threshold ε_i, while ensuring through communication with other robots that the joint constraint for a non-conflicting solution remains satisfied.

A method for solving this problem is the consensus-based bundle algorithm [439]. Although this decomposition makes the problem easier to solve in a distributed fashion, it also introduces the additional complexity of picking the parameters ε_i such that the goal of maximizing the chance-constrained score of the missing distribution

$$y = \sum_{i=1}^{N_a} y_i \tag{1.133}$$

given the mission risk ε is adequately represented.

1.6   CASE STUDIES

1.6.1   RECONNAISSANCE MISSION

Reconnaissance problems can be defined as the **road search problem** also known as the VRP. The cooperative system of aerial robots can considerably improve information superiority in this problem. Road search problems have mainly been handled in the operational research area, and these can be generally classified into two categories: one is the TSP which finds a shortest circular trip through a given number of cities and the other the Chinese postman problem (CPP) on an entire network of roads. The TSP using multiple aerial robots can be considered as a task assignment problem to minimize the cost (time or energy) by assigning each target to the aerial robot, and various methods have been developed such as optimization based on **binary linear programming**, heuristic methods including **iterative network flow** and **tabu search algorithms**. On the other hand, the CPP is normally used for ground vehicle applications. Vehicle routing algorithms usually approximate their path to a line for reduction in the computational load, so the physical constraints imposed on the vehicle are not addressed. Although some algorithms for the TSP consider the physical constraints, they are mostly developed for a single vehicle. For multiple vehicles, only heuristic method is implemented due to the complexity of the problem. Moreover, these constraints have rarely been considered in the CPP. To search the roads identified in the map, a road network should be established which consists of a set of straight lines joining way-points. These way-points are located either on road junctions or are located along the roads at sufficient separation to allow accurate representation of the curved roads by a set of straight lines[168].

1.6.1.1   General Vehicle Routing Problem

The **General Vehicle Routing Problem** (GVRP) is a combined load acceptance and routing problem which generalizes the well-known VRP and **Pickup and Delivery Problem** (PDP). Among the real-life requirements are time-window restrictions, a heterogeneous vehicle fleet with different travel times, travel costs and capacity, multi-dimensional capacity constraints, order/vehicle compatibility constraints, order with multiple pickup delivery and service locations, different start and end locations for vehicles and route restrictions for vehicles. The most widely studied VRPs are the **capacitated VRP** and the **vehicle routing problem with time windows** (VRPTW). **Variable neighborhood search** is meta-heuristic based on the idea of systematically changing the neighborhood structure during the search. VNS systematically exploits the following observations:

1. A **local optimum** with respect to one neighborhood structure is not necessary the same for another.
2. A **global optimum** is a local optimum with respect to all possible neighborhood structures.
3. For many problems, the **local optima** with respect to one or several neighborhoods are relatively close to each other.

In the GVRP, a transportation request is specified by a non-empty set of a pickup, delivery and/or service location which has to be visited in a particular sequence by the same vehicle; the time windows in which these locations have to be visited and the revenue gained when the transportation request is served. Furthermore, some characteristics can be specified which constrain the possibility of assigning the transportation requests to certain vehicles due to compatibility constraints and capacity constraints. At each of the locations, some shipments with several describing attributes can be loaded or unloaded. In contrast to many other commonly known routing problems, not all transportation requests have to be assigned to a vehicle. Instead, a so-called **make-or-buy** decision is necessary to determine whether a transportation request should be assigned to a self-operated vehicle (make) or not (buy). A **tour** of a vehicle is a journey starting at the vehicle's start location and ending at its final location, passing all other locations the vehicle has to visit in the correct sequence and passing all locations belonging to each transportation request assigned to the vehicle in the correct respective sequence.

Definition 48. *Feasibility A tour is feasible if and only if, for all orders assigned to the tour, compatibility constraints hold at each point in the tour time window and capacity restrictions hold.*

The objective is to find distinct feasible tours maximizing the profit which is determined by the accumulated revenue of all served transportation requests reduced by the accumulated costs for operating these tours. The GVRP is the problem of finding distinct feasible tours maximizing the profit determined by the accumulated revenue of all orders served by a vehicle reduced by the cost for operating the tours. More information on this implementation can be found in [151].

1.6.1.2 Chinese Postman Problem

This section focuses on the CPP and its variation which involves constructing a tour of the road network traveling along each road with the shortest distance. Typically, the road network is mapped to an undirected graph $G = (V, E)$ and edge weights $w : E \to \mathbb{R}_0^+$, where the roads are represented by the edge set E and road crossings are represented by the node set V. Each edge is weighted with the length of the road or the amount of time needed to pass it. The CPP algorithm involves first constructing an even graph from the road network graph. This even graph has a set of vertices with an even number of edges attached to them. This is required as any traverse of the junction by approaching on one road and leaving on another, which means that only an even number of edges will produce an entry and exit pair for the tour. As roads in the road network graph may have junctions with an odd number of edges, some roads are chosen for duplication in the graph. The technique chooses a set of roads with the shortest combined length to minimize duplication. The tour of the even graph is calculated by determining the Euler tour of the graph, which visits every edge exactly once or twice in the case of duplicated edge. The k-**CPP** deals with the deployment of several postmen. **Min–Max k-CPP** (MM k-CPP) algorithms are described for multiagent road network search. **MM k-CPP** is a variation of k-CPP which considers the route of a similar length. This objective can be required if the aerial robot should finish road search mission with the minimum mission completion time.

1.6.1.3 Cluster Algorithm

This algorithm is based on cluster first, route second. In the first step, the edge set E is divided into k clusters, and then a tour for each cluster is computed. This algorithm pseudocode can be represented as a constructive heuristic method and described by Algorithm 8.

Algorithm 8 Cluster Algorithm

---

1. Determine the set of representative edges. First of all, k representative edges $f_1, \ldots, f_k$ of cluster F_i for each vehicle are determined. Let f_1 be the edge having the maximum distance from the depot and f_2 be the edge having maximum distance from f_1. The rest of successive edges are successively determined by maximizing the minimum distance to the already existing representatives. Then, remaining edges are assigned to the cluster according to the weighted distance between e and f_i. Consider the distance between representative edges and depot, the number of assigned edges to the cluster F_i, and the cost of the cluster.
2. Include edges for connectivity. Add edges between every vertex and depot, and determine minimum spanning tree which includes original edges in each cluster for connection between the edges.
3. The rural CPP: Compute CPP route of required subset of edges out of total edges by using the conventional CPP.

1.6.1.4 The Rural CPP

Unlike the cluster algorithm, the first route algorithm follows route first, cluster second. In a first step, the postman tour which covers all edges is computed, and then this tour is divided by k tour segments which have similar length. This algorithm (Algorithm 9) is described as follows.

Algorithm 9 Rural Chinese Postman Algorithm

---

1. Compute an optimal postman C^* using the conventional CPP.
2. Compute splitting nodes: $(k-1)$ splitting nodes $v_{p_1}, \ldots, v_{p_{k-1}}$ on C^* are determined in such a way that they mark tour segments of C^* approximately having the same length. Approximated tour segment length L_j is computed by using lower bounded $s_{\max}$ of the shortest-path tour:

$$s_{\max} = \tfrac{1}{2}\max_{e=u,v\in E} w(SP(v_1, u)) + w(e) + w(SP(v_1, v_1)) \qquad (1.134)$$

$$L_j = \frac{j}{k}\left(w(C^* - 2s_{\max})\right) + s_{\max}, \, 1 \le k \le k-1 \qquad (1.135)$$

where k denotes the number of aerial robots, $w(\alpha)$ represents the distance of the sub-tour α and SP represents the shortest path between nodes considering road network. Then, the splitting node v_{p_j} is determined as being the last node such that $w(C^*_{v_{p_j}}) \le L_j, C^*_{v_n}$ is the sub-tour of C^* starting at the depot node and ending at v_n.
3. k-postmen tours: Construct k tours $C = (C_1, \ldots, C_k)$ by connecting tour segments with shortest paths to the depot node.

For the road network search using multiple aerial robot, a variation of the typical CPP algorithm is required so that it can consider the operational and physical characteristics of the aerial robot in the search problem. Since the aerial robot cannot change its heading angle instantaneously due to the physical constraint, the trajectory has to meet the speed and turn limits of the aerial robot. Moreover, differently from ground vehicles, the aerial robot has to fly along the road only to cover a certain edge which is not connected. This modified search problem is formulated as a **multi-choice multi-dimensional Knapsack problem** (MMKP) which is to find an optimal solution minimizing flight time. Classical MMKP is to pick up items for knapsacks for maximum total values so that the total resource required, does not exceed the resource constraint of knapsack. For applying MMKP to the road network search, aerial robots are assumed as the knapsacks, the roads to be searched are resources and limited flight time or energy of each aerial robot is capacity of knapsacks. MMKP formulation allows us to consider the limitations of each aerial robot flight time and different types of roads, vehicles and minimum turning radius and get the sub-optimal solution of the coordinated road search assignment. Moreover, the Dubins path planning produces the shortest and flyable paths taking into consideration their dynamical constraints; thus the Dubins path is used to calculate the cost function of the modified search problem [12].

1.6.2   EXPANDING GRID COVERAGE

In this section, an approach is considered to coordinate a group of aerial robots without a central supervision, by using only local interactions between the robots [17]. When this decentralized approach is used, much of the communication overhead (characteristic to centralized system) is saved, the hardware of the robots can be simple and better modularity is achieved. A properly designed system should be readily scalable, achieving **reliability** and **robustness** through **redundancy**. A group must cover an unknown region in the grid that possibly expands over time. This problem is strongly related to the problem of distributed search after mobile or evading targets. In general, most of the techniques used for the task of a distributed coverage use some sort of cellular decomposition. In the dynamic cooperative cleaners problem, the time is discrete. Let the undirected graph $G(V, E)$ denote a $2D$ integer grid Z^2 whose vertices have a binary property called contamination. Let $\text{Cont}_t(v)$ state the contamination state of the vertex v at time t, taking either the value "1" or "0". Let F_t be the contamination state of the vertex at time t, i.e.,

$$F_t = v \in G | \text{Cont}_t(v) = 1 \qquad (1.136)$$

F_0 is assumed to be a single connected component. This algorithm will preserve this property along its evolution. Let a group of k aerial robots that can move on the grid G (moving from one vertex to its neighbor in one time step) be placed at time t_0 on F_0 at point $P_0 \in F_t$. Each aerial robot is equipped with a sensor capable of telling the contamination status of all vertices in the digital sphere of diameter 7.

Namely, in all the vertices, the **Manhattan distance** from the aerial robot is equal or smaller than 3. An aerial robot is also aware of other aerial robots which are located in these vertices, and all of them agree on a common direction. Each vertex may contain any number of aerial robots simultaneously. Each aerial robot is equipped with a memory of size O(1) bits. When an aerial robot moves to a vertex v, it has the possibility of cleaning this tile (i.e., causing $\text{Cont}(v)$ to become 0). The aerial robots do not have any prior knowledge of the shape or size of the subgraph F_0 except that it is a single and simply connected component. The contaminated region F_t is assumed to be surrounded at its boundary by a rubber-like elastic barrier, dynamically reshaping itself to fit the evolution of the contaminated region over time. This barrier is intended to guarantee the preservation of the simple connectivity of F_t crucial for the operation of the aerial robots due to their limited memory. When an aerial robot cleans a contaminated vertex, the barrier retreats in order to fit the void previously occupied by the cleaned vertex. At every time step, the contamination spreads; that is, if $t = nd$ for some positive integer n, then

$$\forall v \in F_t, \forall u \in \{4-\text{Neighbors}(v)\}, \text{Cont}_{t+1}(u) = 1 \qquad (1.137)$$

Here the term "4-Neighbors(v)" simply means the four vertices adjacent to vertex v. While the contamination spreads, the elastic barrier stretches while preserving the simple connectivity of the region. For the aerial robots traveling along the vertices of F, the barrier signals the boundary of the contaminated region. The aerial robot's goal is to clean G by eliminating the contamination entirely. It is important to note that no central control is allowed and that the system is fully decentralized (i.e., all the aerial robots are identical, and no explicit communication between the aerial robots is allowed). An important advantage of this approach in addition to the simplicity of the aerial robots is **fault tolerance**; even if all the aerial robots evaporate before completion, the remaining ones will eventually complete the mission if possible. A cleaning algorithm is proposed in [17] for exploring and cleaning an unknown contaminated subgrid F, expanding every d time steps. This algorithm is based on a constant traversal of the contaminated region, preserving the connectivity until the region is cleaned entirely. Until the conditions of completion of mission are met, each aerial robot goes through the following sequence of commands. The pseudocode is presented in Algorithm 10.

Algorithm 10 Cluster Algorithm

1. First, each aerial robot calculates its desired destination at the current time.
2. Then each aerial robot calculates whether it should give priority to another aerial robot located at the same vertex that wishes to move to the same destination.
3. When two or more aerial robots are located at the same vertex and wish to move toward the same direction, the aerial robot that had entered the vertex first gets to leave the vertex, while the other aerial robots wait.
4. Before actually moving, each aerial robot who had obtained permission to move must now locally synchronize its movement with its neighbors in order to avoid simultaneous movements which may damage the connectivity of the region.
5. When an aerial robot is not delayed by any other agent, it executes its desired movement.

It is important to notice that, at any given time, waiting aerial robots may become active again if the conditions which made them become inactive in the first place had changed. More details on this implementation can be found in [17].

1.6.3 OPTIMIZATION OF PERIMETER PATROL OPERATIONS

This section addresses the following base perimeter patrol problem: a team of aerial robots equipped with cameras perform the task of perimeter surveillance. There are m alert stations/sites on the perimeter where a nearby breaching of the perimeter by an intruder is flagged by an unattended ground sensor (UGS). Upon detection of an incursion in its sector, an alert is flagged by the **UGS**. The statistics of the alert's arrival process are known, and in queuing theory, the process is a **Poisson process**. Camera-equipped aerial robots are on a continuous patrol along the perimeter, and their task is to inspect UGSs with alerts. To determine whether an incursion flagged by an UGS is a false alarm or a real threat, a patrolling aerial robot flies to the alert site to investigate the alert. The longer an aerial robot loiters at an alert site, the more information it gathers; however, this also increases the delay in responding to other alerts. The decision problem for the aerial robot is to determine the dwell time so as to maximize the expected payoff. This perimeter patrol problem falls in the domain of discrete time-controlled queuing systems. In general, a queuing system includes arriving customers, servers and waiting lines/buffers or queues for the customers awaiting service.

A **stochastic dynamic programming**approach is employed to obtain optimal policies for the patrolling aerial robot. Theoretical performance bounds from queuing system literature can be used to benchmark the optimal controller [376]. In the context of perimeter patrol, the customers are the flagged UGSs/alerts waiting to be serviced, and the aerial robots are the servers. In queuing theory, the queues/buffers could be of finite or infinite capacity. Here, only unit/single buffer queuing is considered, for the UGS either flags an alert or it does not. Once it flags an alert, its state does not change even if additional triggering events were to occur, until the flag is reset by a loitering aerial robot. Hence there is at most only one alert waiting at an alert site. Thus, this perimeter patrol problem constitutes a **multi-queue multi-server** unit buffer queuing system with deterministic interstate travel and service times. Because the aerial robots are constantly on patrol or are servicing a triggered UGS, the framework considered here is analogous to a cyclic polling system. The basic model of a **cyclic polling system** consists of separate queues with independent Poisson arrivals served by a single server in cyclic order. A related problem is the **dynamic traveling repairmen problem**, where the stations are not restricted to being on a line segment or a closed path. Traditionally, in queuing theorem, including the system, the server's action, i.e., which station to move toward next, is considered a control variable. But the service time itself is either considered to be a random variable with a known probability density function (PDF), for example, an exponential distribution, or a deterministic constant. One is interested from a patrolling perspective in the optimal service time in addition to the dynamic scheduling of the server's movement. The basic question then would be to decide how long the server/aerial robot should dwell at a triggered alert station/UGS as well as in which direction a bidirectional server is.

In patrol missions, the status of some sites must be monitored for events. This paragraph is based mostly on [428]. If an aerial robot must be close to a location to monitor it correctly and the number of aerial robots does not allow covering each site simultaneously, a path planning problem arises.

Problem 49. *How should the aerial robots visit the locations in order to make sure that the information about all locations is as accurate as possible?*

Some patrolling algorithms were developed in the last decade. One of the first algorithms was based on a graph patrolling formulation on which agents use reinforcement learning on a particular MDP. The MDP was defined over a countably infinite state space, assuming that aerial robots communicate by leaving messages on the nodes of the graph, leading to an unrealistic communication model. Reactive algorithms such as the **ant-colony** approach have been shown to perform well; however, this approach relies on the simplistic communication models relying on pheromones. When all locations are equally important, the shortest **Hamiltonian circuit** is an optimal solution for a single aerial robot. Multiagent strategies using a unique cycle are the best whatever the graph is. However, as some locations may be more important than others, not visiting the less important ones from time to time may be advantageous. The patrolling problem has a graphical structure. V is the vertex set of that graph and E the edges set. Let L be an $|V| \times |V|$ matrix in which L_{ij} is a real number that represents the time required to travel from i to j if $[i, j] \in E$ and is infinite otherwise. Each vertex i has a non-negative importance weight w_i. **Idleness** can be used as a performance measure. The idleness of vertex i, denoted by τ_i, represents the time since the last visit of an aerial robot to that vertex. The idleness is 0 if and only if an aerial robot is currently at vertex i, and $\tau_{i+1} = \tau_i + \Delta t$ if there are no visits to i in the time interval $(t, t + \Delta t)$. As idleness is an unbounded quantity, exponential idleness is used: $k_i^t = b\tau_i^t$ with $0 < b < 1$. It can be seen as the expected value of a **Bernoulli random variable**, and k_i^t is the probability that this random variable is 1 at time t. b is the rate at which k_i decays over time. The probability evolves as $k_i^{t+\Delta t} = k_i^t b\Delta t$ if there are no visits to i during time interval $(t, t + \Delta t)$. If an aerial robot with noisy observations visits i at time t, idleness becomes 0 with probability $b < (1 - a) \leq 1$, where a is the probability that the idleness does not become 0 when an aerial robot visits a vertex. If n aerial robots visit vertex i at time $t + \Delta t$ and there were no visits since time t,

$$k_i^{t+\Delta t} = k_i^t b\Delta t + 1 - a^n \qquad (1.138)$$

An instance of the patrolling problem is a tuple $\langle L, w, a, b \rangle$ consisting, respectively, of the matrix L of edge lengths, the vector w of importance weights and parameters a, b.

1.6.3.1 Multiagent Markov Decision Process

The problem is assumed to be fully observable; i.e., each aerial robot has the same complete information to make its decision [37]. In the patrolling problem, however, the actions of each aerial robot have a concurrent effect on the environment. These actions also have different durations. Concurrency in decision processes is modeled with a **generalized MDP**. Such decision processes generalize multiagent Markov decision processes (MMDPs) to continuous time with asynchronous events. The state variables for this problem describe the position of each aerial robot and the idleness of each vertex. If the total number of agents is N, the state space is

$$S = V^N \times [0, 1]^{|V|} \tag{1.139}$$

Given some states $s = (v, k) \in S, v_i$ is the position of the ith aerial robot, and k_i the idleness of the ith vertex. At various time points, called **decision epochs**, the aerial robot must choose an action. The actions from which an aerial robot can choose from depend on the structure of the graph and its position: if an aerial robot is at vertex v, it can choose its action from $A_v = \{u : [v, u] \in E\}$. If an aerial robot occurs at time $t^{i+1} = t^i + L_{vu}$ and $v^t = v$ while $t \in [t^i, t^{i+1}]$ and $v^t = u$ as soon as $t = t^{i+1}$, the problem is concurrent, the decision epochs of all aerial robots can be interleaved arbitrarily. Each component k_i of k and the number of agents n.

$$k_i^{t^{j+1}} = k_i^{t^j} a_i^{n_i^{j+1}} b^{\Delta t^j} + 1 - a_i^{n_i^{j+1}} \tag{1.140}$$

Let $\{y^j\}_j$ be the non-decreasing sequence of decision epochs and n_i^j be the number of agents arriving at vertex i at time t^j, $\Delta t^j = \{t^{j+1} - t^j\}$. The reward R is defined in terms of k; the rate at which reward is gained is given by

$$dR = w^T k^t dt \tag{1.141}$$

The discounted value function for a generalized MDP is defined as

$$V^\pi(s) = E\left[\int_0^\infty \gamma^t dR\right] = E\left[\gamma^{t^j} \int_0^{\Delta t^j} \gamma^t w^T k^t dt\right] = E\left[\gamma^{t^j} w^T k^{t^j} \frac{(b\gamma)^{\Delta t^j} - 1}{ln(b\gamma)}\right] \tag{1.142}$$

where $\gamma \in (0, 1]$ is the discount factor.

1.6.3.2 Anytime Error Minimization Search

Online planning has the advantage that it solves equation (4.45) only for the current state, in contrast with offline algorithms that do so for all states. The patrolling problem is simpler to solve online than offline. **Anytime error minimization search** (AEMS) performs a heuristic search in the state space. The search procedure proceeds using a typical **branch-and-bound** scheme. Since the exact long-term value of any state is not exactly known, it is approximated using lower and upper bounds, guiding the expansion of the search tree by greedily reducing the error on the estimated value of the root node. In the patrolling problem, actions have the same interpretation as in a partially observable setting, whereas observations are the travel durations. In AEMS, the error is defined using the upper bound and the lower bound on the value of some state. Let $s \in S$ be a state, $L(s) \le V(s) \le U(s)$, where $L(s), U(s)$ are, respectively, the lower and upper bounds and $V(s)$ the actual value of s. Given some search tree T, whose set of leaf nodes is noted $F(T)$, the bounds for the root node are estimated recursively according to

$$L(s) = \left\{ \begin{array}{lr} \hat{L}(s) & \text{if } s \in F(T) \\ L(s, a) = \max_{a \in A} R(s, a) + \gamma L(\tau(s, a)) & \text{otherwise} \end{array} \right\} \tag{1.143}$$

and

$$U(s) = \left\{ \begin{array}{ll} \hat{U}(s) & \text{if } s \in F(T) \\ U(s,a) = \max_{a \in A} R(s,a) + \gamma U(\tau(s,a)) & \text{otherwise} \end{array} \right\} \quad (1.144)$$

where $\tau(s,a)$ is the next state if action a is taken in state s. $\hat{U}(s), \hat{L}(s)$ are problem-dependent heuristics. An estimation of the error on the value of s is given by

$$\hat{e}(s) = U(s) - L(s) \quad (1.145)$$

Let s^0 be the root state of search tree T. Since all states are not reachable with equal probability (depending on the policy), the contribution of any state s to the error on s^0 is approximated by

$$\hat{E}\left(s^0, s^t, T\right) = \gamma^t Pr\left(h_0^t|s^0, \hat{\pi}\right) \hat{e}(s^t) \quad (1.146)$$

where t is the depth of s in T and $Pr\left(h_0^t|s^0, \hat{\pi}\right)$ denotes the probability of having history h_0^t (the sequence of joint actions that lead from s^0 to s^t) while following policy $\hat{\pi}$.

If $h_0^t = a^0, o^0, a^1, o^1, ..., a^t, o^t$ is the joint action history for some sequence of states $s^0, s^1, ..., s^t$, then

$$Prob\left(h_0^t|s^0, \hat{\pi}\right) = \prod_{i=0}^{t} Prob\left(a^i = \hat{\pi}(s^i)|s^i\right) Prob\left(o^i|s^i, a^i\right) \quad (1.147)$$

As the optimal policy is unknown, a good approximation is to use

$$\text{Prob}(a|s) = \left\{ \begin{array}{ll} 1 & \text{if } U(s,a) = \max_{a' \in A} U(s,a') \\ 0 & \text{otherwise} \end{array} \right\} \quad (1.148)$$

Given a search tree T, rooted at s^0, the next state to expand is

$$\tilde{s}(T) = \arg\max_{s \in F(T)} \hat{E}\left(s, s^0, T\right) \quad (1.149)$$

Each time a node $\tilde{s}$ is expanded, it is removed from $F(T)$, its children are added to $F(T)$ and the bounds of $\tilde{s}$ and its parents are updated. When an agent must choose an action, the action of maximum lower bound is chosen. A lower bound for the value of any state is the value of following any policy from that state. A **greedy policy** is arguably a good simple policy. It is defined to always choose the action with arrival state for which $w^T k$ is maximal. Equation (4.45) defines the value of such a policy. An upper bound is usually obtained by relaxing problem constraints, assuming that aerial robots are ubiquitous. (They can be in more than one location at the same time.) Whenever an aerial robot reaches a vertex, it instantaneously multiplies itself and starts heading to adjacent unvisited locations. This bound estimates the shortest time that a team of aerial robots would take to cover the entire graph and estimates through the discount factor and upper bound on the maximum reward obtainable. This bound implicitly assumes that the optimal policy does not require having more than one aerial robot at any vertex.

Extending AEMS to asynchronous multiagent is simple: whenever a node is expanded, there is a branch for every joint action and observation. Asynchronicity is handled with state augmentation. The state is now (s, η) where η_i is the time remaining before the next decision epoch of aerial robot i. At any time t, the next decision epoch happens at time $t + min_i \{\eta_i\}$. The expand operation adds actions and observations for any aerial robot for which $\eta = 0$. Whenever aerial robot i performs an action of duration Δt, η_i is assigned Δt. Otherwise, η_i is updated according to its depth in the search tree. AEMS can be used to perform online planning for any subset of aerial robots. However, it is unlikely that any aerial robot has the computational capacity to compute a joint policy, because the complexity is exponential in the number of aerial robots. These aerial robots are thus coordinated locally, and a partial order is defined among aerial robots. An aerial robot is said to be greater than (respectively less than) another aerial robot if it must choose its policy before (respectively after). The aerial robots compute their policy according to that order. Once an aerial robot knows the policies of all greater aerial robots, it proceeds to compute its policy and then communicates it to the lesser aerial robots. Whenever an aerial robot selects its policy, it chooses the best policy given the policy of greater aerial robots. A useful property of this coordination algorithm is that if the aerial robots use an online anytime planner, then it is also anytime and online. A **fallback strategy** is to ignore the presence of the greater aerial robots until their policy has been received.

1.6.4 STOCHASTIC STRATEGIES FOR SURVEILLANCE

A team of aerial robots are engaged in a surveillance mission, moving in a unpredictable fashion. The analysis assumes that these aerial robots can move forward and backward as rotary wings vehicles do. Stochastic rules are used to guide the motions of the aerial robots, minimizing centralized computation and communication requirements by focusing on local rules for each agent. The surveillance problem is formulated abstractly as a random walk on a hypergraph, where each node on a hypergraph corresponds to a section of the environment and where each edge of the graph is labeled with the probability of transition between the nodes. The problem of **parallel Markov chains** and **fastest mixing** when a team of agent moves among states on a graph is considered [154]. A hypergraph is associated to a general class of search environments which are modeled as sets of line segments in a bounded domain in the plane. Since the states are directional segments between the way-points, the transition between states can be viewed as a set of turning probabilities. Given this representation, a Markov chain has the following transition probability matrix:

$$P = [P_{ij}] \tag{1.150}$$

$$\sum_i P_{ij} = 1 \tag{1.151}$$

$$0 \le P_{ij} \le 1 \tag{1.152}$$

where P_{ij} is a stochastic matrix representing the probability of the agent going to state i from state j. Constraints (equations (1.151) and (1.152)) must hold since the sum of the probabilities must be 1 and all the probabilities must be non-negative.

The problem of defining surveillance strategies is parameterized. The following questions arise:

1. What **types of surveillance coverage** can the stochastic aerial robots provide (i.e., what is the steady state distribution for the system of aerial robots with specified turning probabilities)?
2. What is the **rate of convergence** of the system of aerial robots to this invariant distribution?
3. What are the appropriate **measures for comparing** different surveillance strategies?
4. How can **randomness** be captured in the motions of the aerial robots?
5. How is the **trade-off between randomness and speed of convergence**?

1.6.4.1 Analysis Methods

The aim of this paragraph is to analyze the type of surveillance coverage, invariant distribution, that the stochastic aerial robots can provide. The probability distribution at each time $k+1$ is determined according to

$$\vec{p}_i^{k+1} = P_i \vec{p}_i^k \tag{1.153}$$

where $\vec{p}_i^k$ is the probability distribution for aerial robot i at time k and P_i is the transition probability matrix for aerial robot i. There is a unique invariant distribution for an **irreducible, aperiodic Markov chain,** which is the eigenvector associated with the eigenvalue 1. This invariant distribution represents the steady state probability of the aerial robot being at any state. The eigenvalues of the Markov chain can be sorted by magnitude:

$$1 = |\lambda_1|(P) \geq |\lambda_2|(P) \geq \ldots |\lambda_n|(P) \tag{1.154}$$

The **mixing rate** for the Markov chain is given by

$$\mu(P) = |\lambda_2|(P) \tag{1.155}$$

where $|\lambda_2|(P)$ is the eigenvalue which is the second largest in magnitude. The smaller the mixing rate, the faster the Markov chain converges to its steady state distribution. The expected composite distribution of a can be explicitly determined by

$$\vec{p} = \frac{\sum_{i=1}^{a} \vec{p}_i^k}{a} \tag{1.156}$$

The aerial robots can move randomly and independently.

1.6.4.2   Problems in 1D

1.6.4.2.1   Single Aerial Robot Investigations

The aim of this paragraph is to show how tuning problem parameters can affect the invariant distribution. The case of a probabilistic aerial robot walking on an $n-$node, 1D lattice taking steps to the right with probability ρ and steps to the right with probability $1 - \rho$ is considered. For a given initial probability distribution $p_1^0, \ldots, p_n^0$, the probability distribution at each time step evolves according to

$$\vec{p}^{k+1} = P\vec{p}^k \tag{1.157}$$

where

$$P = \begin{pmatrix} 0 & 1-\rho & 0 & \cdots & 0 \\ 1 & 0 & 1-\rho & \cdots & 0 \\ 0 & \rho & 0 & \cdots & 0 \\ \vdots & \vdots & \ddots & \ddots & \vdots \\ 0 & 0 & \cdots & \rho & 0 \end{pmatrix} \tag{1.158}$$

The steady state, invariant distribution satisfies

$$\vec{p} = P\vec{p} \tag{1.159}$$

The components of this steady state distribution can be found explicitly as solutions of recursive equations. While this approach can be taken from simple environments for a single aerial robot, it becomes cumbersome as the environments become even slightly more complex and there are more sentries performing the surveillance. An empirical measure of non-uniformity can also be defined

Definition 50. *Non-Uniformity: NU*

$$NU(k) = \frac{\sum_i (\tilde{x}_i - \tilde{\pi}_i)^2}{n^2} \tag{1.160}$$

where

$$\tilde{x}_i = \frac{x(i) \times n}{k \times a} \qquad \tilde{\pi}_i = \pi_i \times n \qquad \lim_{k \to \infty} NU(k) \to 0 \tag{1.161}$$

where $\tilde{x}_i$ is a history of visitation frequency for state i, x_i is the visitation history, $\tilde{\pi}_i$ is a normalized invariant distribution for state i, n is the number of states, a is the number of agents and k is the number of steps each agent has taken.

This non-uniformity measure essentially quantifies how quickly the surveillance team covers the environment. Mathematically, it is the mean distance the team of aerial robots are from the composite invariant distribution. Normalization of the visitation history and the invariant distribution is important because the measure must be applicable to both small and large environments. As k increases, the normalized visitation history will approach the normalized steady state distribution.

1.6.4.2.2 Multi-Aerial-Robot Investigations

A strategy for 1D, n-node lattice is presented in this paragraph. By appropriate choice of parameters, it is possible to implement a probabilistic strategy in which aerial robots disperse in the lattice domain as fast as possible. The turning parameters of this strategy are assigned by the following relation:

$$\rho_i = \left\{ \begin{array}{ll} 0.9 & \text{for } k \leq \frac{a+1-i}{a+1} \times n \\ 0.5 & \text{Otherwise} \end{array} \right\} \tag{1.162}$$

where ρ_i is the probability that aerial robot i turns right, k is the number of steps, a is the number of aerial robots and n is the number of nodes in the lattice. This strategy disperses the aerial robots along the graph before switching to equal turning probabilities. The aerial robots do not have a uniform steady state distribution prior to switching to the equal turning probabilities. After the aerial robots switch to the equal turning probabilities, their steady state distribution approach the uniform distribution as the initial distributions are suppressed with time.

1.6.4.3 Complete Graphs

Complete graphs are considered, and the associated turning parameters that yield the fastest mixing and approach the uniform steady state distribution are investigated.

Theorem 51. *For a complete graph with n vertices, the probabilistic random walk having transition probability matrix*

$$p_{ij} = \left\{ \begin{array}{ll} \frac{1}{n-1} & \text{if } i \neq j \\ 0 & \text{Otherwise} \end{array} \right\} \tag{1.163}$$

has eigenvalues 1 of multiplicity 1 and $\frac{-1}{n-1}$ of multiplicity $n-1$. The invariant distribution for this Markov chain is uniform, and the eigenvalue $\frac{-1}{n-1}$ is smaller in magnitude than the eigenvalue of second largest magnitude corresponding to any other set of transition probabilities.

In the general case, the structure of the problem can be considered as line segments on a bounded plane. A complete graph on its vertices is associated to each edge of rank ≥ 3 in $H(X, \varepsilon)$ representing the set of segments in the search environment. The edges of these complete graphs represent the possible choices of transitions from one segment to another. Any general graph can be decomposed into a system of interconnected complete subgraphs, cliques. Cliques having two vertices are distinguished from those having more than two vertices. Those having two vertices correspond to transitions in which the only choices available to the aerial robot are to move ahead or to move backward. The intersections of a general graph can be thought of as a complete graph where the number of nodes in the complete graph is equivalent to the number of choices an aerial robot can make. With no restrictions on the movement of an aerial robot, the number of nodes in the complete graph is equal to the number of edges incident to the intersection in the graph representation. A hybrid strategy is obtained by combining the strategies for both linear graphs and complete graphs. It will provide uniform coverage of a general graph while achieving this coverage quickly without a large sacrifice in the randomness in the behavior of the aerial robot. More details on this implementation can be found in [154].

1.7 CONCLUSIONS

The evolution of multiagent systems has revolutionized the services that distributed computation can offer. In this chapter, team approach is introduced; then deterministic decision-making is presented followed by information about association with limited communications. Then, decision-making under uncertainty is analyzed. Finally, some case studies are discussed.

Some future challenges are the discovery of adaptation to new environments, cooperation and communication between agents, autonomous decision-taking and resource management and collaboration in the realization of activities with common objectives. In-depth research is still required in the areas of modeling, control and guidance relevant in coordination and control problems. The distributed nature of information processing, sensing and actuation makes the teams of vehicles a significant departure from the traditional centralized control system paradigm. The following is an open question: "Given a team of locally interacting robots, and a high-level (global) specification over some environment, how can we automatically generate provably correct (local) control strategies?" What global (expressive) specifications can be efficiently distributed? How should local interactions (for example, message passing versus synchronization on common events) be modeled?

2 Flight Planning

2.1 INTRODUCTION

Flight planning generates paths that are consistent with the physical constraints of the autonomous aircraft, the obstacle and collision avoidance and weighed regions. Weighed regions are regions with abnormally low or high pressure, wind speeds or any other factor affecting flight. Three-dimensional mission planning involves creating a path generation system which helps the aircraft to reach the mission goal but also creates a path to satisfy different constraints during the mission. This path generation system generates the path from the initial point to the mission goal and navigates the aircraft. Flight planning requires an awareness of the environment in which it is operating [56]. The position, orientation and velocity of the aircraft are known from the sensors, and the flight management system has information about the meteorological conditions and probable obstacles to avoid. In this chapter, the assumption is made that the information required will be available. More information about situation awareness can be found in the following chapter.

The human approach to navigation is to make maps and erect sign post or use landmarks. Robust navigation in natural environments is an essential capability of smart autonomous aircraft. In general, they need a map of their surroundings and the ability to locate themselves within that map, in order to plan their motion and successfully navigate [402,404]. Map-based navigation requires that the aircraft's position is always known.

Smart autonomous aircraft should be able to make decisions for performing tasks or additional actions or for changing current task. They should have the capacity to perceive their environment and consequently update their activity. The autonomy of these systems is increased as high-level decisions, such as aircraft way-point assignment and collision avoidance, are incorporated in the embedded software [451].

Within the autonomy area, automated guidance and trajectory design play an important role. On-board maneuver planning and execution monitoring increase the aircraft maneuverability, enabling new mission capabilities and reducing costs [65,332].

The main tasks of the flight planning system are

1. Given a mission in terms of way-points, generate a series of paths while providing a minimum clearance.
2. Generate the reference trajectories while satisfying aircraft constraints.

The route planning problem is about finding an optimum path between a start point and a destination point considering the dynamics of the aircraft, the environment and specific constraints implied by operational reasons. The calculation of a flight plan involves the consideration of multiple elements. They can be classified as either continuous or discrete, and they can include nonlinear aircraft performance, atmospheric conditions, wind forecasts, airspace structure, amount of departure fuel and operational constraints [380]. Moreover, multiple differently characterized flight phases must be considered in-flight planning. The multiphase motion of an aircraft can be modeled by a set of differential algebraic dynamic subsystems:

$$\Upsilon = \{\Upsilon_0, \Upsilon_1, \ldots, \Upsilon_{N-1}\}$$

so that for $k \in \{0, \ldots, N-1\}$,

$$\Upsilon_k = \left\{ f_k : \mathbb{X}_k \times \mathbb{U}_k \times \mathbb{R}^{n_{l_k}} \longrightarrow \mathbb{R}^{n_{X_k}}, g_k : \mathbb{X}_k \times \mathbb{U}_k \times \mathbb{R}^{n_{l_k}} \longrightarrow \mathbb{R}^{n_{Z_k}} \right\}$$

where f_k represents the differential equation

$$\dot{X} = f_k(X, U, p)$$

for the kth subsystem, g_k describes the algebraic constraints, and k represents the index for phases. The state set has the following property: $\mathbb{X}_k \subset \mathbb{R}^{n_{X_k}} \subseteq \mathbb{R}^{n_X}$, and the control set is such that $\mathbb{U}_k \subset \mathbb{R}^{n_{U_k}} \subseteq \mathbb{R}^{n_U}$. A vector of parameter is $p \in \mathbb{R}^{n_p}$. Let the switch times between phases defined as

$$t_I = t_0 \leq t_1 \leq \cdots \leq t_N = t_f$$

That is, at time t_k, the dynamic subsystem changes from Υ_{k-1} to Υ_k. As a consequence, in the time sub-interval $[t_k, t_{k+1}]$, the system evolution is governed by the dynamic subsystem Υ_k. In the sub-interval $[t_{N-1}, t_N]$, the active dynamic subsystem is Υ_{N-1}. The switch is triggered by a sequence of switch conditions in the set $\mathbb{S} = \{\mathbb{S}_1, \mathbb{S}_2, \ldots, \mathbb{S}_{N-1}\}$, $\mathbb{S} = \mathbb{S}_A \cup \mathbb{S}_c$ provides logic constraints that relate the continuous state and mode switch. $\mathbb{S}_A$ corresponds to the set of autonomous switch and $\mathbb{S}_c$ to the set of controlled switch. For instance, for an autonomous switch, when the state trajectory intersects a certain set of the state space at subsystem $k - 1$, the system is forced to switch to subsystem k. For a controlled switch, only when the state belongs to a certain set, the transition from $k - 1$ to k is possible. This controlled switch might take place in response to the control law. Key parameters depend on the mission. There is no universal way of picking them.

The configuration space may be altered if the aircraft properties or the characteristics of the environment change. The scenarios in which the Unmanned Aircraft Systems (UAS) missions are executed are dynamic and can change continuously. Part of the mission or in the worst case, the initial mission objectives could thus be modified. Based on the particular situation, the aircraft could fly toward a destination point, to monitor a set of objectives or a specific rectangular area or to survey a target of interest. When the aircraft must monitor an area or survey a target, the payload parameters and the altitude of the maneuver must be set. The planning algorithm is composed of three sub-algorithms:

1. **Path planning:** An integral part of Unmanned Aerial Vehicle (UAV) operation is the design of a flight path or mission path that achieves the objectives of the mission. If a monitor or a survey task is commanded, it runs assuming subsequent couples of **primary mission way-points** (PMW) as input data. If a "fly-to" task is commanded, input data consist of the current aircraft position as the starting point and the commanded position as the destination point. The fundamentals of flight are in general straight and level flight (maintenance of selected altitude), ascents and descents, level turns and wind drift correction. The algorithm calculates the path between each couple of primary mission way-points, called a **macro-leg**, generating an appropriate sequence of route way-points and corresponding speed data in a 3D space. Each macro-leg can be composed of climb, cruise and descent phases depending on the relative altitude and position of the primary points and the obstacles. The global route is the connected sequence of the macro-legs. Such a route is safe and efficient and is provided as a set of way-points to pass through at a specified speed. The safe algorithm calculates a set of geometric safe paths from the first to the second primary way-points for each couple of way-points. The cost algorithm manipulates the safe paths and generates the path that fulfills the aircraft performance and mission priority. It generates also the reference trajectories while satisfying aircraft constraints.

2. **Mission planning:** Mission planning ensures that the UAV operates in a safe and efficient manner. It identifies a series of ordered point locations called primary mission way-points, which will be included in the route to accomplish the mission objectives. The mission planning system needs to find motion plans that avoid collision scenarios and maximize mission efficiency and goal attainment. Furthermore, the plans cannot exceed the performance limitations of the aircraft. There are complex trade-offs that must be made with regard to mission goals, mission efficiency and safety objectives. As autonomous aircraft operate in an unpredictable and dynamic outdoor environment, it is necessary to combine pre-flight strategic planning with in-flight tactical replanning. There is significant time pressure on

tactical replanning due to aircraft velocities and the additional constraint for fixed-wing UAV motion, maintaining a minimum stall speed [403].

3. **Mission management:** It provides the translation of mission objectives into quantifiable, scientific descriptions giving a measure to judge the performance of the platform, i.e., the system in which the mission objectives are transformed into system parameters. The mission management functions can be split into two different functional elements:

   a. The payload functions are mission specific and directly relate to the mission.

   b. The aircraft management system is defined as the set of functions that are required for the on-board embedded software to understand, plan, control and monitor the aircraft operations. They usually represent the safety critical functionality required for the safe employment of the platform; hence, they include all the flight critical and safety-related functions.

Remark 52. *Online replanning on-board the aircraft ensures continued conformance with the National Airspace System (NAS) requirements in the event of an outage in the communication link.*

2.2 PATH AND TRAJECTORY PLANNING

The general problem of path planning is to determine a motion for an aircraft allowing it to move between two configurations while respecting a number of constraints and criteria. These arise from several factors of various nature and generally depend on the characteristics of the system, environment and type of task.

Problem 53. *Planning: Given a mechanical system (S), whose motion is governed by a system of differential equations, find aircraft path and trajectory such that its movement is constrained limiting the configuration space and the constraints on the controls.*

The planning problem implies the calculation of a trajectory $(X(t), U(t))$ satisfying the differential equation such that $X(t_0) = X_0$ and $X(t_f) = X_f$.

1. A **path** is a set of configurations reached by the aircraft to go from one configuration to another. Path planning (finding a path connecting two configurations without collision) is a kinematical/geometrical problem. A path is defined as the interpolation of position coordinates. A path does not specify completely the motion of the system in question.

2. A **trajectory** is a path over a law describing the time instants of passage of each system configuration. The path planning is not only a kinematical/geometrical problem but also a dynamical problem. A trajectory refers to timely annotated paths; it is aircraft specific.

In this case, the constraints relate to aircraft geometry, kinematics and dynamics. A solution must optimize a cost function expressed in terms of distance traveled by the aircraft between two extremal configurations, time or energy necessary to the implementation of its motion. Optimization problems are divided into two categories: those with continuous variables and those with discrete variables.

Trajectory planning can find a path expressed in terms of the degrees of freedom of the aircraft and velocity/angle rates. A 4D motion planning comprises a referenced sequence of 3D way-points and the desired track velocities between them. Such tracks are also referred to as trajectory segments. It is necessary to incorporate an approximation of aircraft dynamics to ensure that the generated paths are physically realizable.

A motion plan consists of two classes of motion primitives [187]:

1. the first class is a special class of trajectories: trim trajectories. A trim is a steady-state or quasi-steady flight trajectory.

2. The second class consists of transitions between trims: maneuvers.

Each flight segment is defined by two end flight constraints which together with dynamic model form a system of **differential algebraic equations** (DAE). The resolution of the DAE for the different flight segments is often based on the reduction of the aircraft equations of motion to a system of ordinary differential equations through the explicit utilization of the flight constraints. A continuously differentiable path should be preferred to enable smooth transitions. Typically, way-points and paths are planned when UAS are operated autonomously. After planning way-points, paths are then typically planned joining these way-points. As there are dynamic constraints, the paths are planned by using various geometric curves instead of straight lines. After planning paths, guidance laws are designed for path following.

2.2.1 TRIM TRAJECTORIES

Trimmed flight condition is defined as one in which the rate of change (of magnitude) of the state vector is zero (in the body-fixed frame) and the resultant of the applied forces and moments is zero. In a trimmed trajectory, the autonomous aircraft will be accelerated under the action of non-zero resultant aerodynamic and gravitational forces and moments, and these effects will be balanced by effects such as centrifugal and gyroscopic inertial forces and moments.

$$\dot{u} = \dot{v} = \dot{w} = 0 \qquad \dot{p} = \dot{q} = \dot{r} = 0 \tag{2.1}$$

Under the trim condition, the aircraft motion is uniform in the body-fixed frame. The aerodynamic coefficients which are variable in time and space become stationary under this condition and their identification becomes easier [54,64]. Their geometry depends on the body-fixed linear velocity vector V_e, the roll angle ϕ_e, pitch angle θ_e and the rate of yaw angle $\dot{\psi}_e$. The choice of these quantities should satisfy the dynamic equations, the controls saturation and envelope protection constraints.

For trim trajectories, the flight path angle γ is a constant γ_0, while the angle χ is linearly varying versus time t.

$$\chi(t) = \chi_0 + t\chi_1 \tag{2.2}$$

The parameters γ_0, χ_0, χ_1 are constants, and the following relations can be proposed:

$$x(t) = x_0 + \frac{\cos \gamma_0}{\chi_1} \left(\cos(\chi_0 + \chi_1 t) - \cos \chi_0 \right) \tag{2.3}$$

$$y(t) = y_0 - \frac{\cos \gamma_0}{\chi_1} \left(\sin(\chi_0 + \chi_1 t) - \sin \chi_0 \right) \tag{2.4}$$

$$z(t) = z_0 + \sin(\gamma_0)t \tag{2.5}$$

Trim trajectories are represented in general by helices, with particular cases such as straight motion or circle arcs. For this kind of helix, curvature κ and torsion τ are constant

$$\begin{aligned} \kappa &= \chi_1 \cos(\gamma_0) \\ \tau(s) &= \chi_1 \sin(\gamma_0) \end{aligned} \tag{2.6}$$

The dynamic model allows to compute the following relations:

$$T = \frac{D + mg \sin \gamma_0}{\cos \alpha} \tag{2.7}$$

$$\sigma = 0 \tag{2.8}$$

and

$$(L + T \sin \alpha) - mg \cos \gamma_0 = 0 \tag{2.9}$$

A part of a helix can be used to join two configurations, respectively $X_0 = (x_0, y_0, z_0, \chi_0, \gamma_0)$ and $X_f = (x_f, y_f, z_f, \chi_f, \gamma_f)$. This particular case occurs when $\gamma_0 = \gamma_f$ and the following relationships are verified

$$\chi_1 = \sin(\gamma_0) \frac{\chi_f - \chi_0}{z_f - z_0} \tag{2.10}$$

with the constraint between the initial and final positions:

$$[\chi_1(x_f - x_0) + \cos\gamma_0 \sin\chi_0]^2 + [\chi_1(y_f - y_0) - \cos\gamma_0 \cos\chi_0]^2 = \cos^2\gamma_0 \tag{2.11}$$

The length, L, of the path is given by

$$L = \frac{\chi_f - \chi_0}{\chi_1} \tag{2.12}$$

The trim trajectories have the advantage of facilitating the planning and control problems. The role of the trajectory generator is to generate a feasible time trim trajectory for the aircraft.

2.2.2   TRAJECTORY PLANNING

Aircraft trajectory planning goes from the mathematical optimization of the aircraft trajectory to the automated parsing and understanding of desired trajectory goals, followed by their formulation in terms of a mathematical optimization programming [103,169,357,358].

2.2.2.1   Time Optimal Trajectories

The subject of this section is to formulate the trajectory generation problem in minimum time as this system has bounds on the magnitudes of the inputs and the states [105]. The velocity is assumed to be linearly variable. As the set of allowable inputs is convex, the time optimal paths result from saturating the inputs at all times (or zero for singular control). For a linear time- invariant controllable system with bounded control inputs, the time optimal control solution to a typical two-point boundary value problem (TPBVP) is a bang–bang function with a finite number of switches [48,88,355].

Problem 54. *The Dubins problem is the problem of describing the minimum time trajectories for differential system defined as*

$$\begin{aligned}
\dot{x} &= \cos\chi \\
\dot{y} &= \sin\chi \\
\dot{\chi} &= U \\
|U| &\leq 1
\end{aligned} \tag{2.13}$$

It has been proved by Dubins [110] that the optimal arcs of this problem (2.13) are a concatenation of at most three pieces S (straight), R (turn to the right), L (turn to the left). The shortest path for a Dubins vehicle consists of three consecutive path segments, each of which is a circle of minimum turn radius C or a straight line L. So the Dubins set $\mathbb{D}$ includes six paths: $\mathbb{D} = \{LSL, RSR, RSL, LSR, RLR, LRL\}$. If the paths are sufficiently far apart, the shortest path will always be of CSC type. In the case of an airplane, this is always the case.

In [329], the problem of finding a fastest path between an initial configuration $(0, 0, \chi_s)$ and a final configuration (x_t, y_t, χ_t) is considered. The direction-dependent model represents an extension of the original Dubins aircraft model which assumes isotropic speed and minimum turning radius $\frac{V}{R(\theta)} U$.

In [387], the optimal 3D curves are helicoidal arcs. This direction-dependent framework generalizes some of the previous work, in particular Dubins-like vehicles moving in constant and uniform wind [464].

Problem 55. *The Markov–Dubins problem is the problem of describing the minimum time trajectories for the system:*

$$\dot{x} = \cos\chi$$
$$\dot{y} = \sin\chi$$
$$\dot{\chi} = \omega \tag{2.14}$$
$$\dot{\omega} = U$$
$$|U| \leq 1$$

This system is a dynamic extension of Dubins system equation (2.13).

Flight planning aim is leg-based navigation. A leg specifies the flight path to get to a given way-point. From the crossing points that define the rough paths of aircraft, the goal is to refine these paths to generate trajectories parametrized function of time and satisfying the kinematic constraints of the aircraft... The principle of maximum of Pontryagin (PMP) provides necessary conditions for excluding certain types of trajectories. Most often, the conclusions derived by applying the PMP characterize a family of sufficient controls containing the optimal control between two points. An application of PMP combined with the tools of control theory, nonlinear Lie bracket and Lie algebra, led, in the case of mobile robots, to refine the set of optimal trajectories, initially obtained by Reed and Sheep [81,315].

2.2.2.2   Nonholonomic Motion Planning

Nonholonomic motion planning relies on finding a trajectory in the state space between given initial and final configurations subject to nonholonomic constraints [61,63,361]. The Lie algebraic method relies on a series of local planning around consecutive current states. Global trajectory results from joining local trajectories. At a current state, a direction of motion toward the goal state is established. Then, a rich enough space of controls is taken [366]. As the system is controllable, via some vector fields, the controls are able to generate the fields [123]. The steering method for affine drift-less systems exploits different properties of such a system, namely, nilpotence, chained form and differential flatness [23,56].

The methods involving sinusoid at integrally related frequencies can be modified using some elementary **Fourier analysis** to steer system.

The kinematic equations of motion of an aircraft in 3D are given by

$$\dot{x} = V\cos\gamma\cos\chi$$
$$\dot{y} = V\cos\gamma\sin\chi$$
$$\dot{z} = -V\sin\gamma \tag{2.15}$$
$$\dot{\chi} = \omega_1$$
$$\dot{\gamma} = \omega_2$$

If the angles are assumed to be small, an approximation to this system is obtained by setting $\cos\gamma \approx 1$, $\sin\gamma \approx \gamma$, $\cos\chi \approx 1$, $\sin\chi \approx \chi$. Relabeling the variables, the preceding system can be written under a chained form:

$$\dot{X}_1 = U_1$$
$$\dot{X}_2 = U_2$$
$$\dot{X}_3 = U_3 \tag{2.16}$$
$$\dot{X}_4 = X_2 U_1$$
$$\dot{X}_5 = -X_3 U_1$$

where $X = (x, \chi, \gamma, y, z)^T$, $U = (\dot{x}, \dot{\chi}, \dot{\gamma})^T$. This system can also be written under the following form:

$$
X = \begin{pmatrix} 1 \\ 0 \\ 0 \\ X_2 \\ -X_3 \end{pmatrix} U_1 + \begin{pmatrix} 0 \\ 1 \\ 0 \\ 0 \\ 0 \end{pmatrix} U_2 + \begin{pmatrix} 0 \\ 0 \\ 1 \\ 0 \\ 0 \end{pmatrix} U_3 = g_1 U_1 + g_2 U_2 + g_3 U_3 \qquad (2.17)
$$

Using Lie brackets, the following vectors are calculated:

$$
g_4 = [g_1, g_2] = \begin{pmatrix} 0 \\ 0 \\ 0 \\ 1 \\ 0 \end{pmatrix} \qquad g_5 = [g_1, g_3] = \begin{pmatrix} 0 \\ 0 \\ 0 \\ 0 \\ -1 \end{pmatrix} \qquad (2.18)
$$

The determinant of the matrix $(g_1, g_2, g_3, g_4, g_5)$ is different from zero, and the controllability rank condition is satisfied. However, it should be noticed that for a fixed-wing aircraft, U_1 is not symmetric as $0 < V_{stall} \le U_1 \le V_{max}$, while U_2 and U_3 are symmetric. Hence, small time local controllability cannot be ensured.

This multi-chained form system can be steered using sinusoid at integrally related frequencies. To steer this system, first, the controls U_1, U_2, U_3 are used to steer x, χ, γ to their desired locations.

$$
U_1 = \delta_1 \cos(\omega t) \qquad U_2 = \delta_2 \cos(k_2 \omega t) \qquad U_3 = \delta_3 \cos(k_3 \omega t) \qquad (2.19)
$$

where k_2 and k_3 are positive integers.

By integration, x, χ, γ are all periodic and return to their initial values

$$
x = \frac{\delta_1}{\omega} \sin(\omega t) + X_{10} \qquad (2.20)
$$

$$
\chi = \frac{\delta_2}{k_2 \omega} \sin(k_2 \omega t) + X_{20} \qquad (2.21)
$$

$$
\gamma = \frac{\delta_3}{k_3 \omega} \sin(k_3 \omega t) + X_{30} \qquad (2.22)
$$

$$
y = -\frac{\delta_1 \delta_2}{2 k_2 \omega} \left(\frac{\cos((k_2 + 1)\omega t)}{(k_2 + 1)\omega} + \frac{\cos((k_2 - 1)\omega t)}{(k_2 - 1)\omega} \right) + X_{20} \frac{\delta_1}{\omega} \sin(\omega t) + X_{40} \qquad (2.23)
$$

and

$$
z = \frac{\delta_1 \delta_3}{2 k_3 \omega} \left(\frac{\cos((k_3 + 1)\omega t)}{(k_3 + 1)\omega} + \frac{\cos((k_3 - 1)\omega t)}{(k_3 - 1)\omega} \right) - X_{30} \frac{\delta_1}{\omega} \sin(\omega t) + X_{50} \qquad (2.24)
$$

where X_{10}, X_{20}, X_{30}, X_{40} and X_{50} are integration constants.

The problem of steering the approximate model from $X_0 \in \mathbb{R}^5$ at $t = 0$ to $X_f \in \mathbb{R}^5$ at $t = 1$ is considered. The initial conditions allow the calculation of the integration constants:

$$
X_{10} = x_0 \quad X_{20} = \chi_0 \quad X_{30} = \gamma_0 \qquad (2.25)
$$

$$
X_{40} = y_0 + \frac{\delta_1 \delta_2}{2 k_2 \omega} \left(\frac{1}{(k_2 + 1)\omega} + \frac{1}{(k_2 - 1)\omega} \right) \qquad (2.26)
$$

and

$$X_{50} = z_0 - \frac{\delta_1 \delta_3}{2k_3 \omega} \left(\frac{1}{(k_3 + 1)\omega} + \frac{1}{(k_3 - 1)\omega} \right) \tag{2.27}$$

The final conditions allow to write the following:

$$\delta_1 = \frac{\omega}{\sin \omega}(x_f - x_0) \tag{2.28}$$

$$\delta_2 = \frac{k_2 \omega}{\sin(k_2 \omega)}(\chi_f - \chi_0) \tag{2.29}$$

$$\delta_3 = \frac{k_3 \omega}{\sin(k_3 \omega)}(\gamma_f - \gamma_0) \tag{2.30}$$

while the following nonlinear equations must be solved in k_2, k_3, ω to characterize entirely the system:

$$y_f - y_0 - \chi_0(x_f - x_0) = -\frac{\delta_1 \delta_2}{2k_2 \omega} \left(\frac{\cos((k_2 + 1)\omega) - 1}{(k_2 + 1)\omega} + \frac{\cos((k_2 - 1)\omega) - 1}{(k_2 - 1)\omega} \right) \tag{2.31}$$

$$z_f - z_0 - \gamma_0(x_f - x_0) = -\frac{\delta_1 \delta_3}{2k_3 \omega} \left(\frac{\cos((k_3 + 1)\omega) - 1}{(k_3 + 1)\omega} + \frac{\cos((k_3 - 1)\omega) - 1}{(k_3 - 1)\omega} \right) \tag{2.32}$$

Once this has been done, all the reference trajectories are characterized.

In the general case, motion planning may be facilitated by a preliminary change of state coordinates which transforms the kinematics equations of the aircraft into a simpler canonical form.

2.2.3 PATH PLANNING

Path planning problem can be formulated as follows:

Problem 56. *Given a C-space Ω, the path planning problem is to find a curve:*

$$C : [0, 1] \longrightarrow C_{\text{free}} \qquad s \to C(s)$$

where s is the arc-length parameter of C.

C_{free} represents the set of configurations free of obstacles. An optimal path is a curve C that minimizes a set of internal and external constraints (time, fuel consumption or risk). The complete set of constraints is described in a cost function τ, which can be isotropic or anisotropic [351]:

1. **Isotropic case:** The cost function τ depends only on the configuration X.
2. **Anisotropic case:** The cost function τ depends on the configuration X and a vector of field force.

The aircraft needs to move smoothly along a path through one or more way-points. This might be to avoid obstacles in the environment or to perform a task that involves following a piecewise continuous trajectory. Several path models, such as straight lines segments, Lagrange interpolation, Hermite interpolation, piecewise linear (quadratic, cubic) interpolation and spline interpolation (cubic, Bezier, etc.), can be used [103,353,398]. Other techniques can also be used such as wavefront algorithm and Pythagorean hodograph [102,179,183].

The unmanned aircraft minimum **curvature radius** and pitch angle constraints should be satisfied because the aircraft's curvature radius is highly related to the geometry, kinematics and dynamics of the aircraft [50,51]. In addition, the pitch angle is generally limited in a certain range for the sake of the aircraft safety in 3D space, and the pitch angle at all points on the trajectory must be constrained between the assigned lower and upper bounds.

The optimal path is essential for path planning. Path planning should produce not only a feasible path, but also the optimal one connecting the initial configuration to the final configuration. In [399], a real-time dynamic **Dubins helix** method for trajectory smoothing is presented. The projection of 3D trajectory on the horizontal plane is partially generated by Dubins path planner such that the curvature radius constraint is satisfied. The helix curve is constructed to satisfy the pitch angle constraint, even in the case where the initial and final configurations are close.

2.2.3.1 B-Spline Formulation

Splines are a set of special parametric curves with certain desirable properties. They are piecewise polynomial functions, expressed by a set of control points. There are many different forms of splines, each with their own attributes. However, there are two desirable properties:

1. Continuity: the generated curve smoothly connects its points.
2. Locality of the control points: the influence of a control point is limited to a neighborhood region.

Different alternate paths can be represented by B-spline curves to minimize computation, because a simple curve can easily be defined by three control points. **A parametric B-spline curve** $p(s)$ of the order k or degree $k-1$ is defined by $(n+1)$ control points p_i, knot vector X and the relationship

$$p(s) = \sum_{i=0}^{n} p_i N_{i,k}(s) \tag{2.33}$$

where $N_{i,k}(s)$ are the **Bernstein basis functions** and are generated recursively using

$$N_{i,k}(s) = \frac{(s - X_i)N_{i,k-1}(s)}{X_{i+k-1} - X_i} + \frac{(X_{i+k} - s)N_{i+1,k-1}(s)}{X_{i+k-1} - X_{i+1}} \tag{2.34}$$

and

$$N_{i,1} = \left\{ \begin{array}{cc} 1 & \text{If } X_i \leq s \leq X_{i+1} \\ 0 & \text{Otherwise} \end{array} \right\} \tag{2.35}$$

The control points define the shape of the curve. By definition, a low-degree B-spline will be closer and more similar to the control polyline (the line formed by connecting the control points in order). The B-splines used can be third-degree B-splines to ensure that the generated curves stay as close to the control points as possible [127].

2.2.3.2 Cubic Hermite Spline

The Hermite spline is a special spline with the unique property that the curve generated from the spline passes through the control points that define the spline. Thus, a set of predetermined points can be smoothly interpolated by simply setting these points as control points for the Hermite spline.

Cubic Hermite spline interpolation passes through all the way-points, and it is possible to assign the derivative values at the control points and also obtain local control over the paths. A solution to the path generation problem is to use a cubic Hermite spline for each pair of successive way-points [321].

Given a non-negative integer n, P_n denotes the set of all real-valued polynomials. The partition of the interval $I = [a,b]$ is given as $a = X_1 < X_2 < \cdots < X_n = b$ and $f_i, i = 1, \ldots, n$ the corresponding set of monotone data at the partition points:

$$p(s) = f_i H_1(s) + f_{i+1} H_2(s) + h_i H_3(s) + h_{i+1} H_4(s) \tag{2.36}$$

where $H_k(s)$ are the cubic Hermite basis functions for the interval I_i:

$$H_1(s) = \varphi \frac{X_{i+1}-s}{h_i} \qquad H_2(s) = \varphi \frac{s-X_i}{h_i}$$
$$H_3(s) = -h_i \eta \frac{X_{i+1}-s}{h_i} \qquad H_4(s) = h_i \eta \frac{s-X_i}{h_i} \tag{2.37}$$

where $h_i = X_{i+1} - X_i$, $\varphi = 3t^2 - 2t^3$, $\eta = t^3 - t^2$.

This methodology can be extended to parametric splines. This entails the introduction of the independent variable ϑ and the formulation of one separate equation for each one of the data variable:

$$x_d(\vartheta) = C_{x_3}(\vartheta - \vartheta_i)^3 + C_{x_2}(\vartheta - \vartheta_i)^2 + C_{x_1}(\vartheta - \vartheta_i) + C_{x_0} \tag{2.38}$$

where

$$C_{x_0} = X_i \qquad C_{x_1} = X_i'$$
$$C_{x_2} = \frac{3S_i^x - x_{i+1}' - 2x_i'}{\Delta\vartheta_i} \qquad C_{x_3} = \frac{-2S_i^x + x_{i+1}' + x_i'}{\Delta\vartheta_i} \tag{2.39}$$

where $(.)'$ denotes differentiation with respect to parameter ϑ, $\Delta\vartheta_i = \vartheta_{i+1} - \vartheta_i$ is the local mesh spacing, and $S_i^x = \frac{x_{i+1}+x_i}{\Delta\vartheta_i}$ is the slope of the linear interpolant.

2.2.3.3 Quintic Hermite Spline

The task is to find a trajectory: a parametrized curve, $\eta(t) = \begin{pmatrix} x(t) \\ y(t) \\ z(t) \end{pmatrix}$ with $t \in [0, T_f]$ from a start point $\eta(0)$ with specified velocity $\dot{\eta}(0)$ and acceleration $\ddot{\eta}(0) = 0$ to a destination point $\eta(T_f)$ with specified velocity $\dot{\eta}(T_f)$ and acceleration $\ddot{\eta}(T_f) = 0$, taking into account the kinematics of the aircraft, the operational requirements of the mission and the no-fly areas.

The objective function to be minimized can be the flight time T_f, the length of the trajectory $\int_0^{T_f} |\dot{\eta}(t)| dt$, the mean flight height above ground $\frac{1}{T_f} \int_0^{T_f} (z(t) - h_{\text{terrain}}(x(t), y(t)))$ with $h_{\text{terrain}}(x(t), y(t))$ denoting the terrain elevation of point $(x(t), y(t))$,...

The approach is based on a discretization of the airspace by a 3D network [9]. Its topology depends on the kinematic properties of the aircraft, the operational requirements of the mission and the relief of the terrain. Each directed path in the network corresponds to a trajectory which is both flyable and feasible [129]. Generation of flight path segment is split into two subsystems:

1. A twice continuously differentiable parametrized curve with appropriate conditions is determined using quintic Hermite interpolation

$$x(t) = a_{51}t^5 + a_{41}t^4 + a_{31}t^3 + a_{21}t^2 + a_{11}t + a_{01}$$
$$y(t) = a_{52}t^5 + a_{42}t^4 + a_{32}t^3 + a_{22}t^2 + a_{12}t + a_{02}$$
$$z(t) = a_{53}t^5 + a_{43}t^4 + a_{33}t^3 + a_{23}t^2 + a_{13}t + a_{03} \tag{2.40}$$

The parameters a_{ij} can be determined by the endpoint conditions.

2. The flyability and feasibility of the trajectories are checked using a simplified model of the aircraft.

2.2.3.4 Pythagorean Hodographs

The Pythagorean hodograph condition in $\mathbb{R}^3$ is given by

$$x'^2(t) + y'^2(t) + z'^2(t) = \tilde{\sigma}^2(t) \tag{2.41}$$

where $\tilde{\sigma}(t)$ represents the parametric speed. The problem lies in finding an appropriate characterization for polynomial solutions [124].

Theorem 57. *If relatively prime real polynomials $a(t), b(t), c(t), d(t)$ satisfy the Pythagorean condition:*

$$a^2(t) + b^2(t) + c^2(t) = d^2(t) \tag{2.42}$$

they must be expressible in terms of other real polynomials $\tilde{u}(t), \tilde{v}(t), \tilde{p}(t), \tilde{q}(t)$ in the form:

$$
\begin{aligned}
a(t) &= \tilde{u}^2(t) + \tilde{v}^2(t) - \tilde{p}^2(t) - \tilde{q}^2(t) = x'(t) \\
b(t) &= 2\left(\tilde{u}(t)\tilde{q}(t) + \tilde{v}(t)\tilde{p}(t)\right) = y'(t) \\
c(t) &= 2\left(\tilde{v}(t)\tilde{q}(t) - \tilde{u}(t)\tilde{p}(t)\right) = z'(t) \\
d(t) &= \tilde{u}^2(t) + \tilde{v}^2(t) + \tilde{p}^2(t) + \tilde{q}^2(t) = \sigma(t)
\end{aligned}
\tag{2.43}
$$

This form can be written in several different ways corresponding to permutations of $a(t), b(t), c(t)$ and $\tilde{u}(t), \tilde{v}(t), \tilde{p}(t), \tilde{q}(t)$.

If the polynomials $\tilde{u}(t), \tilde{v}(t), \tilde{p}(t), \tilde{q}(t)$ are specified in terms of Bernstein coefficients on $t \in [0,1]$, the Bezier control points of the spatial PH curves they define can be expressed in terms of these coefficients. For example for $\tilde{u}(t) = u_0(1-t) + u_1 t$ and similarly for $\tilde{v}(t), \tilde{p}(t), \tilde{q}(t)$, the control points of spatial Pythagorean hodograph cubic are found to be of the form:

$$
P_1 = P_0 + \frac{1}{3}\begin{pmatrix} u_0^2 + v_0^2 - p_0^2 - q_0^2 \\ 2\left(u_0 q_0 + v_0 p_0\right) \\ 2\left(v_0 q_0 - u_0 p_0\right) \end{pmatrix}
\tag{2.44}
$$

$$
P_2 = P_1 + \frac{1}{3}\begin{pmatrix} u_0 u_1 + v_0 v_1 - p_0 p_1 - q_0 q_1 \\ \left(u_0 q_1 + u_1 q_0 + v_0 p_1 + v_1 p_0\right) \\ \left(v_0 q_1 + v_1 q_0 - u_0 p_1 - u_1 p_0\right) \end{pmatrix}
\tag{2.45}
$$

$$
P_3 = P_2 + \frac{1}{3}\begin{pmatrix} u_1^2 + v_1^2 - p_1^2 - q_1^2 \\ 2\left(u_1 q_1 + v_1 p_1\right) \\ 2\left(v_1 q_1 - u_1 p_1\right) \end{pmatrix}
\tag{2.46}
$$

The point P_0 corresponds to the integration constants.

2.2.4 THE ZERMELO PROBLEM: AIRCRAFT IN THE WIND

2.2.4.1 Initial Zermelo's Problem

Zermelo's problem was originally formulated to find the quickest nautical path for a ship at sea in the presence of currents, from a given departure point in $\mathbb{R}^2$ to a given destination point. It can also be applied to the particular case of an aircraft with a constant altitude and a zero flight path angle and the wind velocity represented by $W = (W_N, W_E)$.

A UAV has to travel through a region of strong constant wind at a constant altitude. The wind is assumed to have a constant wind velocity W in the y direction. The autopilot modulates the aircraft's heading χ to minimize travel time to the origin.

2.2.4.1.1 First Case Study

In the first case study, the UAV is assumed to have a constant velocity V and its heading χ is chosen as an input. The goal is to minimize time with the following boundary conditions: $x_0 = y_0 = 0$ and $x_f = 1; y_f = 0$. The control is assumed to be unconstrained. The minimal time problem can be formulated as follows:

$$\min \int_0^{T_f} dt \tag{2.47}$$

subject to

$$\dot{x} = V \cos \chi$$
$$\dot{y} = V \sin \chi + W_E$$
$$x(0) = y(0) = 0 \qquad (2.48)$$
$$x(T_f) = 1 \quad y(T_f) = 0$$

Using the Pontryagin minimum principle, the following optimal control can be calculated:

$$\chi^* = -\arcsin \left(\frac{W_E}{V} \right) \qquad (2.49)$$

while the optimal trajectories are

$$x^*(t) = tV \cos \chi$$
$$y^*(t) = t(V \sin \chi + W_E) \qquad (2.50)$$

The final time is:

$$T_f = \frac{1}{\sqrt{V^2 - W_E^2}}$$

This resolution is only possible if $|W_E| \leq V$.

2.2.4.1.2   Second Case Study

The second case study is an attempt to be more realistic. The input now is the rate of the heading $\dot{\chi}$, constrained to belong to the interval $[-U_{max}, U_{max}]$. The boundary conditions are slightly different from the first case: $x_0 = y_0 = \chi_0 = 0$ and $\chi_f = 0; y_f = 0$. The minimal time problem can be formulated as follows:

$$\min \int_0^{T_f} dt \qquad (2.51)$$

subject to

$$\dot{x} = V \cos \chi$$
$$\dot{y} = V \sin \chi + W_E$$
$$\dot{\chi} = U \qquad (2.52)$$
$$x(0) = y(0) = \chi(0) = 0$$
$$\chi(T_f) = 0 \quad y(T_f) = 0$$

Using the Pontryagin minimum principle, the following optimal control can be calculated:

$$U^* = \left\{ \begin{array}{ll} U_{max} & 0 \leq t \leq t_1 \\ -U_{max} & t_1 \leq t \leq T_f \end{array} \right\} \qquad (2.53)$$

with $t_1 = \frac{T_f}{2}$ while the optimal trajectories are

$$\chi^* = \left\{ \begin{array}{ll} U_{max} t & 0 \leq t \leq t_1 \\ U_{max}(T_f - t) & t_1 \leq t \leq T_f \end{array} \right\} \qquad (2.54)$$

$$x^* = \left\{ \begin{array}{ll} \frac{V}{U_{max}} \sin (U_{max} t) & 0 \leq t \leq t_1 \\ \frac{V}{U_{max}} \sin (U_{max} (t - T_f)) + 2 \frac{V}{U_{max}} \sin \left(\frac{T_f U_{max}}{2} \right) & t_1 \leq t \leq T_f \end{array} \right\} \qquad (2.55)$$

$$y^* = \left\{ \begin{array}{ll} -\frac{V}{U_{max}} \cos (U_{max} t) + W_E t + \frac{V}{U_{max}} & 0 \leq t \leq t_1 \\ \frac{V}{U_{max}} \cos (U_{max} (t - T_f)) + W_E (t - T_f) - \frac{V}{U_{max}} & t_1 \leq t \leq T_f \end{array} \right\} \qquad (2.56)$$

The final time can be found from the resolution of the following equation:

$$\frac{U_{\max}}{2} \frac{W_E}{V} T_f - \cos\left(\frac{U_{\max} T_f}{2}\right) = 1 \tag{2.57}$$

Depending on the values of $U_{\max}, V, W$, this equation may have or not a real positive solution.

Another case study is treated next. The equations describing the optimal path for the case of linearly varying wind velocity are

$$\begin{aligned} \dot{x} &= V\cos\chi + W_N(y) \\ \dot{y} &= V\sin\chi \end{aligned} \tag{2.58}$$

where (x,y) are its coordinates, and $W_N = \frac{V_y}{h}$ is the velocity of the wind. The initial value of χ is chosen so that the path passes through the origin. For the linearly varying wind strength considered here, the optimal steering angle can be related to the aircraft position through a system of implicit feedback equation [117]

$$\dot{\chi} = -\cos^2\chi \frac{dW_N}{dy} \tag{2.59}$$

If $W_N = \frac{W}{a}y$, a being a constant then

$$\chi = \arctan\left(\frac{W}{a}t + \tan\chi_0\right) \tag{2.60}$$

The optimal trajectory is then given by

$$y = a\left(\frac{1}{\cos\chi_f} - \frac{1}{\cos\chi}\right) \tag{2.61}$$

and

$$\begin{aligned} x &= \frac{a}{2}\left(\frac{1}{\cos\chi_f}\left(\tan\chi_f - \tan\chi\right) - \tan\chi\left(\frac{1}{\cos\chi_f} - \frac{1}{\cos\chi}\right)\right) \\ &+ \frac{a}{2}\ln\left(\frac{\tan\chi_f + \frac{1}{\cos\chi_f}}{\tan\chi + \frac{1}{\cos\chi}}\right) \end{aligned} \tag{2.62}$$

If $W_N(x,y) = -Wy$, it has been proved in [184] that the time to go is given by

$$T_f = \frac{1}{W}\left(\tan\chi_f - \tan\chi_0\right) \tag{2.63}$$

Remark 58. *The following approximate relations can be implemented, for N way-points:*

$$\dot{\chi}_k = \frac{y_k - y_{k-1}}{x_k - x_{k-1}}\cos^2\chi_k \tag{2.64}$$

$$\dot{\gamma}_k = -\frac{(z_k - z_{k-1})\cos^2\gamma_k}{\cos\chi_k(x_k - x_{k-1}) + \sin\chi_k(y_k - y_{k-1})} \tag{2.65}$$

for $k = 1, \ldots, N$.

2.2.4.2 2D Zermelo's Problem on a Flat Earth

The following problem can be solved in the same way using the Pontryagin maximum principle:

Problem 59. *2D Zermelo's Problem on a flat Earth Time optimal trajectory generation can be formulated as follows:*

$$min \int_0^{T_f} dt \tag{2.66}$$

subject to

$$\dot{x} = U_1(t) + W_N(x,y)$$
$$\dot{y} = U_2(t) + W_E(x,y) \tag{2.67}$$

with:

$$U_1^2(t) + U_2^2(t) \leq V_{max}^2 \tag{2.68}$$

The heading angle is the control available for achieving the minimum time objective [184].

Zermelo's navigation formula consists of a differential equation for $U^*(t)$ expressed in terms of only the drift vector and its derivatives. The derivation can be explained as follows. Let the angle be $\chi(t)$ given by $U_1(t) = V_{max} \cos \chi(t)$ and $U_2(t) = V_{max} \sin \chi(t)$, the following ordinary differential equation must be solved:

$$\frac{d\chi}{dt} = -\cos^2 \chi \frac{\partial W_N}{\partial y} + \sin \chi \cos \chi \left(\frac{\partial W_N}{\partial x} - \frac{\partial W_E}{\partial y} \right) + \sin^2 \chi \frac{\partial W_E}{\partial x} \tag{2.69}$$

Remark 60. *Flying exact trajectories relative to the ground may be good for optimizing operations (fixed approach trajectories), but the autopilot has to control the aircraft that drifts with the air mass. Relative to the air mass to adhere to the ground trajectory, the autopilot must continuously vary the bank angle during turns.*

The problem is generating optimal path from an initial position and orientation to a final position and orientation in the 2D plane for an aircraft with bounded turning radius in the presence of a constant known wind. Some researchers have addressed the problem of optimal path planning of an aircraft at a constant altitude and velocity in the presence of wind with known magnitude and direction. A dynamic programming method to find the minimum time way-point path for an aircraft flying in known wind was proposed by [186,187]. A target trajectory in the horizontal plane can be described by connecting straight lines and arcs. Clearly, this type of description forces the aircraft to attain a bank angle instantaneously at the point of transition between the straight and curved segments. In [137], the curved part of a trajectory is designed assuming a certain steady bank angle and initial speed in a no-wind condition. If a strong wind exists, the autopilot has to change the bank angle depending on the relative direction of the wind during the curved flight phase or continuously fly with a steeper than nominal bank angle. A steeper bank angle and changing relative wind direction both affect thrust control.

In the absence of wind, this is the Dubins problem [130]. The original problem of finding the optimal path with wind to a final orientation is transformed over a moving virtual target whose velocity is equal and opposite to the velocity of the wind. An UAV approaching a straight line under a steady wind can be considered as a virtual target straight line moving with an equal and opposite velocity to the wind acting on it in a situation where the wind is absent. So in the moving frame, the path will be a Dubins path. The ground path is indeed the time optimal path [170].

Time optimal navigation for aircraft in a planar time-varying flow field has also been considered in [389]. The objective is to find the fastest trajectory between initial and final points. It has been shown that in a point symmetric time-varying flow field, the optimal steering policy necessarily has to be such that the rate of the steering angle equals the angular rotation rate of the fluid particles.

In many real scenarios, the direction of wind is not known a priori or it changes from time to time [336]. A receding Horizon Controller was used in [198] to generate trajectories for an aircraft operating in an environment with disturbances. The proposed algorithm modifies the online receding horizon optimization constraints (such as turn radius and speed limits) to ensure that it remains feasible even when the aircraft is acted by unknown but bounded disturbances [335].

2.2.4.3 3D Zermelo's Problem on a Flat Earth

Now, the wind optimal time trajectory planning problem for an aircraft in a 3D space is considered [184,185]. An aircraft must travel through a windy region. The magnitude and the direction of the winds are known to be functions of position, i.e., $W_N = W_N(x,y,z)$, $W_E = W_E(x,y,z)$ and $W_D = W_D(x,y,z)$ where (x,y,z) are 3D coordinates, and (W_N, W_E, W_D) are the velocity components of the wind. The aircraft velocity relative to the air V is constant. The minimum time path from point A to point B is sought. The kinematic model of the aircraft is

$$\begin{aligned}
\dot{x} &= V\cos\chi\cos\gamma + W_N \\
\dot{y} &= V\sin\chi\cos\gamma + W_E \\
\dot{z} &= -V\sin\gamma + W_D
\end{aligned} \tag{2.70}$$

where χ is the heading angle of the aircraft relative to the inertial frame, γ is the flight path angle.

Using the PMP, the evolution of the heading is obtained as a nonlinear ordinary differential equation:

$$\begin{aligned}
\dot{\chi} &= \sin^2\chi\frac{\partial W_E}{\partial x} + \sin\chi\cos\chi\left(\frac{\partial W_N}{\partial x} - \frac{\partial W_E}{\partial y}\right) \\
&\quad - \tan\gamma\left(\sin\chi\frac{\partial W_D}{\partial x} - \cos\chi\frac{\partial W_D}{\partial y}\right) - \cos^2\chi\frac{\partial W_N}{\partial y}
\end{aligned} \tag{2.71}$$

The evolution of the flight path angle is given by a nonlinear ordinary differential equation:

$$\begin{aligned}
\dot{\gamma} &= \cos^2\gamma\cos\chi\frac{\partial W_N}{\partial z} + \cos^2\gamma\frac{\partial W_E}{\partial z} + \sin\gamma\cos\gamma\left(\frac{\partial W_N}{\partial x} - \frac{\partial W_D}{\partial z}\right) \\
&\quad + \sin^2\gamma\frac{\partial W_E}{\partial x} - \sin^2\gamma\sec\chi\frac{\partial W_D}{\partial x} - \sin\gamma\cos\gamma\tan\chi\sin^2\chi\frac{\partial W_E}{\partial x} \\
&\quad - \sin\gamma\cos\gamma\sin^2\chi\left(\frac{\partial W_N}{\partial x} - \frac{\partial W_E}{\partial y}\right) + \tan\chi\sin^2\gamma\left(\sin\chi\frac{\partial W_D}{\partial x} - \cos\chi\frac{\partial W_D}{\partial y}\right) \\
&\quad + \sin\gamma\cos\gamma\sin\chi\cos\chi\frac{\partial W_N}{\partial y}
\end{aligned} \tag{2.72}$$

More information about the derivation of these equations can be found in [56,156].

Remark 61. *When it is assumed that there is no wind, the velocity of the aircraft is constant and that the available control inputs are the flight path angle and the heading angle, the following optimal trajectory can be calculated:*

$$\begin{aligned}
x &= Vt\cos\gamma\cos\chi + x_0 \\
y &= Vt\cos\gamma\sin\chi + y_0 \\
z &= -Vt\sin\gamma + z_0
\end{aligned} \tag{2.73}$$

where (x_0, y_0, z_0) is the initial position of the aircraft. If the final position is given by (x_f, y_f, z_f) then the predicted arrival time is

$$T = \frac{1}{V}\sqrt{(x_f - x_0)^2 + (y_f - y_0)^2 + (z_f - z_0)^2} \tag{2.74}$$

while the heading angle is given by

$$\chi = \arctan\left(\frac{y_f - y_0}{x_f - x_0}\right) \tag{2.75}$$

and the flight path angle by

$$\gamma = \arctan\left(\frac{z_f - z_0}{\cos\chi(x_f - x_0) + \sin\chi(y_f - y_0)}\right) \tag{2.76}$$

When the considered control inputs are the flight path angle rate $\dot{\gamma}$, the heading angle rate $\dot{\chi}$ and the derivative of the velocity $\dot{V}$, the idea is to use the structure and to apply simple bang–bang controls in the planning. The amount of control available is a concern in the planning for this system due to the drift term. The class of bang–bang controls is often a sufficiently rich class of controls for analysis of nonlinear systems. This simple class of controls makes it possible to integrate the equations forward.

2.2.4.4   3D Zermelo's Problem on a Spherical Surface

This paragraph presents the practical trajectory optimization algorithm that approximates the minimization of the total cost of travel time and fuel consumption for aircraft on a spherical surface. A typical aircraft trajectory consists of an initial climb, a steady-state cruise and a final descent. Here, aircraft performance is optimized for the cruise phase only. The cruise trajectory is divided into segments on several altitudes as the optimal cruise altitude increases due to the reduction in aircraft weight as fuel is used. The aircraft optimal heading during cruise is the solution of the **Zermelo's problem** derived on a spherical Earth surface in the absence of constraints. The horizontal trajectory segments are optimized based on the cost-to-go associated with extremal trajectories generated by forward–backward integrating the dynamical equations for optimal heading and aircraft motion from various points in the airspace [60,342].

The direct operating cost for a cruising aircraft can be written as

$$J = \int_{t_0}^{T_f} \left(C_t + C_f \tilde{f}(m, z, V)\right) dt \tag{2.77}$$

where C_t and C_f are the cost coefficient of time and fuel. The **fuel flow rate** $\tilde{f}$ can be approximated by a function of aircraft mass m, altitude z and airspeed V. The fuel burn for aircraft during cruise $\tilde{F}$ is calculated as

$$\tilde{F} = t\tilde{f} \tag{2.78}$$

where t is the elapsed time. The fuel burn rate f for jets and turboprops is determined by the specific fuel consumption (SFC) and thrust T

$$\tilde{f} = \frac{C_{fcr}}{1,000} \cdot \text{SFC} \cdot T_f \qquad \text{SFC} = C_{f_1}\left(1 + \frac{V_{TAS}}{C_{f_2}}\right) \tag{2.79}$$

where $C_{fcr}, C_{f_1}, C_{f_2}$ are the thrust SFC coefficients and V_{TAS} is the true airspeed.

During cruise, thrust equals the aerodynamic drag forces and lift the weight.

$$T = D = \frac{1}{2}C_D(V, \alpha)\rho S V^2 \qquad C_D(V, \alpha) = C_{D_0} + KC_L^2 \qquad C_L(V, \alpha) = \frac{2mg}{\rho S V^2} \tag{2.80}$$

Under the **international standard atmosphere** (ISA), the tropopause altitude is at $11,000m$, and the optimal cruise altitude z_{opt} at or below the tropopause can be calculated as

$$z_{opt} = \left(1 - \exp\left(-f(m, V)K_T R_{gas}/2(g + K_T R_{gas})\rho_{0ISA}^2\right)\right)\left(\frac{1,000 T_{0ISA}}{6.5}\right) \tag{2.81}$$

Above the tropopause, it is

$$z_{opt} = \frac{-f(m, V)R_{gas}T_{propISA}}{2g\rho_{tropISA}} + 11,000m \tag{2.82}$$

where

$$f(m,V) = \ln\left(\frac{4m^2g^2K}{S^2V^4C_{D_0}}\right) \tag{2.83}$$

R_{gas} is the real gas constant for air, the temperature gradient K_T, the sea level density ρ_{0ISA} and the sea level temperature T_{0ISA} considered constant under the ISA. The air density $\rho_{tropISA}$ and temperature $T_{rmtropISA}$ are all constant at the troposphere.

Optimal cruise altitudes are computed from relations (2.81) and (2.82) based on the atmospheric constants and aerodynamic drag coefficients that are aircraft-type dependent. They vary also with aircraft mass and airspeed.

The aircraft equations of motion at a constant altitude above the spherical Earth's surface are

$$\dot{\ell} = \frac{V\cos\chi + W_E(\ell,\lambda,z)}{R\cos\lambda} \tag{2.84}$$

$$\dot{\lambda} = \frac{V\sin\chi + W_N(\ell,\lambda,z)}{R} \tag{2.85}$$

$$\dot{m} = -f \tag{2.86}$$

subject to the conditions that the thrust equals the drag, ℓ is the longitude, λ is the latitude, χ is the heading angle and R is the Earth radius $R \gg z$.

The dynamical equation for the optimal aircraft heading is

$$\dot{\chi} = -\frac{F_{wind}(\chi,\ell,\lambda,W_E,W_N,V)}{R\cos\lambda} \tag{2.87}$$

where $F_{wind}(\chi,\ell,\lambda,W_E,W_N,V)$ is aircraft heading dynamics in response to winds and is expressed as

$$\begin{aligned}F_{wind} = &-\sin\chi\cos\chi\frac{\partial W_E}{\partial \ell} + \cos^2\chi\sin\lambda W_E + \cos^2\chi\cos\lambda\left(\frac{\partial W_E}{\partial \lambda} - \frac{\partial W_N}{\partial \ell}\right)\\ &+\sin\chi\cos\chi\sin\lambda W_N + \cos\chi\sin\chi\cos\lambda\frac{\partial W_N}{\partial \lambda} + V\cos\chi\sin\lambda + \cos^2\chi\frac{\partial W_N}{\partial \ell}\end{aligned} \tag{2.88}$$

The minimum time trajectory is the combination of wind optimal extremals on several different altitudes, each solved using conditions on that altitude. The optimal virtual profile provides the initial and subsequent optimal cruise altitudes as well as the transition time between the altitudes.

2.2.4.5 Virtual Goal

The problem is to determine the optimal path in 3D space between the initial configuration: position X_1 and orientation e_1 and the final configuration: position X_2 and orientation e_2 for a constant speed aircraft and with turn rate constraints. The unit orientation vectors at the initial and final points are $e_1 = (\cos\gamma_1\cos\chi_1,\cos\gamma_1\sin\chi_1,-\sin\gamma_1)^T$ and $e_2 = (\cos\gamma_2\cos\chi_2,\cos\gamma_2\sin\chi_2,-\sin\gamma_2)^T$. The proposed path planning algorithm is based on the following kinematic equations of motion:

$$\begin{aligned}\dot{x} &= V\cos\gamma\cos\chi + W_N\\ \dot{y} &= V\cos\gamma\sin\chi + W_E\\ \dot{z} &= -V\sin\gamma + W_D\\ \dot{\chi} &= \omega_1\\ \dot{\gamma} &= \omega_2\end{aligned} \tag{2.89}$$

where the state vector is defined as $X = (x,y,z,\chi,\gamma)$, the velocity V is assumed to be constant and ω_1,ω_2 are the control inputs. The trajectory must satisfy a maximum turn rate constraint or the curvature bound $\pm\kappa_{max}$ and torsion $\pm\tau_{max}$ for the aircraft.

The original problem of computing an optimal path in the presence of wind can be expressed as one of computing the optimal path from an initial position X_1 and orientation e_1 with no wind to a final orientation e_2 and a virtual goal that moves with a velocity equal and opposite to the velocity of the wind [364]. The air path defined as the path traveled by the aircraft with respect to the moving frame can be of CSC type or of helicoidal type. Let the minimum time required to reach the final point be T_f. At T_f, the virtual final point moves from the given final position $X_2 = (x_2, y_2, z_2)$ to a new position $X_{2v} = (x_{2v}, y_{2v}, z_{2v})$ which can be expressed as a function of T_f.

$$X_{2v} = X_2 - (W \cos \gamma_1 \cos \chi_1, W \cos \gamma_1 \sin \chi_1, -W \sin \gamma_1)^T T_f \tag{2.90}$$

or equivalently

$$X_{2v} = X_2 - \int_0^{T_f} W(t)dt \tag{2.91}$$

with $W = \sqrt{W_N^2 + W_E^2 + W_D^2}$.

Finally, the position to be reached is X_{2v}. The heading angle and the flight path angle at the virtual goal are χ_2, γ_2. The reformulated problem (2.90) is similar to the kinematic model of a 3D Dubins vehicle in the absence of wind except that the final point X_{2v} is also dependent on T_f. Thus, a CSC or helicoidal path can be computed, the variables X_{2v}, T_f can be obtained using nonlinear equations solving algorithm and the ground path can be computed in the inertial frame using the state equation in (2.89).

2.3 GUIDANCE AND COLLISION/OBSTACLE AVOIDANCE

Aircraft collision is a serious concern as the number of aircraft in operation increases. In the future, they will be expected to carry sophisticated avoidance systems when flying together with conventional aircraft. **On-board sensor systems** combined with self-operating algorithms will ensure collision avoidance with little intervention from ground stations. On-board sensors can detect other aircraft nearby. Information related to the other aircraft such as position, velocity and heading angle can be used to build an avoidance command. In order for an aircraft to maneuver successfully in such a dynamic environment, a feasible and collision-free trajectory needs to be planned in the physical configuration space. The avoidance law should be generated in real-time and simple to implement. The ability to sense and avoid natural and man-made obstacles and to rebuild its flight path is an important feature that a smart autonomous aircraft must possess [391]. Guidance, trajectory generation, flight and mission planning are the core of the flight management system of a smart autonomous aircraft [52,70]. The computational abilities provide the strongest constraints on the autonomous aircraft, although advances in the hardware mechanisms are to be expected. Improvements in software are essential.

Collision avoidance is of vital importance for small UAV flying on low heights which usually encounter a large number of mostly static obstacles [87]. Due to a limited knowledge of the environment and small computing power, collision avoidance needs to rely on little information while being computationally efficient. Guidance laws are one way of tackling such difficulties [373]. For autonomous missions, sense and avoid capability is a critical requirement.

Collision avoidance can be broadly classified into global and local path planning algorithms, to be addressed in a successful mission. Whereas global path planning broadly lays out a path that reaches the goal point, local collision avoidance algorithms, which are usually fast, reactive and carried out online, ensure safety of the aircraft from unexpected and unforeseen obstacles collisions. The algorithm in [340] first plans a path to the goal avoiding the obstacles known a priori. If a collision is predicted to occur, the path is replanned so that the obstacle is avoided. However, the objective is to always move toward the goal point after collision is avoided. Hence, these algorithms are considered to be global path planning algorithm with local collision avoidance features embedded into it.

The problem of trajectory prediction is encountered whenever it is necessary to provide the control system with a projection of the future position of the aircraft, given the current flight conditions, together with an envelope of feasible trajectory. The prediction can be integrated in a ground proximity warning system, to reveal a possible ground collision with a time margin sufficient for undertaking an appropriate control action for avoiding obstacles on the prescribed flight path. In this case, the knowledge of an envelope of feasible future position is sought. In [29], an algorithm is presented for determining future positions of the aircraft center of gravity inside a given prediction horizon from measurements of acceleration and angular velocity. The technique is based on the weighted combination of two estimates:

1. The projection in the **Frenet frame** of a helix with vertical axis for long-term prediction in steady and quasi-steady maneuver segments.
2. A third-order accurate power series expansion of the trajectory in the Frenet frame, useful for short-term trajectory prediction during transient maneuvering phases.

2.3.1 GUIDANCE

Guidance is a dynamic process of directing an object toward a given point that may be stationary or moving [168,407]. Inertia of the aircraft is ignored in most approaches, and their dynamics are ignored [197]. In this section, three approaches are presented: two conventional and the last one based on fuzzy techniques.

Definition 62. *Guidance is the logic that issues steering commands to the aircraft to accomplish certain flight objectives. The guidance algorithm generates the autopilot commands that steer the autonomous aircraft. A guidance system is defined as a group of components that measures the position of the guided aircraft with respect to the target and change its flight path according to a guidance law to achieve the flight mission goal.*

A guidance law is defined as an algorithm that determines the required commanded aircraft accelerations. In guidance studies, only local information on the wind flow field is assumed to be available and a near optimal trajectory, namely, a trajectory that approximates the behavior of the optimal trajectory, is determined [337].

There are various approaches to both static and moving obstacles detection that are mostly based on the collision cone approach. Often, collision avoidance is achieved by tracking an aim-point at a safe distance from the obstacle, using a homing guidance law. Explicit avoidance laws can be derived by limiting the consideration to a plane, defined by the relative geometry between the aircraft and the obstacle. The main issue is that primary mission objectives are not fully considered which may cause trajectories to be far from optimal [377].

Geometric techniques including pure pursuit and variants of pursuit and **line of sight** (LOS) guidance laws are mainly found in [407]. The path following algorithms based on pure pursuit and LOS-based guidance laws uses a **virtual target point** (VTP) on the path. The guidance laws direct the aircraft to chase the virtual target point which eventually drives the aircraft onto the path. The distance between the virtual target point and the UAV position projected on the path is often called virtual distance. The stability of LOS guidance laws for path following depends significantly on the selection of the virtual distance parameter. Pure pursuit and LOS guidance laws can be combined to create a new guidance law for path following [384].

3D guidance refers to following the mission path in both the horizontal and vertical planes. It covers both 2D ground track following and altitude profile following of the desired trajectory. In the 3D case, guidance commands are generated for the lateral directional control system in terms of reference roll or heading angles, and for the longitudinal control system in terms of pitch or altitude commands [377]. A subset of the general 3D problem is the 2D lateral guidance problem in which the guidance objective is to ensure accurate ground track following of the aircraft. Thus, it must exactly fly over lines and arcs joining mission way-points as projected on plane ground, with minimum cross-track or lateral deviation.

2.3.1.1 Proportional Navigation

Line of sight (LOS) and its variations are still the simplest and most popular guidance law in use today [74]. The **Proportional Navigation** (PN) guidance law is a strategy that can be applied to any situation where the nominal trajectory is well known. According to this law, the maneuver of the aircraft is proportional to the LOS rate. It is based on the fact that if the target and the aircraft have constant velocities, then on collision course, the LOS rate is zero. There are two basic disturbances that may influence the guidance loop: the target and the initial conditions. The guidance law is designed to steer the aircraft in a flight path toward the boundary of a safety bound. The safety bound can be a minimal radius circle and/or cylinder to prevent collision.

The center of mass of the aircraft is instantaneously located at $R(t)$ and its desired value $R_T(t)$ with respective velocities $V(t)$ and $V_T(t)$ relative to a stationary frame of reference. The instantaneous position of the target relative to the aircraft is given by

$$e(t) = R_T(t) - R(t) \tag{2.92}$$

with

$$V_e(t) = \frac{de}{dt} = V_T(t) - V(t) \tag{2.93}$$

The following control law for guidance is used:

$$V_e(t) = \mathbf{K}(t)e(t) \tag{2.94}$$

where $\mathbf{K}(t)$ is a time-varying gain matrix.

The required acceleration control $U(t)$ to be applied to the aircraft is

$$U(t) = \frac{dV_e}{dt} = \mathbf{K}(t)V(t) + \dot{\mathbf{K}}(t)e(t) \tag{2.95}$$

A choice of state vector for the guidance problem is

$$X(t) = (e(t), \dot{e}(t))^T \tag{2.96}$$

which yields the linear feedback law

$$U(t) = \left(\dot{\mathbf{K}}(t), \mathbf{K}(t)\right) X(t) \tag{2.97}$$

This proportional navigation guidance can be easily implemented.

A suitable navigation strategy for a rendezvous would be to achieve simultaneously a zero miss distance and a zero relative speed.

A simple approach is to have the instantaneous velocity error $V(t)$ become aligned with the acceleration error $a = \dot{V}_T(t) - \dot{V}(t)$. This implies that the cross product of velocity and acceleration errors must vanish:

$$V \times a = 0 \tag{2.98}$$

Such a navigation law is termed cross product steering [407].

2.3.1.2 Method of Adjoints

A **terminal guidance** is considered where two objects, having constant speeds, move in a plane toward a collision point. Small perturbations are assumed with respect to the nominal collision triangle. That is, the line of sight deviation λ is small. Let R be the range and $V_c = -\dot{R}$ the closing speed. The usual small perturbation is assumed with the reference line, in that case the closing speed is assumed to be constant. Then along the LOS, the final time T_f and time to go t_{go} are defined as follows:

$$T_f = \frac{R_0}{-\dot{R}} = \frac{R_0}{V_c} \tag{2.99}$$

$$t_{go} = \frac{R}{-\dot{R}}$$

(2.100)

The miss distance m is defined as

$$m = y(T_f)$$

(2.101)

where y is the relative separation perpendicular to the initial LOS. Proportional navigation guidance law states that the command acceleration of the aircraft normal to the LOS n_c is proportional to the LOS rate:

$$n_c = N'V_c\dot{\lambda}$$

(2.102)

where N' is the navigation constant [407].

2.3.1.3 Fuzzy Guidance Scheme

The overall guidance scheme has two components: a way-point generator and a fuzzy guidance system (FGS). The desired trajectory is specified in terms of a sequence of way-points without any requirements on the path between two successive way-points [180]. The way-point generator holds a list of way-points in 5D, checks aircraft position and updates the desired way-point when the previous one has been reached with a given tolerance. The way-point generator only task is to present the actual way-point to the FGS [463,466].

A tolerance ball is included around the way-point, defining that as actual target reached. Between the way-point generator and the FGS, a coordinate transformation is performed to convert Earth-fixed frame position errors into way-point frame components. Each way-point defines a coordinate frame centered at the way-point position (x_w, y_w, z_w) and rotated by its heading angle χ_w around the $z-$axis. γ_w is the flight path angle of the way-point. The coordinate transformation allows the synthesis of a fuzzy rule-set valid in the way-point fixed coordinate frame, which is invariant with respect to the desired approach direction χ_w. When a way-point is reached, the next one is selected, the actual reference value is changed and the rotation matrix is updated to transform position and orientation errors into the new way-point coordinate frame.

The aircraft autopilots are designed to track desired airspeed, heading and flight path angles V_d, γ_d, χ_d using decoupled closed-loop inertial dynamics and so three independent Takagi–Sugeno controllers were synthesized to constitute the FGS.

1. The first controller generates the desired flight path angle γ_d for the autopilot using altitude error:

$$e_z = z_w - z_A \qquad \gamma_d = f_\gamma(e_z)$$

(2.103)

The state vector $(V_A, \gamma_A, \chi_A, z_A)^T$ represents aircraft speed, flight path angle, heading and altitude, respectively.

2. The second controller computes desired aircraft velocity

$$e_V = V_w - V_A \qquad V_d = V_w + f_V(e_V)$$

(2.104)

3. The third one is responsible for the generation of the desired heading angle χ_d using the position error along the X, Y axes and the heading error e_χ.

A fuzzy rule-set designed at a specific trim airspeed value could yield insufficient tracking performance when the desired way-point crossing speed V_w differs significantly from V [180]. To accommodate large values of e_V and to investigate the effect of disturbance, modeled as aircraft's speed differential with respect to crossing speed V_w, a speed-correlated scale coefficient of position error is introduced. The rotation matrix is defined as

$$\mathbf{R}(\chi_w) = \begin{pmatrix} \cos(\chi_w + \frac{\pi}{2}) & \sin(\chi_w + \frac{\pi}{2}) \\ -\sin(\chi_w + \frac{\pi}{2}) & \cos(\chi_w + \frac{\pi}{2}) \end{pmatrix}$$

(2.105)

The position errors in the fixed way-point coordinate frame are given by

$$
\begin{pmatrix} e_x^w \\ e_y^w \end{pmatrix} = \mathbf{R}(\chi_w) \begin{pmatrix} x - x_w \\ y - y_w \end{pmatrix}
\tag{2.106}
$$

The velocity compensated position errors $e_{x_c}^w, e_{y_c}^w$ are defined by

$$
\begin{pmatrix} e_{x_c}^w \\ e_{y_c}^w \end{pmatrix} = \frac{V^*}{V^w} \begin{pmatrix} x - x_w \\ y - y_w \end{pmatrix}
\tag{2.107}
$$

where V^* represents the airspeed value used during FGS membership rules design. In this way, position errors used by the FGS to guide the aircraft toward the way-point with desired approach direction are magnified when V^w requested way-point crossing speed is larger than V^* or reduced otherwise.

Remark 63. *Equation (2.107) may diverge if $V^* \to 0$. However, this is not an operationally relevant condition because the requested way-point crossing speed should be defined according to aircraft flight parameters and the stall velocity must be avoided.*

Finally, the desired heading angle produced by fuzzy controller is

$$
\chi_d = \chi_w + f_\chi(e_{X_c}^w, e_{Y_c}^w)
\tag{2.108}
$$

The fuzzy guidance system is based on a Takagi–Sugeno system model described by a blending of IF-THEN rules. Using a weighted average defuzzifier layer, each fuzzy controller output is defined as follows:

$$
Y = \frac{\sum_{k=1}^m \mu_k(X) U_k}{\sum_{k=1}^m \mu_k(X)}
\tag{2.109}
$$

where $\mu_i(X)U_i$ is the ith membership function of input X to the ith zone. The membership functions are a combination of Gaussian curves of the form:

$$
f(X, \sigma, c) = \exp\left(-\frac{(X - c)^2}{\sigma^2} \right)
\tag{2.110}
$$

The fuzzy rules are defined according to the desired approach direction and the angular rate limitations of the aircraft. The fuzzy knowledge base is designed to generate flyable trajectories using the max linear and angular velocities and accelerations. The fuzzy guidance system provides different desired flight path and heading angle commands for different way-points. The altitude and velocity controllers are implemented using a Takagi–Sugeno model directly. For the altitude, the input is the altitude error e_z, and the output is the desired flight path angle γ_d. Inputs and outputs are mapped with four fuzzy sets each:

1. If e_z is N_∞ then γ_d is P_∞, for big negative errors
2. If e_z is N_s then γ_d is P_s, for small negative errors
3. If e_z is P_s then γ_d is N_s, for small positive errors
4. If e_z is P_∞ then γ_d is N_∞, for big positive errors

Here, the generic output constant P_s represents the output value s, and the constant N_s represents the output value $-s$.

The velocity controller is similar to the altitude controller. Three input fuzzy sets are used for the velocity error e_V and three for the resulting ΔV_d output:

1. If e_V is N_∞ then ΔV_d is P_s, for negative errors,
2. If e_V is ZE then ΔV_d is P_0, for near to zero errors,
3. If e_V is P_∞ then ΔV_d is N_s, for positive errors.

Guidance in the horizontal (x, y) plane is more complex. The horizontal plane fuzzy controller takes its input from scaled position errors $e_{x_c}^w, e_{y_c}^w$ and heading error e_χ. The error along the X axis is coded into five fuzzy sets:

1. N_∞ for big negative lateral errors
2. N_s for small negative lateral errors
3. ZE for near exact alignment
4. P_s for small positive lateral errors
5. P_∞ for big positive lateral errors.

Three sets are also defined over the Y^w axis error:

1. ZE for aircraft over the way-point
2. N_s for way-point behind the aircraft
3. P_s for way-point in front of the aircraft.

Finally, the heading error may be coded in seven fuzzy sets. In the application of equation (2.109), the m fuzzy rules are grouped into S groups, each with K rules: $m = SK$. The S groups correspond to S areas on the xy plane. From the preceding,

$$Y = \frac{1}{C(X)} \sum_{i=1}^{s} \sum_{j=1}^{K} \mu_i^{xy} \left(e_{X_c}^w, e_{Y_c}^w \right) \mu_{ij}^x (e_x) U_{ij} \tag{2.111}$$

or

$$Y = \frac{1}{C(X)} \sum_{i=1}^{s} \mu_i^{xy} \left(e_{X_c}^w, e_{Y_c}^w \right) \delta_i^x (e_x) \tag{2.112}$$

where

$$Y = \sum_{i=1}^{s} \frac{\mu_i^{xy} \left(e_{X_c}^w, e_{Y_c}^w \right)}{C(x)} \delta_i^x (e_x) = \sum_{i=1}^{s} \bar{\mu}_i^{xy} \left(e_{X_c}^w, e_{Y_c}^w \right) \delta_i^x (e_x) \tag{2.113}$$

Fixing $\left(e_{X_c}^w, e_{Y_c}^w \right)$ in the middle of the pth zone under the assumption that the contribution from the other zones is near zero yields:

$$Y \left(e_{X_c}^{wP}, e_{Y_c}^{wP} \right) = \bar{\mu}_p^{xy} \left(e_{X_c}^{wP}, e_{Y_c}^{wP} \right) \delta_p^x (e_x) + \sum_{i=1; i \neq p}^{s} \bar{\mu}_i^{xy} \left(e_{X_c}^w, e_{Y_c}^w \right) \delta_i^x (e_x) \tag{2.114}$$

or

$$Y \left(e_{X_c}^{wP}, e_{Y_c}^{wP} \right) \approx \bar{\mu}_p^{xy} \left(e_{X_c}^{wP}, e_{Y_c}^{wP} \right) \delta_p^x (e_x) \tag{2.115}$$

Equation (2.115) shows that once the fuzzy sets for the position errors $\left(e_{X_c}^{wP}, e_{Y_c}^{wP} \right)$ are fixed, the definition of fuzzy sets for e_x should be computed by looking first at each area on the XY plane and then adding the cumulative results. Under this assumption, seven fuzzy sets are defined for the heading error $e_x \in \{N_b, N_m, N_s, ZE, P_s, P_m, P_b\}$, b is for big, m for medium and s for small.

The design goals of this FGS are thus

1. Capability of reaching a set of way-points in a prescribed order,
2. Possibility of specifying the speed and heading of the aircraft at a way-point crossing,
3. Capability of quickly reconfiguring the way-point set in response to changes in the mission scenario,
4. Reaching fixed way-points as well as tracking and reaching moving way-points.

In an operational scenario, the way-point generator may be interfaced with a mission management system that updates the way-points when needed.

2.3.2 STATIC OBSTACLES AVOIDANCE

The basic representations of an environment are configuration space and occupancy grid. In configuration space, the dimensions of the environment plus the coordinates of all obstacles are given. In an occupancy grid, the environment is specified at a certain resolution with individual voxels representing either free space or obstacles [65]. There are many ways to represent a map and the position of the aircraft within the map. The free regions and obstacles may be represented as polyhedral, each comprising a list of vertices or edges. This is potentially a very compact form, but determining potential collisions between the aircraft and obstacles may involve testing against long lists of edges. A simpler representation is the occupancy grid. The environment is treated as a grid of cells and each cell is marked as occupied or unoccupied.

Motion planning is realized by an integrated set of algorithms such as collision checking [167], configuration sampling [191,333] and path planning [328]. It can be categorized as static, in which all obstacle configuration is known prior to planning or dynamic in which environment (obstacles) information becomes known to the planner only through real-time sensing of its local environment.

Given the nature of these aspects as diverse and difficult, most of the proposed work in the field of motion planning has focused on the consideration of some versions of the general problem. The ability to perform autonomous mission planning is considered one of the key enabling technologies for UAV. The development of algorithms capable of computing safe and efficient routes in terms of distance, time and fuel is very important [349].

The uniform framework to study the path planning problem among static obstacles is the **configuration space** (C-space). The main idea of the C-space is to represent the aircraft as a point called a configuration. The C-space is the set of all possible configurations, and C-free are the regions of C-space which are free of obstacles. Obstacles in the environment become C-obstacles in the C-space.

The main idea in sampling-based path planning method is to avoid an exhaustive construction of C-obstacles by sampling the C-space. The sampling scheme may be deterministic [352,354] or probabilistic [338,350]. The key issue is then to use an efficient grid-search algorithm to find an optimal path in the sense of a metric.

Definition 64. *A metric defines the distance between two configurations in the C-space, which becomes a metric space. This metric can be seen as the "cost-to-go" for a specific aircraft to reach a configuration from another one.*

A grid-search algorithm is an optimization technique that successively performs an exploration and an exploitation processes:

1. The **exploration process** builds a minimum cost-to-go map, called distance function, from the start to the goal configuration.
2. The **exploitation process** is a backtracking from the goal to the start configuration.

Accurate environmental mapping is essential to the path planning process [74]:

1. **Qualitative or topological mapping** represents features without reference to numerical data and is therefore not geometrically exact. It consists of nodes and arcs, with vertices representing features or landmarks.
2. **Quantitative or metric mapping** adopts a data structure which is feasible for path planning based on way-points or sub-goals: meadow maps, Voronoi diagrams, regular occupancy grid and quad-tree mapping.

Planning can be formulated as either a continuous or a discrete task [363,395]. Continuous methods such as potential fields, vector field histogram and bug algorithms and discrete methods such as visibility graph planning or Voronoi diagram or A* algorithm [28,30,56,86,323].

2.3.2.1 Discrete Methods

In discrete planning, the system is typically modeled as a graph and is called a transition system. **Nodes** represent states, and **edges** represent transitions between states and are labeled by **actions** that enable these transitions. An important feature of sampling-based approaches is that the required controllers for feasible trajectories are automatically constructed as a result of the exploration process.

While many navigation algorithms work directly on the environment description, some algorithms such as Dijkstra or A* requires a **distance graph** as an input. A distance graph is an environment description at a higher level. It does not contain the full environment information, but it allows for an efficient path planning step. While the **quadtree method** identifies points with some free space around them, the **visibility graph** method uses corner points of obstacles instead.

Other approaches can abstract a geometric environment into a topological map based on landmarks. Planning is carried out on this topological map. A planned route has to be converted back to the geometric space for continuous motion control. After obtaining discrete routing information, an admissible and safe path has to be generated for an aircraft to travel from one node to another, compliant with all kinematic and dynamic constraints. Many planning algorithms address obstacle avoidance while planning a path to reach a destination point using A*, D*, **Voronoi diagrams**, **probabilistic roadmap** (PRM) or rapidly exploring random tree (RRT) methods. Goerzen in [152] reviewed deterministic motion planning algorithms in the literature from the autonomous aircraft guidance point of view.

Definition 65. *A **directed graph** (N, E) is a structure where N is a set of nodes or vertices, and E is a set of edges connecting the nodes $(E \subseteq N \times N)$. In path planning, nodes usually stand for positions in the space, and edges determine whether it is possible to transit directly between these positions.*

UAV has strict payloads and power constraints which limit the number and variety of sensors available to gather information about knowledge of its surroundings. In [196], the paths are planned to maximize collected amount of information from desired regions while avoiding forbidden regions.

Graph search algorithms look for a feasible path in a given environment. Examples of graph search are deterministic graph search techniques such as A* search algorithm, Voronoi graph search and probabilistic sampling-based planners. Although these algorithms are mainly used for global path planning, reactive planners have been achieved by modifying the algorithms in order to reduce the computation time.

Differential constraints naturally arise from the kinematics and dynamics of the smart autonomous aircraft. When incorporated into the planning process, a path is produced that already satisfies the constraints [318]. Because of the difficulty of planning under differential constraints, nearly all planning algorithms are sampling-based as opposed to combinatorial. From an initial state X_0, a reachability tree can be formed by applying all sequences of discretized actions. Sampling-based algorithms proceed by exploring one or more reachability trees that are derived from discretization. In some cases, it is possible to trap the trees onto a regular lattice structures. In this case, planning becomes similar to grid search. A solution trajectory can be found by applying standard graph search algorithms to the lattice. If a solution is not found, then the resolution may need to be increased.

A planning algorithm that allows the aircraft to autonomously and rapidly calculates 3D routes can also follow a heuristic approach [405]. This algorithm is developed and verified in a virtual world that replicates the real world in which the aircraft is flying. The elements of the real world such as the terrain and the obstacles are represented as mesh-based models. Such an algorithm is also based on a graph but aims at increasing the portion of space explored identifying several nodes in the 3D space with an iterative approach.

2.3.2.1.1 Deterministic Methods

Several methods have been proposed and are mainly based on the construction of **visibility graphs**. Following this approach, the nodes of the graph are candidate path points that the aircraft will fly through and each arc represents an approximate path between them. To build the visibility graph, the set of nodes and the arcs that join them without intersecting obstacles have to be determined.

2.3.2.1.1.1 Visibility graph   The visibility graph uses corner points of obstacles. If the environment is represented as a configuration space, the polygon description of all obstacles is already available. The list of all start and end points of obstacle border lines plus the autonomous aircraft's start and goal position is available. A complete graph is then constructed by linking every node position of every other one. Finally, all the lines that intersect an obstacle are deleted, leaving only the lines that allow the flight from one node to another in a direct line. The characteristic of Algorithm 11 are as follows. Let $V = \{v_1, ..., v_n\}$ be the set of vertices of the polygons in the configuration space as well as the start and goal configurations. To construct the visibility graph, other vertices visible to $v \in V$, must be determined. The most obvious way to make this determination is to test all line segments $vv_i, v \neq v_i$, to see if they intersect an edge of any polygon. A more efficient way is the rotational sweep algorithm. For the problem of computing the set of vertices visible from v, the sweep line I that is a half-line emanating from v and a rotational sweep rotating I from 0 to 2π are used.

Algorithm 11 Visibility Algorithm

1. **Input:** A set of vertices $\{v_i\}$ (whose edges do not interset) and a vertex v.
2. **Output:** A subset of vertices from $\{v_i\}$ that are within the LOS of v.
3. For each vertex v_i, calculate α_i, the angle from the horizontal axis to the line segment vv_i
4. Create the vertex list ε, containing the α_i sorted in increasing order.
5. Create the active list S, containing the sorted list of edges that intersect the horizontal half-line emanating from v.
6. for all α_i do
7. if v_i is visible to v then
8. add the edge (v, v_i) to the visibility graph
9. endif
10. if v_i is the beginning of an edge E, not in S, then
11. insert the edge E into S
12. endif
13. if v_i is the end of an edge in S then
14. Delete the edge from S
15. endif
16. endfor

This algorithm incrementally maintains the set of edges that intersect I, sorted in order of increasing distance from v. If a vertex v_i is visible to v, then it should be added to the visibility graph. It is straight forward to determine if v_i is visible to v. Let S be the sorted list of edges that intersects the half-line emanating from v. The set is incrementally constructed as the algorithm runs. If the line segment vv_i does not intersect the closed edge in S and if I does not lie between the two edges incident on v (the sweep line does not intersect the interior of the obstacles at v.), then v_i is visible from v [86].

2.3.2.1.1.2 Voronoi algorithm   The **Voronoi tessellation** of a set of planar points, known as sites, is a set of Voronoi cells. Each cell corresponds to a side and consists of all points that are closer to its site than to any other sites. The edges of the cells are the points that are equidistant to two nearest sites. A generalized Voronoi diagram comprises cells defined by measuring distances to objects rather than points.

Definition 66. *Planar ordinary Voronoi diagram: Given a set of a finite number of distinct points in the Euclidean plane, all locations in that space are associated with the closest members of the point set with respect to the Euclidean distance. The result is a tessellation of the plane into a set of the regions associated with members of the point set. This tessellation is called a **Planar ordinary Voronoi diagram** generated by the point set and the regions constituting the Voronoi diagram, ordinary Voronoi polygons [240].*

In hostile environments, a Voronoi diagram can decompose a space defined by random scattered points into separate cells in such a manner that each cell will include one point that is closer to all elements in this cell than any other points [97]. The graph is constructed by using Delaunay triangulation and its dual graph Voronoi diagrams. The procedure of this cell decomposition starts with a priori knowledge of the location and of the number of the scattered points. Formation of triangles formed by three of the Voronoi sites without including any other sites in this circumcircle is called Delaunay triangulation. By connecting all the edges together, polygons are formed and these polygons construct the Voronoi graph.

Considering the formulation of 3D networks, **3D Delaunay method** is used in [360] to partition the space because of uniqueness: regional, nearest and convex polyhedron.

As it is difficult to control aircraft precisely enough to follow the minimum-distance path without risk of colliding with obstacles, many skeleton-based road map approaches have been taken. The Voronoi approach builds a skeleton that is maximally distant from the obstacles and finds the minimum-distance path that follows this skeleton. This algorithm is a 2D algorithm, complete but not optimal. Voronoi diagram is a special kind of decomposition of a metric space determined by distances to a specified discrete set of objects in the space [240]. Given a set of points S, the corresponding Voronoi diagram is generated, and each point P has its own Voronoi cell which consists of all points closer to P than any other points. The border points between polygons are the collection of the points with the distance to shared generators [28].

The Voronoi algorithm works by constructing a skeleton of points with minimal distances to obstacles. A free space F in environment (white voxels) as well as an occupied space F' (black voxels) is defined, and $b' \in F'$ is a basis point for $p \in F$ if and only if b has minimal distance to p, compared with all other points in F', Voronoi diagram = $\{p \in F | p$ has at least two basis points$\}$. If the logic of the algorithm requires the set of nodes to be deleted during an incremental update to form a tree, then supplying the subsequently executed portions of the code, with anything but a tree, will likely cause the code to loop or crash. Similarly if a new node is to be inserted into the Voronoi Algorithm 12, then the code may subsequently need to access the coordinates of this node.

Typical computational units are given in Algorithm 12 in below section. Line 1 sets ε to maximum precision, while line 3 computes some data. Line 4 checks topological and conditions, while line 6 relaxes the ε threshold and line 7 makes sure to reset ε. Line 10 checks locally for soundness of input, and line 13 fixes the problem in the input data. Finally, line 14 replaces "correct" by "best possible." If the relaxation of the ε threshold and the heuristics of the multi-level recovery process have not helped to compute data that meets the topological conditions then the code finally enters "desperate mode," because the "optimum" is replaced by "best possible."

One problem of this approach is that it allows lines to pass very closely to an obstacle, and so this would only work for a theoretical point robot. However, this problem can be easily solved by virtually enlarging each obstacle by at least half of the autonomous aircraft largest dimension before applying the algorithm.

Algorithm 12 Voronoi algorithm

---

1. ε = lower-bound
2. Repeat
3. x = ComputeData(ε)
4. success = CheckConditions (x, ε)
5. If (not success) then
6. ε=10ε
7. reset data structure appropriatedly
8. until (success OR ε > upper-bound
9. ε = lower-bound
10. If (not success) then
11. illegal = CheckInput()
12. If (illegal) then
13. clean data locally
14. restart computation from scratch
15. else
16. x = DesperateMode()

2.3.2.1.1.3 Dijkstra algorithm is a method for computing all shortest paths from a given starting node in a fully connected graph. Relative distance information between all nodes is required. In the loop, nodes are selected with the shortest distance in every step, then distance is computed to all of its neighbors and path predecessors are stored. The cost associated with each edge is often the distance but other criteria such as safety and clearance can be incorporated into the cost function as well.

The algorithm works by maintaining for each vertex the shortest-path distance from the start vertex. Further, a back-pointer is maintained for each vertex indicating from which neighboring vertex the shortest path from the start comes. Hence, a shortest path to some vertex can be read out by following the back-pointers back to the start vertex. From the start vertex, the shortest-path distances are propagated through the graph until all vertices have received their actual shortest-path distance.

Dijkstra's algorithm solves the single source shortest-path problem on a weighted directed graph $G = (V, E)$ for the case in which all edge weights are non-negative, $w(u, v) \geq 0$ for each edge $(u, v) \in E$. This algorithm maintains a set S of vertices whose final shortest-path weights from the source s have already been determined. The algorithm repeatedly selects the vertex $u \in V - S$, with the minimum shortest-path estimate, adds u to S and relaxes all edges leaving u.

Algorithm 13 Dijkstra Algorithm: DIJKSTRA

---

1. INITIALIZE-SINGLE-SOURCE(G, s)
2. S = $\emptyset$
3. Q = G.V
4. while $Q \neq \emptyset$
5. u = EXTRACT-MIN(Q)
6. S = S $\cup \{u\}$
7. for each vertex $v \in G.Adj[u]$
8. RELAX(u, v, w)

The Dijkstra algorithm (Algorithm 13) works as follows. Line 1 initializes the d and π values, and line 2 initializes the set S to the empty set. The algorithm maintains the invariant that $Q = V - S$

at the start of each iteration of the while loop of lines 4–8. Line 3 initializes the min-priority queue Q to contain all the vertices in V. Since $S = \emptyset$ at that time, the invariant is true after line 3. Each time through the while loop of lines 4–8, line 5 extracts a vertex u from $Q = V - S$ and line 6 adds it to set S, thereby maintaining the invariant. Vertex u, therefore, has the smallest shortest-path estimate of any vertex in $V - S$. Then, lines 7–8 relax each edge (u,v) leaving u, thus updating the estimate $v \cdot d$ and the predecessor $v \cdot \pi$ if the shortest path to v can be improved.

Because Dijkstra's algorithm always chooses the lightest or closest vertex in $V - S$ to add to set S, it is said to be a greedy strategy. Greedy algorithms do not always yield optimal results in general, but Dijkstra's algorithm computes shortest paths.

Some hybrid methods can be proposed. One method consists of building the visibility graph using Dijkstra algorithm to find the approximate shortest paths from each node to the goal and finally implementing a mixed integer linear programming (MILP) receding horizon control to calculate the final trajectory.

2.3.2.1.1.4 A* algorithm The **A* algorithm** is based on Dijkstra's algorithm. It focuses the search in the graph toward the goal. A heuristic value is added to the objective function. It must be a lower-bound estimate of the distance from the current vertex to the goal vertex. A* heuristic algorithm computes the shortest path from one given start node to one given goal node. Distance graph with relative distance information between all nodes plus lower bounds of distance to goal from each node are required. In every step, it expands only the currently shortest path by adding adjacent node with shortest distance including estimate of remaining distance to goal. The algorithm stops when the goal vertex has been treated in the priority queue.

Algorithm 14 A* Algorithm

---

1. Input: A graph
2. Output: A path between start and goal nodes
3. Repeat
   a. Pick n_{best} from 0 such that $f(n_{\text{best}}) < f(n)$
   b. Remove n_{best} from O and add to C
   c. If $n_{\text{best}} = q_{\text{goal}}$, EXIT
   d. expand n_{best}: for all $x \in Star(n_{\text{best}})$ that are not in C
   e. if $x \notin O$ then
   f. add x to O
   g. else if $g(n_{\text{best}}) + C(n_{\text{best}}, x) < g(x)$ then
   h. update x's back-pointer to point to n_{best}
   i. end if
4. Until O is empty

The pseudocode for this approach can be formulated as Algorithm 14. If the heuristic is admissible, i.e., if $h(s) \leq c^*(s, s_{\text{goal}})$ for all s, A* is guaranteed to find the optimal solution in optimal running time. If the heuristic is also consistent, i.e., if $h(s) \leq c^*(s, s') + h(s')$, it can be proven that no state is expanded more than once by the A* algorithm. The A* algorithm has a priority queue which contains a list of nodes sorted by priority, determined by the sum of the distance traveled in the graph thus far from the start node and the heuristic. The first node to be put into the priority queue is naturally the start node. Next, the start node is expanded by putting all adjacent nodes to the start node into the priority queue sorted by their corresponding priorities. These nodes can naturally be embedded into the autonomous aircraft free space and thus have values corresponding to the cost required to traverse between the adjacent nodes. The output of the A* algorithm is a back-pointer path, which is a sequence of nodes starting from the goal and going back to the start. The difference from Dijkstra is that A* expands the state s in OPEN with a minimal value of $g(s) + h(s)$ where $h(s)$

is the heuristic that estimates the cost of moving from s to s_{goal}. Let $c^*(s, s')$ denote the cost of the optimal solution between s and s'. The "Open List" saves the information about the parental nodes found as a candidate solution. The 3D cells in the grid not only have elements in the neighborhood on the same height level used but also have cell nodes with locations above and below. The output of the A* algorithm is a back-pointer path, which is a sequence of nodes starting from the goal and going back to the start. Two additional structures are used: an open set O and a closed set C. The open set O is the priority queue, and the closed set C contains all processed nodes [86]. The Euclidean distance between the current point and the destination goal, divided by the maximum possible nominal speed, can be employed as a heuristic function. This choice ensures that the heuristic cost will always be lower than the actual cost to reach the goal from a given node, and thus, the optimum solution is guaranteed.

Trajectory primitives provide a useful local solution method within sampling-based path planning algorithm to produce feasible but sub-optimal trajectories through complicated environments with relatively low computational cost. Extensions to the traditional trajectories primitives allow 3D maneuvers with continuous heading and flight path angles through the entire path. These extensions which include simple transformations as well as additional maneuvers can maintain closed form solutions to the local planning problem and therefore maintain low computational cost [188].

2.3.2.1.2 Probabilistic Methods

Sampling-based motion planning algorithms have been designed and successfully used to compute a probabilistic complete solution for a variety of environments. Two of the most successful algorithms include the PRM and RRTs. At a high level, as more samples are generated randomly, these techniques provide collision-free paths by capturing a larger portion of the connectivity of the free space [320].

2.3.2.1.2.1 Probabilistic Roadmap Method **Probabilistic roadmap method** (PRM) sparsely samples the word map, creating the path in two-phase process: planning and query. The **query** phase uses the result of the planning phase to find a path from the initial configuration to the final one. The planning phase finds N random points that lie in free space. Each point is connected to its nearest neighbors by a straight line path that does not cross any obstacles, so as to create a network or graph, with a minimal number of disjoint components and no cycles. Each edge of the graph has an associated cost which is the distance between its nodes. An advantage of this planner is that once the roadmap is created by the planning phase, the goal and starting points can be changed easily. Only the query phase needs to be repeated.

PRM algorithm combines an offline construction of roadmap with a randomized online selection of an appropriate path from the roadmap. However, this algorithm cannot be applied in rapidly changing environment due to offline construction of the roadmap.

The roadmap methods described above are able to find a shortest path on a given path. The issue most path planning methods are dealing with is how to create such a graph. To be useful for path planning applications, the roadmap should represent the connectivity of the free configuration space well and cover the space such that any query configuration can be easily connected to the roadmap. Probabilistic roadmap approach is a probabilistic complete method that is able to solve complicated path planning problems in arbitrarily high-dimension configuration spaces. The basic concept in PRM is that rather than attempt to sample all of C-space, one instead samples it probabilistically. This algorithm operates in two phases, a roadmap construction phase in which a roadmap is constructed within the C-space and a query phase, in which probabilistic searches are conducted using the roadmap to speed the search:

1. **Roadmap Construction Phase:** it tries to capture the connectivity of free configuration space. An undirected, acyclic graph is constructed in the autonomous aircraft C-space

in which edges connect nodes if and only if a path can be found between the nodes corresponding to way-points. The graph is grown by randomly choosing new locations in C-space and attempting to find a path from the new location to one of the nodes already in the graph while maintaining the acyclic nature of the graph. This relies on a local path planner to identify possible paths from the randomly chosen location and one or more of the nodes in the graph. The choice of when to stop building the graph and the design of the local path planner are application specific, although performance guarantees are sometimes possible. **Local planning** (m milestones, e edges) connects nearby milestones using local planner and form a roadmap. Local planning checks whether there is a local path between two milestones, which corresponds to an edge on the roadmap. Many methods are available for local planning. The most common way is to discretize the path between two milestones into n_i steps and the local path exists when all the intermediate samples are collision free, performing discrete collision queries at those steps. It is the most expensive part of the PRM algorithm.

2. **Query Phase:** When a path is required between two configurations s and g, paths are first found from s to node $\bar{s}$ in the roadmap and from g to some node $\bar{g}$ in the roadmap. The roadmap is then used to navigate between $\bar{g}$ and $\bar{s}$. After every query, the nodes s and g and the edges connecting them to the graph can be added to the roadmap. As in the learning phase, the query phase relies on a heuristic path planner to find local paths in the configuration space.

The local planner should be able to find a path between two configurations in simple cases in a small amount of time. Given a configuration and a local planner, one can define the set of configurations to which a local planning attempt will succeed. This set is called the visibility region of a node under a certain local planner. The larger the visibility region is, the more powerful the local planner. The most straightforward sampling scheme shown in Algorithm 15 is to sample configurations uniformly randomly over the configuration space.

Algorithm 15 Roadmap Algorithm

---

1. nodes ← sample N nodes random configuration
2. for all nodes
3. find $k_{nearest}$ nearest neighbors
4. if collision check and $\gamma \leq \gamma_{max}$ then roadmap ← edge
5. end

For every node, a nearest-neighbor search is conducted. Several constraints have to be satisfied during the construction phase before an edge connection between two nodes is possible.

2.3.2.1.2.2 Rapidly Exploring Random Tree Method The method (RRT) is able to take into account the motion model of the aircraft. A graph of aircraft configurations is maintained, and each node is a configuration. The first node in the graph is the initial configuration of the aircraft. A random configuration is chosen, and the node with the closest configuration is found. This point is near in terms of a cost function that includes distance and orientation. A control is computed that moves the aircraft from the "near" configuration to the random configuration over a fixed period of time. The point that it reaches is a new point, and this is added to the graph. The distance measure must account for a difference in position and orientation, and requires appropriate weighting of these quantities. The random point is discarded if it lies within an obstacle. The result is a set of paths or roadmap. The trees consist of feasible trajectories that are built online by extending branches toward randomly generated target states.

The rapidly expanding random tree approach is suited for quickly searching high-dimensional spaces that have both algebraic and differential constraints. The key idea is to bias the exploration toward unexplored portions of the space by sampling points in the state space and incrementally pulling the search tree toward them, leading to quick and uniform exploration of even high-dimensional state spaces. A graph structure must be built with nodes at explored positions and with edges describing the control inputs needed to move from node to node. Since a vertex with a larger Voronoi region has a higher probability to be chosen as (x_{near}) and it is pulled to the randomly chosen state as close as possible, the size of larger Voronoi regions is reduced as the tree grows. Therefore, the graph explores the state space uniformly and quickly. The basic RRT Algorithm 16 operates as follows: the overall strategy is to incrementally grow a tree from the initial state to the goal state. The root of this tree is the initial state; at each iteration, a random sample is taken and its nearest neighbor in the tree computed. A new node is then created by growing the nearest neighbor toward the random sample.

Algorithm 16 RRT Basic Algorithm

1. Build RRT (x_{init})
2. G_{sub}, init(x_{init})
3. for k = 1 to maxIterations do
4. $x_{rand} \leftarrow$ RANDOM-STATE()
5. $x_{near} \leftarrow$ NEAREST-NEIGHBOR(x_{rand}, G_{sub})
6. u_{best}, x_{new}, success $\leftarrow$ CONTROL$(x_{near}, x_{rand}, G_{sub})$;
7. if success
8. G_{sub}.add-vertex x_{new}
9. G_{sub}.add-edge $x_{near}, x_{new}, u_{best}$
10. Return G_{sub}
11. RRT-EXTEND
12. $V \leftarrow \{x_{init}\}, E \leftarrow \emptyset, i \leftarrow 0$;
13. While $i < N$, do
14. $G \leftarrow (V, E)$
15. $x_{rand} \leftarrow$ Sample(i); $i \leftarrow i + 1$
16. $(V, E) \leftarrow$ Extend(G, x_{rand})
17. end

For each step, a random state (x_{rand}) is chosen in the state space. Then, (x_{near}) in the tree that is the closest to the (x_{rand}) in metric ρ is selected. Inputs $u \in \mathbf{U}$, the input set, are applied for Δt, making motions toward (x_{rand}) from (x_{near}). Among the potential new states, the state that is as close as possible to (x_{rand}) is selected as a new state (x_{new}). The new state is added to the tree as a new vertex. This process is continued until (x_{new}) reaches (x_{goal}).

To improve the performance of the RRT, several techniques have been proposed such as biased sampling and reducing metric sensitivity. **Hybrid systems** models sometimes help by switching controllers over cells during a decomposition. Another possibility is to track space-filling trees, grown backwards from the goal.

2.3.2.2 Continuous Methods

2.3.2.2.1 Receding Horizon Control

Receding horizon control, a variant of model predictive control, repeatedly solves online a constrained optimization problem over a finite planning horizon. At each iteration, a segment of the total path is computed using a dynamic model of the aircraft that predicts its future behavior. A sequence of control inputs and resulting states are generated that meet the kino-dynamic and environmental constraints and that optimize some performance objectives. Only a subset of these inputs is actually implemented, however, and the optimization is repeated as the aircraft maneuvers and new measurements are available. The approach is specially useful when the environment is explored online.

2.3.2.2.2 Mixed Integer Linear Programming

Mixed integer linear programming (MILP) approaches the problem of collision avoidance as an optimization problem with a series of constraints. The goal or objective function is to minimize the time needed to traverse several way-points. The constraints are derived from the problem constraints (flight speed, turning radius) and the fact that the aircraft must maintain a safer distance from obstacles and other aircraft.

Autonomous aircraft trajectory optimization including collision avoidance can be expressed as a list of linear constraints, involving integer and continuous variables known as mixed integer linear program. The mixed integer linear programming approach in [362] uses indirect branch-and-bound optimization, reformulating the problem in a linearized form and using commercial software to solve the MILP problem. A single aircraft collision avoidance application was demonstrated, and then, this approach was generalized to allow for visiting a set of way-points in a given order. Mixed integer linear programming can extend continuous linear programming to include binary or integer decision variables to encode logical constraints and discrete decisions together with the continuous autonomous aircraft dynamics. The approach to optimal path planning based on MILP was introduced in [374]. The autonomous aircraft trajectory generation is formulated as a 3D optimization problem under certain conditions in the Euclidean space, characterized by a set of decision variables, a set of constraints and the objective function. The decision variables are the autonomous aircraft state variables, i.e., position and speed. The constraints are derived from a simplified model of the autonomous aircraft and its environment. These constraints include

1. Dynamics constraints, such as a maximum turning force which causes a minimum turning radius, as well as a maximum climbing rate.
2. Obstacles avoidance constraints like no-flight zones
3. Target reaching constraints of a specific way-point or target.

The objective function includes different measures of the quality in the solution of this problem, although the most important criterion is the minimization of the total flying time to reach the target. As MILP can be considered as a geometric optimization approach, there is usually a protected airspace setup around the autonomous aircraft in the MILP formulation. The stochasticity that stems from uncertainties in observations and unexpected aircraft dynamics could be handled by increasing the size of protected airspaces. An advantage of the MILP formulation is its ability to plan with non-uniform time steps between way-points. A disadvantage of this approach is that it requires all aspects of the problem (dynamics, ordering of all way-points in time and collision avoidance geometry) to be specified as a carefully designed and a usually long list of many linear constraints, and then, the solver's task is basically to find a solution that satisfies all of those constraints simultaneously [390].

Then, a MILP solver takes the objective and constraints and attempts to find the optimal path by manipulating the force effecting how much a single aircraft turns at each time step. Although MILP is an elegant method, it suffers from exponential growth of the computations [362,374].

2.3.2.2.3 Classical Potential Field Method

The **artificial potential field method** is a collision avoidance algorithm based on electrical fields. Obstacles are modeled as repulsive charges, and destinations (way-points) are modeled as attractive charges. The summation of these charges is then used to determine the safest direction to travel.

Let $X = (x, y, z)^T$ denote the UAV current position in airspace. The usual choice for the **attractive potential** is the standard parabolic that grows quadratically with the distance to the goal such that

$$U_{\text{att}} = \frac{1}{2} k_a d_{\text{goal}}^2(X) \tag{2.116}$$

where $d_{\text{goal}} = \left\| X - X_{\text{goal}} \right\|$ is the Euclidean distance of the UAV current position X to the goal X_{goal} and k_a is a scaling factor. The attractive force considered is the negative gradient of the attractive potential:

$$F_{\text{att}}(X) = -k_a \left(X - X_{\text{goal}} \right) \tag{2.117}$$

By setting the aircraft velocity vector proportional to the vector field force, the force $F_{\text{att}}(X)$ drives the aircraft to the goal with a velocity that decreases when the UAV approaches the goal.

The **repulsive potential** keeps the aircraft away from obstacles. This repulsive potential is stronger when the UAV is closer to the obstacles and has a decreasing influence when the UAV is far away. A possible repulsive potential generated by obstacle i is:

$$U_{\text{rep}_i}(X) = \begin{pmatrix} \frac{1}{2} k_{\text{rep}} \left(\frac{1}{d_{\text{obs}_i}(X)} - \frac{1}{d_0} \right)^2 & d_{\text{obs}_i}(X) \le d_0 \\ 0 & \text{Otherwise} \end{pmatrix} \tag{2.118}$$

where i is the number of the obstacle close to the UAV, $d_{\text{obs}_i}(X)$ is the closest distance to the obstacle i, k_{rep} is a scaling constant and d_0 is the obstacle influence threshold.

$$F_{\text{rep}_i}(X) = \begin{pmatrix} k_{\text{rep}} \left(\frac{1}{d_{\text{obs}_i}(X)} - \frac{1}{d_0} \frac{1}{d_{\text{obs}_i}(X)} \right) \hat{e}_i & d_{\text{obs}_i}(X) \le d_0 \\ 0 & \text{Otherwise} \end{pmatrix} \tag{2.119}$$

where $\hat{e}_i = \frac{\partial d_{\text{obs}_i}(X)}{\partial X}$ is a unit vector that indicates the direction of the repulsive force, therefore,

$$\begin{pmatrix} \dot{x}_d \\ \dot{y}_d \\ \dot{z}_d \end{pmatrix} = -\left(F_{\text{att}}(X) + F_{\text{rep}_i}(X) \right) \tag{2.120}$$

After the desired global velocity is calculated by the potential field method, the corresponding desired linear velocity V_d and attitude χ_d, γ_d can also be obtained:

$$V_d = k_u \sqrt{\dot{x}_d^2 + \dot{y}_d^2 + \dot{z}_d^2} \tag{2.121}$$

$$\gamma_d = \text{atan2} \left(-\dot{z}_d, \sqrt{\dot{x}_d^2 + \dot{y}_d^2} \right) \tag{2.122}$$

$$\chi_d = \text{atan2}(\dot{y}_d, \dot{z}_d) \tag{2.123}$$

where the gain k_u is introduced to allow for additional freedom in weighting the velocity commands. The pitch and yaw angle guidance laws are designed so that the aircraft's longitudinal axis steers to align with the gradient of the potential field. The roll angle guidance law is designed to maintain the level flight.

Harmonic field approach is useful in avoiding local minima of the classical potential field methods [56].

2.3.3 MOVING OBSTACLES AVOIDANCE

Smart autonomous aircraft requires a collision avoidance algorithm, also known as **sense and avoid**, to monitor the flight and alert the aircraft to necessary avoidance maneuvers. The challenge is now autonomous navigation in open and dynamic environments, i.e., environments contain moving objects or other aircraft as potential obstacles, and their future behavior is unknown. Taking into account these characteristics requires to solve three main categories of problems [316]:

1. Simultaneous localization and mapping in dynamic environments. This topic will be discussed in the next chapter.
2. Detection, tracking, identification and future behavior prediction of the moving obstacles.
3. Online motion planning and safe navigation.

In such a framework, the smart autonomous aircraft has to continuously characterize with onboard sensors and other means. As far as the moving objects are concerned, the system has to deal with problems such as interpreting appearances, disappearances and temporary occlusions of rapidly maneuvering vehicles. It has to reason about their future behavior and consequently to make prediction. The smart autonomous aircraft has to face a double constraint: constraint on the response time available to compute a safe motion which is a function of the dynamicity of the environment and a constraint on the temporal validity of the motion planned which is a function of the validity duration of the predictions.

Path planning in a priori unknown environment cluttered with dynamic objects and other aircraft is a field of active research. It can be addressed by using explicit time representation to turn the problem into the equivalent static problem, which can then be solved with an existing static planner. However, this increases the dimensionality of the representation and requires exact motion models for surrounding objects. The dimensionality increase raises the computational effort (time and memory) to produce a plan, and motion modeling raises difficult prediction issues.

There are various approaches to both static and moving obstacles that are mostly based on the collision cone approach. Often, collision avoidance is achieved by tracking a way-point at a safe distance from the obstacles using a homing guidance law.

Atmosphere can be very dynamic. The **cloud behavior** is very complex. Typically, in aircraft navigation, clouds and turbulence should be avoided. They are considered as moving obstacles. There are several ways to model their behavior in a complex environment. Physical modeling of the cloud/turbulence can be done using **Gaussian dispersion methods**, which predict the cloud behavior using statistical dispersion techniques. Another modeling approach is to define the points picked up by the UAV as the vertices $\{v_i, i = 1, \ldots, N\}$. The vertices are connected by line segments of constant curvature $\{\kappa_{ij}\}$ with C^2 contact at the vertices. The **splinegon** representation assumes some reasonably uniform distribution of vertices [378]. Each vertex has a curvature and length, and these can be used to determine matrices for each segment.

The work presented in [373] is concerned with developing an algorithm that keeps a certain safety distance while passing an arbitrary number of possibly moving but nonmaneuvering obstacles. Starting from a 3D collision cone condition, input–output linearization is used to design a nonlinear guidance law [79]. The remaining design parameters are determined considering convergence and performance properties of the closed-loop guidance loop. Then, the guidance framework is developed in terms of a constrained optimization problem that can avoid multiple obstacles simultaneously while incorporating other mission objectives.

2.3.3.1 D* Algorithm

The algorithm D* has a number of features that are useful for real-world applications. It is an extension of the A* algorithm for finding minimum cost paths through a graph for environments where the environment changes at a much slower speed than the aircraft. It generalizes the occupancy grid to a cost map $c \in \mathbb{R}$ of traversing each cell in the horizontal and vertical directions. The cost of traversing the cell diagonally is $c\sqrt{2}$. For cells corresponding to obstacles, $c = \infty$. The key features of D* is that it supports incremental replanning. If a route has a higher than expected cost, the algorithm D* can incrementally replan to find a better path. The incremental replanning has a lower computational cost than completely replanning. Even though D* allows the path to be recomputed as the cost map changes, it does not support a changing goal. It repairs the graph allowing for an efficient updated searching in dynamic environments.

Field D*, like D*, is an incremental search algorithm which is suitable for navigation in an unknown environment. It makes an assumption about the unknown space and finds a path with the least cost from its current location to the goal. When a new area is explored, the map information is updated and a new route is replanned, if necessary. This process is repeated until the goal is reached or it turns out that goal cannot be reached (due to obstacles for instance). The algorithm A* can also be used in a similar way [106].

The D* Algorithm 17 is devised to locally repair the graph allowing efficient updated searching in dynamic environments, hence the term D*. D* initially determines a path starting with the goal and working back to the start using a slightly modified Dijkstra's search. The modification involves updating a heuristic function. Each cell contains a heuristic cost h which for D* is an estimate of path length from the particular cell to the goal, not necessarily the shortest-path length to the goal as it was for A*. These h values will be updated during the initial Dijkstra search to reflect the existence of obstacles. The minimum heuristic values h are the estimate of the shortest-path length to the goal. Both the h and the heuristic values will vary as the D* search runs, but they are equal upon initialization [86].

Algorithm 17 D* Algorithm

---

1. **Input:** List of all states L
2. **Output:** The goal state, if it is reachable, and the list of states L are updated so that back-pointer list describes a path from the start to the goal. If the goal state is not reachable, **return NULL**
3. For each $X \in L$ do
4. $t(X) = New$
5. *endfor*
6. $h(G) = 0; 0 = \{G\}; X_c = S$
7. The following loop is Dijkstra's search for an initial path.
8. repeat
9. $k_{min} = $ process-state $(0, L)$
10. until $(k_{min} > h(x_c))$ or $(k_{min} = -1)$
11. $P = $ Get-Pointer-list (L, X_c, G)
12. If $P = $ Null then
13. return (Null)
14. end if
15. end repeat
16. *endfor*
17. X_c is the second element of P Move to the next state in P
18. $P = $ Get-Back-Pointer-List (L, X_c, G)
19. until $X_c = G$
20. return (X_c)

Notation

1. X represents a state
2. O is the priority queue
3. L is the list of all states
4. S is the start state
5. $t(x)$ is the value of state with respect to priority queue
   a. $t(x)$: New if x has never been in O
   b. $t(x)$: Open if x is currently in O
   c. $t(x)$: Closed if x was in O but currently is not

2.3.3.2 Artificial Potential Fields

The harmonic potential field approach works by converting the goal, representation of the environment and constraints on behavior into a reference velocity vector field [331]. A basic setting is

Problem 67. *Solve*

$$\nabla^2 V(P) = 0 \qquad P \in \Omega \qquad (2.124)$$

subject to $V(P) = 1$ *at* $P = \Gamma$ *and* $V(P_r) = 0$

A provably correct path may be generated using the gradient dynamical system

$$\dot{P} = -\nabla V(P) \qquad (2.125)$$

The harmonic potential field approach can incorporate directional constraints along with regional avoidance constraints in a provably correct manner to plan a path to a target point. The navigation potential may be generated using the **boundary value problem** (BVP)

Problem 68. *Solve*

$$\nabla^2 V(P) = 0 \qquad P \in \Omega - \Omega' \qquad (2.126)$$

and

$$\nabla \left(\Sigma(P) V(P) \right) = 0, P \in \Omega' \qquad (2.127)$$

subject to $V(P) = 1$ *at* $P = \Gamma$ *and* $V(P_r) = 0$.

$$\Sigma(P) = \begin{pmatrix} \sigma(P) & 0 & \cdots & 0 \\ 0 & \sigma(P) & 0 & 0 \\ 0 & \cdots & 0 & \sigma(P) \end{pmatrix} \qquad (2.128)$$

A provably correct trajectory to the target that enforces both the regional avoidance and directional constraints may be simply obtained using the gradient dynamical system in equation (2.125). The approach can be modified to take into account ambiguity that prevents the partitioning of an environment into admissible and forbidden regions [331].

Dynamic force fields can be calculated based on aircraft position and velocity. The force field uses scalar modifications to generate a larger and stronger field in front of the aircraft. Therefore, the force exerted by one aircraft on another can be calculated using the difference between the bearing of the aircraft exerting the force and the one feeling that force. Furthermore, a secondary calculation occurs that scales the force exerted into a force belt. This scaling is computed by determining the difference between the exerted force and the bearing of the aircraft feeling the force. Finally, the repulsive forces are summed, scaled and added to an attractive force of constant magnitude. This resultant vector is bound by the maximum turning rate and then used to inform the aircraft of its next maneuvers: a new target way-point is sent to the aircraft.

When an aircraft is close to its destination and because it begins to ignore the other aircraft, it is necessary to extend its force field such that the forces are exerted on nonpriority aircraft at a larger distance. This combination allows the prioritized aircraft to go directly to its goal while providing other aircraft with an early alert through the expanded force field [365].

The artificial potential method is modified in [365] to handle special cases by including a **priority system** and techniques to prevent an aircraft from circling its final destination. Two-dimensional constraints can be imposed because many UAS must operate in a limited range of altitude. Each UAS has a maximum operating altitude and the minimum altitude may be imposed by stealth or the application requirements.

The design and implementation of a potential field obstacle algorithm based on fluid mechanics panel methods is presented in [95]. Obstacles and the UAV goal positions are modeled by harmonic functions, thus avoiding the presence of local minima. Adaptations are made to apply the method to the automatic control of a fixed-wing aircraft, relying only on a local map of the environment that is updated from sensors on-board the aircraft. To avoid the possibility of collision due to the dimension of the UAV, the detected obstacles are expanded. Considering that the detected obstacles are approximated by rectangular prisms, the expansion is carried out by moving the faces outwards by an amount equal to the wingspan. The minimum value that assured clearance to the obstacle is half-span. Then, obstacles are further extended creating prisms.

2.3.3.3 Online Motion Planner

The virtual net comprises a finite set of points $X_e(R)$ corresponding to a finite set of prescribed relative positions [400]:

$$R \in \mathbb{M} = \{R_1, R_2, \ldots, R_n\} \subset \mathbb{R}^3 \tag{2.129}$$

$$X_e(R_k) = (R_k, 0)^T = (R_{x,k}, R_{y,k}, R_{z,k}, 0, 0, 0)^T \qquad k = 1, \ldots, n \tag{2.130}$$

where velocity states are zero, n is the number of points in the virtual net. The obstacle position and uncertainty is represented by an ellipsoid. The set $O(q, Q)$ centered around the position $q \in \mathbb{R}^3$ is used to over-bound the position of the obstacle, i.e.,

$$\mathbb{O}(q, Q) = \left\{ X \in \mathbb{R}^6, (\mathbf{S}X - q)^T \mathbf{Q} (\mathbf{S}X - q) \leq 1 \right\} \tag{2.131}$$

where $\mathbf{Q} = \mathbf{Q}^T$ and $\mathbf{S} = \begin{bmatrix} 1 & 0 & 0 & 0 & 0 & 0 \\ 0 & 1 & 0 & 0 & 0 & 0 \\ 0 & 0 & 1 & 0 & 0 & 0 \end{bmatrix}$

The set $\mathbb{O}(q, Q)$ can account for the obstacle and aircraft physical sizes and for the uncertainties in the estimation of the obstacle/aircraft positions. The set $\mathbb{O}(q, Q)$ has an ellipsoidal shape in the position directions and is unbounded in the velocity directions. Ellipsoidal sets, rather than polyhedral sets, can be used to over-bound the obstacle because ellipsoidal bounds are typically produced by position estimation algorithms, such as the extended Kalman filter. This filter will be presented in the next chapter.

The online motion planning with obstacle avoidance is performed according to Algorithm 18.

Algorithm 18 Online Motion Planner

1. Determine the obstacle location and shape (i.e. q and Q).
2. Determine the growth distance.
3. Construct a graph connectivity matrix between all $R_i, R_j \in \mathbb{M}$. In the graph connectivity matrix, if two vertices are not connected, the corresponding matrix element is $+\infty$. If they are connected, the corresponding matrix element is 1. The graph connectivity matrix is multiplied element-wise to produce a constrained cost of transition matrix.
4. Perform graph search using any standard graph search algorithm to determine a sequence of connected vertices, $R(k) \in \mathbb{M}$ such that $R[1]$ satisfies the initial constraints, $R[l_p]$ satisfies the final constraints and the cumulative transition cost computed from the constrained cost of transition matrix is minimized.
5. After the path has been determined as a sequence of the way-points, the execution of the path proceeds by checking if the current state $X(t)$ is in the safe positively invariant set corresponding to the next reference R^+.

2.3.3.4   Zermelo–Voronoi Diagram

In many applications of autonomous aircraft, ranging from surveillance, optimal pursuit of multiple targets, environmental monitoring and aircraft routing problems, significant insight can be gleaned from data structures associated with **Voronoi-like partitioning** [33,34]. A typical application can be the following: given a number of landing sites, divide the area into distinct non-overlapping cells (one for each landing site) such that the corresponding site in the cell is the closest one (in terms of time) to land for any aircraft flying over this cell in the presence of winds. A similar application that fits in the same framework is the task of subdividing the plane into **guard/safety** zones such that a **guard/rescue** aircraft residing within each particular zone can reach all points in its assigned zone faster than any other guard/rescuer outside its zone. This is the generalized minimum-distance problems where the relevant metric is the minimum intercept or arrival time. Area surveillance missions can also be addressed using a frequency-based approach where the objective implies to optimize the elapsed time between two consecutive visits to any position known as the refresh time [5].

Recent work in patrolling can be classified as [272]

1. **Offline versus Online:** Offline computes patrols before sensors are deployed, while online algorithm controls the sensor's motion during operation and is able to revise patrols after the environment has changed.
2. **Finite versus Infinite:** Finite planning horizon algorithm computes patrols that maximize reward over finite horizon, while infinite horizon maximizes an expected sum of rewards over an infinite horizon.
3. **Controlling Patrolling versus Single Traversal:** It is dynamic environment monitoring versus a snapshot of an environment.
4. **Strategic versus Nonstrategic Patrolling**
5. **Spatial or Spatio-Temporal Dynamics**.

The construction of **generalized Voronoi diagrams** with time as the distance metric is in general a difficult task for two reasons: first, the distance metric is not symmetric, and it may not be expressible in closed form. Second, such problems fall under the general case of partition problems for which the aircraft's dynamics must be taken into account. The topology of the agent's configuration space may be non-Euclidean; for example, it may be a manifold embedded in an Euclidean space. These problems may not be reducible to **generalized Voronoi diagram** problems for which efficient construction schemes exist in the literature [240].

The following discussion deals with the construction of Voronoi-like partitions that do not belong to the available classes of generalized Voronoi diagrams. In particular, Voronoi-like partitions existing in the plane for a given finite set of generators, such that each element in the partition is uniquely associated with a particular generator in the following sense: an aircraft that resides in a particular set of the partition at a given instant of time can arrive at the generator associated with this set faster than any other aircraft that may be located anywhere outside this set at the same instant of time. It is assumed that the aircraft's motion is affected by the presence of temporally varying winds.

Since the generalized distance of this Voronoi-like partition problem is the minimum time to go of the Zermelo navigation problem, this partition of the configuration space is known as **Zermelo–Voronoi Diagram** (ZVD). This problem deals with a special partition of the Euclidean plane with respect to a generalized distance function. The characterization of this Voronoi-like partition takes into account the proximity relations between an aircraft that travels in the presence of winds and the set of Voronoi generators. The question of determining the generator from a given set which is the closest in terms of arrival time, to the agent at a particular instant of time, reduces the problem of determining the set of the Zermelo–Voronoi partition that the aircraft resides in at the given instant of time. This is the **point location problem**.

The **dynamic Voronoi diagram problem** associates the standard Voronoi diagram with a time-varying transformation as in the case of a time-varying winds. The **dual Zermelo–Voronoi diagram** problem leads to a partition problem similar to the ZVD with the difference that the generalized distance of the dual Zermelo–Voronoi diagram is the minimum time of the Zermelo navigation problem from a Voronoi generator to a point in the plane. The minimum time of the Zermelo navigation problem is not a symmetric function with respect to the initial and final configurations. The case of nonstationary spatially varying winds is much more complex.

The problem formulation deals with the motion of an autonomous aircraft. It is assumed that the aircraft's motion is described by the following equation:

$$\dot{X} = U + W(t) \tag{2.132}$$

$X = (x, y, z)^T \in \mathbb{R}^3, U \in \mathbb{R}^3, W = (W_N, W_E, W_D)^T \in \mathbb{R}^3$ is the wind which is assumed to vary uniformly with time, known a priori. In addition, it is assumed that $|W(t)| < 1, \forall t \geq 0$ which implies that system equation (2.132) is controllable. Furthermore, the set of admissible control inputs is given by $\mathbb{U} = \{U \in \mathbf{U}, \forall t \in [0, T], T > 0\}$ where $\mathbf{U} = \{(U_1, U_2, U_3) \in \mathbb{U} | U_1^2 + U_2^2 + U_3^2 = 1\}$ the closed unit ball, and U is a measurable function on $[0, T]$.

The Zermelo navigation problem solution when $W = 0$ is the control $U^*(\chi^*, \gamma^*) = (\cos \gamma^* \cos \chi^*, \cos \gamma^* \sin \chi^*, -\sin(\gamma^*))$ where χ^*, γ^* are constants, as shown in the previous section. Furthermore, the Zermelo navigation problem is reduced to the shortest-path problem in 3D.

Next, the Zermelo–Voronoi Diagram problem is formulated:

Problem 69. Zermelo–Voronoi Diagram problem: *Given the system described by* (2.132), *a collection of goal destination* $\mathbb{P} = \{p_i \in \mathbb{R}^3, i \in \ell\}$ *where ℓ is a finite index set and a transition cost*

$$C(X_0, p_i) = T_f(X_0, p_i) \tag{2.133}$$

determine a partition $\mathbb{B} = \{\mathbb{B}_i : i \in \ell\}$ *such that*

1. $\mathbb{R}^3 = \cup_{i \in \ell} \mathbb{B}_i$
2. $\bar{\mathbb{B}}_i = \mathbb{B}_i, \forall i \in \ell$
3. *For each* $X \in int(\mathbb{B}_i), C(X, p_i) < C(X, P_j), \forall j \neq i$

It is assumed that the wind $W(t)$ induced by the winds is known in advance over a sufficiently long (but finite) time horizon. Henceforth, $\mathbb{P}$ is the set of Voronoi generators or sites, $\mathbb{B}_i$ the Dirichlet domain and $\mathbb{B}$ the Zermelo–Voronoi diagram of $\mathbb{R}^3$, respectively. In addition, two Dirichlet domains $\mathbb{B}_i$ and $\mathbb{B}_j$ are characterized as neighboring if they have a non-empty and non-trivial (i.e., single point) intersection.

The Zermelo problem can be formulated alternatively as a moving target problem as follows:

Problem 70. Moving Target Problem: *Given the system described by*

$$\dot{\mathbf{X}} = \dot{X} - W(t) = U(t) \quad \mathbf{X}(0) = X_0 \tag{2.134}$$

Determine the control input $U^* \in \mathbb{U}$ such that

1. *The control U^* minimizes the cost functional $J(U) = T_f$ where T_f is the free final time.*
2. *The trajectory $\mathbf{X}^* : [0, T_f] \to \mathbb{R}^3$ generated by the control U^* satisfies the boundary conditions:*

$$\mathbf{X}^*(0) = X_0 \quad \mathbf{X}^*(T_f) = X_f - \int_0^{T_f} W(\tau) d\tau \tag{2.135}$$

The Zermelo problem and problem 70 are equivalent in the sense that a solution of the Zermelo problem is also a solution of Problem 70 and vice versa. Furthermore, an optimal trajectory $\mathbf{X}^*$ of problem 70 is related to an optimal trajectory X^* of the Zermelo problem by means of the time-varying transformation

$$\mathbf{X}^*(t) = X(t) - \int_0^t W(\tau)d\tau \tag{2.136}$$

The Zermelo minimum time problem can be interpreted in turn as an optimal pursuit problem as follows:

Problem 71. *Optimal Pursuit Problem: Given a pursuer and the moving target obeying the following kinematic equations:*

$$\dot{X}_p = \dot{X} = U \quad X_p(0) = X_0 = \mathbf{X}_0 \tag{2.137}$$

$$\dot{X}_T = -W(t) \quad X_T(0) = X_f \tag{2.138}$$

where X_p and X_T are the coordinates of the pursuer and the moving target, respectively, find the optimal pursuit control law U^* such that the pursuer intercepts the moving target in minimum time T_f:

$$X_p(T_f) = \mathbf{X}(T_f) = X_T(T_f) = X_f - \int_0^{T_f} W(\tau)d\tau \tag{2.139}$$

The optimal control of the Zermelo problem is given by

$$U^*(\chi^*, \gamma^*) = (\cos\gamma^* \cos\chi^*, \cos\gamma^* \sin\chi^*, -\sin\gamma^*)$$

The same control is also the optimal control for the moving target in problem 70. Because the angles χ^*, γ^* are necessarily constant, the pursuer is constrained to travel along a ray emanating from X_0 with constant unit speed (constant bearing angle pursuit strategy), whereas the target moves along the time-parametrized curve $X_T : [0, \infty] \to \mathbb{R}^3$ where $X_T(t) = X_f - \int_0^t W(\tau)d\tau$.

2.3.4 TIME OPTIMAL NAVIGATION PROBLEM WITH MOVING AND FIXED OBSTACLES

A fast autonomous aircraft wishes to go through a windy area so as to reach a goal area while needing to avoid n slow-moving aircraft and some very turbulent areas. The fast aircraft is regarded as a fast autonomous agent, and the n slow aircraft is regarded as n slow agent. It is assumed that the trajectories of the n slow agents are known to the fast aircraft in advance. The objective is to find a control such that the mission is accomplished within a minimum time. A time optimal navigation problem with fixed and moving obstacles is considered in this section. This problem can be formulated as an optimal control problem with continuous inequality constraints and terminal state constraints. By using the control parametrization technique together with the time scaling transform, the problem is transformed into a sequence of optimal parameters selection problems with continuous inequality constraints and terminal state constraints. For each problem, an exact penalty function method is used to append all the constraints to the objective function yielding a new unconstrained optimal parameters selection problem. It is solved as a nonlinear optimization problem [55].

An exact penalty function method is applied to construct a constraint violation function for the continuous inequality constraints and the terminal state constraints. It is then appended to the control function, forming a new cost function. In this way, each of the optimal parameter selection is further approximated as an optimal parameter selection subject to a simple non-negativity constraint or a decision parameter. This problem can be solved as a nonlinear optimization by any effective gradient-based optimization technique, such as the sequential quadratic programming (SQP) method [324].

2.3.4.1   Problem Formulation

Given $n+1$ agents in a 3D flow field, where n slow aircraft follow navigated trajectories, while the fastest aircraft is autonomously controllable. Let the trajectories of the n slow agents be denoted as

$$\eta_i = \begin{pmatrix} x_i(t) \\ y_i(t) \\ z_i(t) \end{pmatrix}, i = 1, \ldots, n, \qquad t \geq 0.$$

The flow velocity components at any point (x, y, z) in the 3D flow field can be denoted by $W_N(x, y, z, t)$, $W_E(x, y, z, t)$, $W_D(x, y, z, t)$ respectively. Then, the motion of the autonomous fast aircraft can be modeled as

$$\begin{aligned} \dot{x} &= V \cos \chi \cos \gamma + W_N(x, y, z, t) \\ \dot{y} &= V \sin \chi \cos \gamma + W_E(x, y, z, t) \\ \dot{z} &= -V \sin \gamma + W_D(x, y, z, t) \end{aligned} \qquad (2.140)$$

where V is the velocity of the controlled agent and the angles $\chi(t), \gamma(t)$ are considered as the control variables, subject to limitation constraints:

$$|\chi(t)| \leq \chi_{\max} \qquad |\gamma(t)| \leq \gamma_{\max}.$$

The relations (2.140) are equivalent to

$$\dot{\eta}(t) = f(\eta(t), \chi(t), \gamma(t), t) \quad \eta(0) = \eta_0 \quad t \geq 0 \qquad (2.141)$$

where η_0 is the initial position of the fast autonomous aircraft. The objective of the Zermelo problem is to find an optimal trajectory for the fast agent A_{n+1} such as the shortest route, the fastest route or the least fuel consumption to arrive at its goal area without colliding with the fixed obstacles and the other n slow agents.

Time optimal control problem 72 is formulated as follows:

Problem 72. *Optimal Control Problem*

$$\min_{\chi, \gamma} T_f$$

subject to

$$\dot{\eta}(t) = f(\eta(t), \chi(t), \gamma(t), t) \quad \eta(0) = \eta_0 \quad t \geq 0$$

$$\sqrt{(x(t) - x_i(t))^2 + (y(t) - y_i(t))^2 + (z(t) - z_i(t))^2} \geq \max\{R, R_i\} \qquad (2.142)$$

$$\eta(t) \in \aleph = \left\{ x_{\min} \leq x \leq x_{\max}, y = 2h_y, z = 2h_z \right\}$$

where T_f represents the time instant at which the fast agent reaches the goal area. The terminal time T_f depends implicitly on the control function, which is defined at the first time when the fast autonomous aircraft enters the target set $\aleph$. For each $i = 1, \ldots, n, R_i$ is the safety radius of the ith slow aircraft, and R is the safety radius of the fast autonomous aircraft.

2.3.4.2   Control Parametrization and Time Scaling Transform

Problem 72 is a nonlinear optimal control problem subject to continuous inequality constraints. Control parametrization and time scaling transform are applied to transform this problem into a nonlinear semi-infinite optimization problem to be solved by an exact penalty function method. The control parametrization is achieved as follows:

$$\chi_p(t) = \sum_{k=1}^{p} \vartheta_k^{\chi} \chi_{\tau_{k-1}^{\chi}, \tau_k^{\chi}}^{c}(t) \qquad \gamma_p(t) = \sum_{k=1}^{p} \vartheta_k^{\gamma} \gamma_{\tau_{k-1}^{\gamma}, \tau_k^{\gamma}}^{c}(t) \qquad (2.143)$$

where $\tau_{k-1} \le \tau_k, k = 1, \ldots, p$ and χ^c, γ^c are the characteristic functions defined by

$$\chi^c(t) = \left\{ \begin{array}{cc} 1 & t \in I \\ 0 & \text{otherwise} \end{array} \right\} \quad \gamma^c(t) = \left\{ \begin{array}{cc} 1 & t \in [\tau_{k-1}, \tau_k] \\ 0 & \text{otherwise} \end{array} \right\} \tag{2.144}$$

The switching times $\tau_i^\chi, \tau_i^\gamma, 1 \le i \le p - 1$ are also regarded as decision variables. The time scaling transform these switching times into fixed times $k/p, k = 1, \ldots, p - 1$ on a new time horizon $[0, 1]$. This is achieved by the following differential equation:

$$\dot{t}(s) = \vartheta^t(s) \sum_{k=1}^{p} \vartheta_k^t \chi_{\tau_{k-1}^\chi, \tau_k^\chi}^t(t) \tag{2.145}$$

Observations of weather avoidance maneuvering typically reveal reactive (tactical) deviations around hazardous weather cells. Safety constraints dictate that aircraft must remain separated from one another as from hazardous weather. Because of weather forecasting errors, weather constraints are not usually known with certainty. The uncertainty is smaller for short-range forecasts, but the uncertainty increases and becomes substantial for long-range forecasts. Model weather constraints can be modeled as deterministic constraints varying with time according to a piece-wise constant function that is based on a weather forecast model, and the most recent short-range weather forecast is made. Aircraft are modeled as points in motion. Their dynamics are specified in terms of bounds on the aircraft velocity and magnitude of acceleration. Whereas acceleration bounds give rise to bounds on the radius of curvature of flight, the scale of the solution is assumed to be large enough that aircraft dynamics can be approximated with a single representation of piecewise linear flight legs connected at way-points.

2.3.4.3 RRT Variation

In [326], a path planning algorithm based on 3D Dubins curve for UAV to avoid both static and moving obstacles is presented. A variation of RRT is used as the planner. In tree expansion, branches of the trees are generated by propagating along 3D Dubins curve. The node sequence of the shortest length together with Dubins curves connecting them is selected as the path. When the UAV executes the path, the path is checked for collision with updated obstacles state. A new path is generated if the previous one is predicted to collide with obstacles. Such checking and replanning loop repeats until the UAV reaches the goal. Three-dimensional Dubins curves is used in mode connection because

1. It allows to assign initial and final heading of the UAV as well as position.
2. It is the shortest curve that connects two points with a constraint on the curvature determined by UAV turning radius of the path and prescribed initial and final headings.

To connect node (x_0, y_0, z_0, χ_0) to (x_1, y_1, z_1, χ_1), a 2D Dubins curve C is first created from (x_0, y_0, χ_0) to (x_1, y_1, χ_1), it is then extended to 3D by assigning

$$z = z_0 + \frac{\ell(x, y)}{\ell(x_1, y_1)}(z - z_0) \tag{2.146}$$

to each (x, y) in C where $\ell(x, y)$ stands for the length along C from to (x_0, y_0) to (x, y).

The next step is to propagate the UAV model along the 3D Dubins curve until reaching the end or being blocked by obstacles. If the end is reached, the propagatory trajectory is the connection between the two nodes; otherwise, a connection does not exist due to collision with obstacles.

2.4 MISSION PLANNING

A mission describes the operation of an aircraft in a given region, during a certain period of time while pursuing a specific objective. Way-points are locations to which the autonomous aircraft is required to fly. A flight plan is defined as the ordered set of way-points executed by the aircraft during a mission. Along the way, there may be a set of areas to visit and a set of areas to avoid. In addition, the mission planning strategy should be dynamic as the mission planning problem is to create a path in a dynamic environment. The aim is to replace the human expert with a synthetic one that can be deployed on-board the smart autonomous aircraft [172,175].

A mission is carried out by performing different actions: actions of movements, actions on the environment, information gathering, etc. The resources used for the implementation of actions are available in limited quantities [192]. For the autonomous aircraft, resources are consumable as fuel and electricity levels decline gradually as the mission proceeds. Mission planning adapts flight to mission needs. The mission planning problem is to select and order the best subset among the set of objectives to be achieved and to determine the dates of start and end of each objective, maximizing the rewards obtained during the objectives and criteria for minimizing the consumption of resources while respecting the constraints on resources and mission.

Mission planning can be considered as a selection problem. The objectives are linked to rewards whose values vary depending on the importance of each objective. Planning must choose a subset of objectives to be achieved in time and limited resources. The existing planning systems are mostly unsuitable for solving smart autonomous aircraft problems: they address a problem where the goal is a conjunction of goals and fail if the goal is not reached. Moreover, the selection of targets does not entirely solve the problem of mission planning. Indeed, the selection is often based on a simplified model for the resources of the problem and ignores the various ways to achieve the same goal. In most cases, a practical solution is obtained by combining the selection of objectives, planning and task scheduling in multi-levels: each level defines the problem to the scheduling algorithm of the lower level. Unlike the planning carried out during the mission preparation, planning online is characterized by the fact that the time taken to find a plan is one of the main criteria for judging the quality of a method. Models to formalize a planning problem and associated methods can be classified into three types:

1. Representations of **logical type:** The dynamics of the aircraft translates into a succession of consecutive states indexed by a time parameter. The states are described by a set of logical propositions and action is an operator to move from one state to another. The purpose of planning is to synthesize a trajectory in state space, predict rewards earned for the course, select and organize different types of actions to achieve a goal or optimize the reward functions.
2. Representations of **graph type:** They offer a more structured representation. Among the approaches using graphs include Petri nets and Bayesian networks.
3. Representations of **Object type:** They have spread because of their relevance to the object-oriented programming languages.

The aim is to formulate the art of flying an aircraft into logical tasks with a series of events to maintain control of specific functions of the aircraft. This concept is a key component with systems designed for future use. The initial concept of using an expert system to control an aircraft seems simple but proves difficult to apply. An expert pilot's decision-making processes is difficult to initiate with computers [348]. The dynamic environment and conditions affecting an aircraft are areas that have to be adapted to such an expert system. The many tasks involved in the control of flight must be divided into manageable steps.

The purpose of the **mission planning** is to select the objectives to achieve and find a way to achieve them, taking into account the environment. Among the possible solutions, the planner must choose the one that optimizes a criterion taking into account the rewards for each goal and cost to achieve them, and respects the constraints of time and resources. Rewards and constraints are non-linear functions of time and resources to the different times when the aircraft performs the actions that lead to achieving the objectives. For achieving a goal, there is a beginning of treatment, end of treatment and when the reward associated with the target is obtained. The mission planner must choose and order a subset of targets, to achieve among all mission objectives. It should optimize the choice of its actions, knowing its resources, the environment, the maximum reward associated with each objective and the time constraints associated with them. Techniques were first designed to solve classical problems from combinatorial optimization such as the traveling salesman problem (TSP) or the Chinese postman problem [307], the maximum flow problem and the independent set point [343,345,347]. Most of these problems are closely related to graph theory.

Formalism could be based on the decomposition of the problem into two levels:

1. The top level corresponds to the objectives of the mission.
2. The lowest level describes this achievement as a function of time and resources.

It formalizes a problem with uncertainties where the number of objectives of the plan is not fixed a priori.

2.4.1 TRAVELING SALESMAN PROBLEM

A salesman has to visit several cities (or road junctions). Starting at a certain city, he wants to find a route of minimum length which traverses each of the destination cities exactly once and leads him back to his starting point. Modeling the problem as a complete graph with n vertices, the salesman wishes to make a tour or Hamiltonian cycle, visiting each cycle exactly once and finishing at the city he starts from [99]. A traveling salesman problem (TSP) instance is given by a complete graph G on a node set $V = \{1, 2, \ldots, m\}$ for some integer m and by a cost function assigning a cost c_{ij} to the arc (i, j) for any i, j in V. The salesman wishes to make the tour whose total cost is minimum where the total cost is the sum of the individual costs along the edges of the tour [89].

Problem 73. *The formal language for the corresponding decision problem is*

$$
\text{TSP} = \left\{ (G, c, k) : \begin{bmatrix} G = (V, E) \text{ is a complete graph} \\ c \text{ is a function from } V \times V \to N \\ k \in N \\ G \text{ has a traveling salesman tour with cost at most } k \end{bmatrix} \right\}
$$

The data consist of weights assigned to the edges of a finite complete graph, and the objective is to find a Hamiltonian cycle, a cycle passing through all the vertices, of the graph while having the minimum total weight. $c(A)$ denotes the total cost of the edges in the subset $A \subseteq E$:

$$
c(A) = \sum_{(u,v) \in A} c(u, v) \tag{2.147}
$$

In many practical situations, the least costly way to go from a place u to a place w is to go directly, with no intermediate steps. The cost function c satisfies the triangle inequality if for all the vertices, $u, v, w \in V$

$$
c(u, w) \leq c(u, v) + c(v, w) \tag{2.148}
$$

This triangle inequality is satisfied in many applications; not in all, it depends on the chosen cost. In this case, the minimum spanning tree can be used to create a tour whose cost is no more than twice that of the minimum tree weight, as long as the cost function satisfies the triangle inequality. Thus, the pseudocode of the TSP approach can be presented in Algorithm 19.

Lines 7–11 of Algorithm 19 set the key of each vertex to ∞ (except for the root r, whose key is set to 0 so that it will be the first vertex processed), set the parent of each vertex to NULL and initialize the min-priority queue Q to contain all the vertices. The algorithm maintains the following three-part loop invariant. Prior to each iteration of the while loop of lines 12–17.

Algorithm 19 TSP with Triangle Inequality

---
1. Select a vertex $r \in G,V$ to be a root vertex
2. Compute a minimum spanning tree T for G from root r using MST-PRIM(G,c,r)
3. Let H be a list of vertices, ordered according to when they are first visited in a preorder tree walk of T.
4. A state transition rule is applied to incrementally build a solution.
5. return the Hamiltonian cycle H
6. **Minimum Spanning Trees:** Procedure MST-PRIM(G,c,r)
7. For each $u \in G,V$
8. u·key $= \infty$
9. u·π = NULL
10. r·key $= 0$
11. Q $=$ G·V
12. While Q $\neq 0$
13. u = EXTRACT-MIN(Q)
14. for each $v \in G \cdot Adj[u]$
15. if $v \in Q$ and $w(u,v) < v \cdot key$
16. v ·π = u
17. v·key = w(u,v)

---

1. $A = \{(v,v,\pi) : v \in V - \{r\} - Q\}$
2. The vertices already placed into the minimum spanning tree are those in $V - Q$.
3. For all vertices $v \in Q$, if $v \cdot \pi \neq NULL$, then $v \cdot key < \infty$ and $v \cdot key$ is the weight of a light edge (v,v,π) connecting v to some vertex, already placed into the minimum spanning tree

Line 13 identifies a vertex $u \in Q$ incident on a light edge that crosses the cut $(V - Q, Q)$ (with the exception of the first iteration, in which $u = r$ due to line 4). Removing u from the set Q adds it to the set $V - Q$ of vertices in the tree, thus adding $(u, u \cdot \pi)$ to A. The for loop of lines 14–17 updates the *key* and π attributes of every vertex v adjacent to u but not in the tree, thereby maintaining the third part of the loop invariant.

There are other approximation algorithms that typically perform better in practice. If the cost c does not satisfy the triangle inequality, then good approximate tours cannot be found in polynomial time. There are different approaches for solving the TSP. Classical methods consist of heuristic and exact methods. Heuristic methods like cutting planes and branch and bound can only optimally solve small problems, whereas the heuristic methods, such as Markov chains (MCs), simulated annealing and tabu search are good for large problems [319]. Besides, some algorithms based on greedy principles such as nearest neighbor and spanning tree can be introduced as efficient solving methods. Nevertheless, classical methods for solving the TSP usually result in exponential computational complexities. New methods such as nature-based optimization algorithms, evolutionary computation, neural networks, time adaptive self-organizing maps, ant systems, particle swarm optimization, simulated annealing and bee colony optimization are among solving techniques inspired by observing nature. Other algorithms are intelligent water drops algorithms and artificial immune systems [93].

In an instance of the TSP, the distances between any pair of n points are given. The problem is to find the shortest closed path (tour) visiting every point exactly once. This problem has been traditionally been solved in two steps with the layered controller architecture for mobile robots. The following discussion is mainly based on [223,259].

Problem 74. *Dubins Traveling Salesman Problem (DTSP): Given a set of n points in the plane and a number $L > 0$, DTSP asks if there exists a tour for the Dubins vehicle that visits all these points exactly once, at length at most L*

At the higher decision-making level, the dynamics of the autonomous aircraft are usually not taken into account and the mission planner might typically chose to solve the TSP for the Euclidean metric (ETSP), i.e., using the Euclidean distances between way-points. For this purpose, one can directly exploit many existing results on the ETSP on graphs. The first step determines the order in which the way-points should be visited by the autonomous aircraft. At the lower level, a path planner takes as an input this way-point ordering and designs feasible trajectory between the way-points respecting the dynamics of the aircraft. In this section, the aircraft at a constant altitude is assumed to have a limited turning radius and can be modeled as a Dubins vehicles. Consequently, the path planner could solve a sequence of Dubins shortest-path problems (DSPP) between the successive way-points. Even if each problem is solved optimally, however, the separation into two successive steps can be inefficient since the sequence of points chosen by the TSP algorithm is often hard to follow for the physical system. In order to improve the performance of the autonomous aircraft system, mission planning and path planning steps are integrated. The Dubins vehicle can be considered as an acceptable approximation of a fixed-wing aircraft at a constant altitude. Motivated by the autonomous aircraft applications, the TSP is considered for Dubins vehicle DTSP.

Problem 75. *Given n points on a plane, what is the shortest Dubins tour through these points and what is its length?*

The worst-case length of such a tour grows linearly with n, and an algorithm can be proposed with performance within a constant factor of the optimum for the worst-case point sets. An upper bound on the optimal length is also obtained [259]. A practical motivation to study the DTSP arises naturally for autonomous aircraft monitoring a collection of spatially distributed points of interest. In one scenario, the location of the points of interest might be known and static. Additionally, autonomous aircraft applications motivate the study of the dynamic traveling repairman problem (DTRP) in which the autonomous aircraft is required to visit a dynamically changing set of targets [46]. Such problems are examples of distributed task allocation problems and are currently generating much interest: complexity issues related to autonomous aircraft assignment problems, Dubins vehicles keeping under surveillance multiple mobile targets, missions with dynamic threats, etc. [134]. Exact algorithms, heuristics as well as polynomial time constant factor approximation algorithms, are available for the Euclidean traveling salesman problem. A variation of the TSP is the angular metric problem. Unlike other variations of the TSP, there are not known reductions of the Dubins TSP to a problem on a finite dimensional graph, thus preventing the use of well-established tools in combinatorial optimization.

Definition 76. *Feasible Curve A feasible curve is defined for the Dubins vehicle or a Dubins path as a curve $\gamma: [0,T] \to \mathbb{R}^2$ that is twice differentiable almost everywhere and such that the magnitude of its curvature is bounded above by $1/\rho$ where $\rho > 0$ is the minimum turning radius.*

The autonomous aircraft configuration is represented by the triplet $(x,y,\psi) \in SE(2)$ where (x,y) are the Cartesian coordinates of the vehicle and ψ its heading. Let $P = p_1, \ldots, p_n$ be a set of n points in a compact region $Q \subseteq \mathbb{R}^2$ and P_n be the collection of all point sets $P \subset Q$ with the cardinality n. Let ETSP(P) denote the cost of the Euclidean TSP over P, i.e., the length of the shortest closed path through all points in P. Correspondingly, let DTSP(P) denote the cost of the Dubins path through all points in P, with minimum turning radius ρ. The initial configuration is assumed to be $(x_{\text{init}}, y_{\text{init}}, \psi_{\text{init}}) = (0,0,0)$, let $C_\rho : SE(2) \to \mathbb{R}$ associate to a configuration (x,y,ψ) the length of the shortest Dubins path, define $F_0 :]0,\pi[\times]0,\pi[\to]0,\pi[$, $F_1 :]0,\pi[\to \mathbb{R}$, $F_2 :]0,\pi[\to \mathbb{R}$

$$F_0(\psi,\theta) = 2\arctan\left(\frac{\sin(\psi/2) - 2\sin(\psi/2 - \theta)}{\cos(\psi/2) + 2\cos(\psi/2 - \theta)}\right) \tag{2.149}$$

$$F_1(\psi) = \psi + \sin\left(\frac{F_0(\psi, \psi/2 - \alpha(\psi))}{2}\right) + 4\arccos\left(\sin\left(\frac{0.5(\psi - F_0(\psi, \psi/2 - \alpha(\psi)))}{2}\right)\right)$$
(2.150)

$$F_2(\psi) = 2\pi - \psi + 4\arccos\left(\frac{\sin(\psi/2)}{2}\right)$$
(2.151)

where

$$\alpha(\psi) = \frac{\pi}{2} - \arccos\left(\frac{\sin(0.5\psi)}{2}\right)$$
(2.152)

The objective is the design of an algorithm that provides a provably good approximation to the optimal solution of the Dubins TSP. The alternating algorithm works as follows: compute an optimal ETSP tour of P, and label the edges on the tour in order with consecutive integers. A DTSP tour can be constructed by retaining all odd-numbered (except the nth) edges and replacing all even-numbered edges with minimum length Dubins paths preserving the point ordering. The pseudocode for this approach is presented in Algorithm 20.

Algorithm 20 Dubins Traveling Salesman Problem

---

1. Set $(a_1, \ldots, a_n)$ = optimal ETSP ordering of P
2. Set ψ_1: orientation of segment from a_1 to a_2
3. For $i \in 2, \ldots, n-1$ do
   - if i is even then set $\psi_i = psi_{i-1}$
   - else set ψ_i = orientation of segment from a_i to a_{i+1}
4. if n is even then set $\psi_n = \psi_{n-1}$ else set ψ_n = orientation of segment from a_n to a_1
5. Return the sequence of configurations $(a_i, \psi_i)_{i \in 1, \ldots, n}$

The nearest-neighbor heuristic produces a complete solution for the DTSP, including a way-point ordering and a heading for each point. The heuristic starts with an arbitrary point and chooses its heading arbitrarily, fixing an initial configuration. Then at each step, a point is found which is not yet on the path but close to the last added configuration according to the Dubins metric. This closest point is added to the path with the associated optimal arrival heading. When all nodes have been added to the path, a Dubins path connecting the last obtained configuration and the initial configuration is added. If K headings are chosen for each point, then an a priori finite set of possible headings is chosen for each point and a graph is constructed with n clusters corresponding to the n way-points, and each cluster containing K nodes corresponding to the choice of the headings. Then, the Dubins distances between configurations corresponding to pair of nodes in distinct clusters are computed. Finally, a tour through the n clusters is computed which contains exactly one point in each cluster. This problem is called the generalized asymmetric traveling salesman problem (ATSP) over nK nodes. A path planning problem for a single fixed-wing aircraft performing a reconnaissance mission using one or more cameras is considered in [239]. The aircraft visual reconnaissance problem for static ground targets in terrain is formulated as a polygon-visiting DTSP.

2.4.2   REPLANNING OR TACTICAL AND STRATEGICAL PLANNING

The mission parameters are provided by a higher level automated scheduling system. **Strategic planning**, which occurs before take-off, takes a priori information about the operating environment and the mission goals, and constructs a path that optimizes for the given decision objectives. **Tactical planning** involves re-evaluation and re-generation of a flight plan during flight based on updated information about the goals and operating environment. The generated plan should be as close as possible to the optimal plan given available planning information.

An autonomous aircraft must choose and order a subset of targets, to achieve among all mission objectives. It should optimize the choice of its actions, knowing its resources, the environment, the maximum reward associated with each objective and the time constraints associated with them. Formalism could be based on the decomposition of the problem into two levels: the top level corresponds to the objectives of the mission, and the lowest level describes this achievement as a function of time and resources. It formalizes a problem with uncertainties where the number of objectives of the plan is not fixed a priori. A static algorithm is used offline to produce one or more feasible plans. A dynamic algorithm is then used online to gradually build the right solution to the risks that arise. The terms static and dynamic characterize the environment in which the plan is carried out. Classical planning assumes that the environment is static, meaning that there is no uncertainty. A predictive algorithm is then used offline to produce a single plan which can then be executed online without being questioned. In the case of a dynamic environment, several techniques are possible.

1. Keep a predictive offline, supplemented by one or reactive algorithms that are executed when a hazard line makes incoherent the initial plan, calling it into question and forced most often to replan.
2. Take into account the uncertainties from the construction phase offline: this is called proactive approaches.
3. Plan always predictively but this time online, short term, in a process of moving horizon, in which case, the execution will gradually resolve uncertainties and allow for further planning steps.

The level of decision-making autonomy is referred to the planning board. It requires a calculation of plan online, called replanning. Updating the plan online involves the development of a hybrid architecture, incorporating the outbreak of calculations of new plans in case of hazard and the inclusion of the results of this calculation. The proposed architecture allows to set up an online event planning, with many hierarchical levels of management of the mission. The mission planning problem is to select and order the best subset among the set of objectives to be achieved and to determine the dates of start and end of each objective, maximizing the rewards obtained during the objectives and criteria for minimizing the consumption of resources, while respecting the constraints on resources and mission.

Mission planning can be considered as a selection problem. The objectives are linked to rewards whose values vary depending on the importance of each objective. Planning must choose a subset of objectives to be achieved in time and limited resources. The existing planning systems are mostly unsuitable for solving such problems: they address a problem where the goal is a conjunction of goals and fail if the goal is not reached. The selection of targets is often based on a simplified model for the resources of the problem and ignores the various ways to achieve the same goal. In most cases, a practical solution is obtained by combining the selection of objectives, planning and task scheduling in a multi-levels architecture: each level defines the problem to the scheduling algorithm of the lower level. Unlike the planning carried out during the mission preparation, planning online is characterized by the fact that the time taken to find a plan is one of the main criteria for judging the quality of a method.

The on-board intelligence allows the aircraft to achieve the objectives of the mission and ensuring its survival, taking into account the uncertainties that occur during a mission. The objectives of the planning function are in general: order the passage of the various mission areas, calculate a path between each element of the route, order the realization of the task, etc. The use of hybrid techniques deterministic/random is intended to provide solutions to this problem. A mission planning system must be able to

1. evaluate multiple objectives
2. handle uncertainty
3. be computationally efficient.

The mission planning task is non-trivial due to the need to optimize for multiple decision objectives such as safety and mission objectives. For example, the safety objective might be evaluated according to a midair collision risk criterion and a risk presented to third-party criterion. The degree of satisfaction of the safety objective is obtained by aggregating the constituent criteria. A constraint refers to limits imposed on individual decision criteria (a decision variable) such as the maximum allowable risk.

For some applications, the mission tasks, for example, spray crops or perform surveillance are conducted at the destination point. Another important consideration is online or in-flight replanning. A plan that is optimal when it is generated can become invalidated or sub-optimal by changes to assumptions in the flight plan. For example, the unanticipated wind conditions can increase fuel consumption, it may take an unexpectedly long time to reach a way-point and there may be changes to mission goals as new information becomes available.

As with manned aircraft, the dependability and integrity of a UAV platform can be influenced by the occurrence of endogenous and exogenous events. There are a number of safety-related technical challenges which must be addressed including provision of a safe landing zone detection algorithm which would be executed in the event of a UAV emergency. In the event of such an emergency, a key consideration of any safety algorithm is remaining flight time which can be influenced by battery life or fuel availability and weather conditions. This estimates of the UAV remaining flight time can be used to assist in autonomous decision-making upon occurrence of a safety critical event [346].

2.4.3   ROUTE OPTIMIZATION

In general, the routing approach consists in reducing a general routing problem to the shortest-path problem. The specification of routes is a problem in its own right. If the route network is modeled as a directed graph, then the routing problem is the discrete problem of finding paths in a graph, which must be solved before the speed profile is sought [258].

2.4.3.1   Classical Approach

An automated mission planning can enable a high level of autonomy for a variety of operating scenarios [156,223]. Fully autonomous operation requires the mission planner to be situated onboard the smart autonomous aircraft. The calculation of a flight plan involves the consideration of multiple elements. They can be classified as either continuous or discrete, and they can include nonlinear aircraft performance, atmospheric conditions, wind forecasts, aircraft structure, amount of departure fuel and operational constraints. Moreover, multiple differently characterized flight phases must be considered in-flight planning. The flight planning problem can be regarded as a trajectory optimization problem [380]. The mission planning has to define a series of steps to define a flight route. In the context of mission planning for an autonomous aircraft, the plan is necessarily relative to displacement. The plan then contains a sequence of way-points in geometric space considered. A possible path of research of the current aircraft position to destinations in geometric space, avoiding obstacles in the best way is sought. Scheduling algorithms must be integrated into an embedded architecture to allow the system to adapt its behavior to its state and dynamic environment [155].

The approach reduces the uncertainty inherent in a dynamic environment through online replanning and incorporation of tolerances in the planning process [327]. The motion plan is constrained by aircraft dynamics and environmental/operational constraints. In addition, the planned path must satisfy multiple, possibly conflicting objectives such as fuel efficiency and flight time. It is not computationally feasible to plan in a high-dimensional search space consisting of all the aforementioned decision variables. It is common, instead, to plan the path in the world space (x, y, z, t) by aggregating the decision variables into a single, non-binary cost term.

Integration must take into account that the activation calculations plan is triggered by events. A random event may occur during a mission whose date of occurrence is unpredictable. An autonomous system has two main goals:

1. Make out its mission while remaining operational
2. React to the uncertainties of mission, environment or system.

The embedded architecture must meet these two objectives by organizing the physical tasks of the mission and the tasks of reasoning. This reaction is conditioned by the inclusion of planning during execution in a control architecture. Each controller is composed of a set of algorithms for planning, monitoring, diagnosing and execution. This architecture can be applied to complex problem solving, using hierarchical decomposition. Hierarchical decomposition is employed to break down the problem into smaller parts both in time and function. This approach provides a viable solution to real-time, closed-loop planning and execution problems:

1. The higher levels create plans with the greatest temporal scope, but low level of detail in planned activities.
2. The lower levels temporal scope decreases, but they have an increase in detail of the planned activities.

Remark 77. *Situation awareness includes both monitoring and diagnosis. Plan generation and execution are grouped. A hierarchical planning approach is chosen because it enables a rapid and effective response to dynamic mission events.*

A **functional and temporal analysis** of the mission has to be performed, identifying the activities that need to be performed. As the mission problem is progressively broken down into smaller sub-problems, functional activities emerged. At the lowest level of decomposition, the functional activities are operating on time scales of seconds. These activities are related to each other in a tree structure, with the lowest level (leaf) nodes providing the output commands to the guidance, navigation and control system.

For the autonomous aircraft, the state is given, at least, by three position coordinates, three velocity coordinates, three orientation angles and three orientation rate angles, for a total of twelve variables. The dynamic characteristics of the aircraft determine the dimension of the system, and many systems may use a reduced set of variables that adequately describe the physical state of the aircraft. It is common to consider smaller state space with coupled states, or to extend the state space to include higher order derivatives.

Planning schemes may be classified as explicit or implicit.

1. An **implicit method** is one in which the dynamic behavior of the aircraft is specified, and then, the trajectory and the actuator inputs required to go from the start configuration to the goal configuration are derived from the interaction between the aircraft and the environment. The best-known example of this method is the potential field method [195] and its extensions. Some other examples include the methods that applied randomized approaches [317] or graph theory [86].
2. **Explicit methods** attempt to find solutions for the trajectories and actuators inputs explicitly during the motion. Explicit methods can be discrete or continuous. Discrete approaches focus primarily on the geometric constraints and the problem of finding a set of discrete configurations between the end states that are free from collisions.

Mission planning problems have been considered from the point of view of artificial intelligence, control theory, formal methods and hybrid systems for solving such problems [58]. A class of complex goals impose **temporal constraints** on the trajectories for a given system, also referred to as temporal goals. They can be described using a formal framework such as **linear temporal** logic (LTL), **computation tree logic** and $\mu-$**calculus**. The specification language, the discrete abstraction of the aircraft model and the planning framework depend on the particular problem being solved and the kind of guarantees required. Unfortunately, only linear approximations of the aircraft dynamics can be incorporated. Multi-layered planning is used for safety analysis of hybrid systems with reachability specifications and motion planning involving complex models and environments. The framework introduces a discrete component to the search procedure by utilizing the discrete structure present in the problem.

The framework consists of the following steps:

1. Construction of a discrete abstraction for the system
2. High-level planning for the abstraction using the specifications and exploration information from a low-level planner
3. Low-level sampling-based planning using the physical model and the suggested high-level plans.

There is a two way exchange of information between the high-level and low-level planning layers. The constraints arising due to temporal goals are systematically conveyed to the low-level layer from the high-level layer using synergy. The construction of the discrete abstraction and two-way interaction between the layers are critical issues that affect the overall performance of the approach.

2.4.3.2   Dynamic Multi-Resolution Route Optimization

An approach is described in this section for **dynamic route optimization** for an autonomous aircraft [370]. A multi-resolution representation scheme is presented that uses **B-spline basis functions** of different support and at different locations along the trajectory, parametrized by a dimensionless parameter. A multi-rate receding horizon problem is formulated as an example of online multi-resolution optimization under feedback. The underlying optimization problem is solved with an anytime evolutionary computing algorithm. By selecting particular basis function coefficient as the optimization variables, computing resources can flexibly be devoted to those regions of the trajectory requiring most attention. Representations that can allow a UAV to dynamically re-optimize its route while in-flight is of interest. A popular technique for route optimization is dynamic programming, often in combination with other methods. Way-points generated by dynamic programming serve as input to either an optimal control or a virtual potential fields approach. The potential field method models the route with point masses connected by springs and dampers. Threats and targets are modeled by virtual force fields of repelling and attracting forces. The optimized route then corresponds to the lowest energy state of this mechanical equivalent.

The general requirements are wind optimal routes, avoid regions for several reasons, minimize fuel costs and make allowance for a required time of arrival. The **dynamic programming method** can be used for the intended application. Dynamic programming is however a global optimizer and more flexible methods are preferred.

2.4.3.2.1   Route Optimization Problem Formulation

The route is represented by a sequence of way-points: $(x_k, y_k, z_k, t_k)_{k=1}^K$ where (x_k, y_k, z_k) are Cartesian coordinates of the way-points and t_k is the scheduled time of arrival to the way-points.

Problem 78. *The route optimization problem can be expressed as*

$$X = arg_{X \in D_X} \min J(X) \tag{2.153}$$

where X is a list containing the way-point parameters, J is the route optimality index and D_x is the domain of the allowed routes. The route optimization problems of interest can involve thousands of way-points.

Therefore, direct solution of equation (2.153) is not possible given onboard processing constraints. The optimization set is limited to a parametric family of trajectories represented by spline functions. The trajectory is parametrized by a dimensionless parameter u and is represented by samples of $x(u)$, $y(u)$ and $z(u)$.

Remark 79. *The span of u for the entire route has to be chosen carefully, taking into account the maneuverability and velocity of the aircraft.*

Assuming constant velocity, the distance spanned by a fixed Δu should be approximately the same along the route. Since the optimization will change the position of the way-points when the distance spanned by $\Delta u = 1$ is small, there will be a higher likelihood of generating routes that are not flyable, and the turns required at way-points may exceed the aircraft's maneuvering capability. On the other hand, a large distance spanned by $\Delta u = 1$ will not allow much flexibility to find the best route. The route parametrization uses the following B-spline expansion:

$$x(u) = \sum_{n=0}^{N_{max}} a_n \tilde{\psi}(u - n) + x_0(u) \tag{2.154}$$

$$y(u) = \sum_{n=0}^{N_{max}} b_n \tilde{\psi}(u - n) + y_0(u) \tag{2.155}$$

$$z(u) = \sum_{n=0}^{N_{max}} c_n \tilde{\psi}(u - n) + z_0(u) \tag{2.156}$$

where $\tilde{\psi}(u)$ is a basis function, and $(x_0(u), y_0(u), z_0(u))$ is the initial approximation of the route (from offline mission planning). The following second-order B-spline basis function is used:

$$\tilde{\psi}(w) = \begin{Bmatrix} 0 & w < 0 \\ w^2 & 0 \leq w \leq 1 \\ -2w^2 + 6w - 3 & 1 \leq w \leq 2 \\ (3 - w)^2 & 2 \leq w \leq 3 \\ 0 & w \geq 3 \end{Bmatrix} \tag{2.157}$$

This basis function has support on the interval $w \in [0, 3]$. The representation equations (2.154) and (2.155) defines a 3D trajectory. This has to be complemented by the time dependence of the aircraft position (for evaluating turn constraints). Assuming a constant speed V,

$$\dot{s} = V \tag{2.158}$$

the path from the trajectory start at $u = 0$ to a point parametrized by $u = w$ on the route is given by:

$$s(w) = \int_{u=0}^{w} \sqrt{\left(\left(\frac{dx}{du}\right)^2 + \left(\frac{dy}{du}\right)^2 + \left(\frac{dz}{du}\right)^2 \right)} du \tag{2.159}$$

By equations (2.158), (2.159) and solving for w, it is possible to generate a route, represented by K time-stamped way-points $(x_k, y_k, z_k, t_k)_{k=1}^{K}$.

2.4.3.2.2 *Online Optimization under Feedback*

Most existing aircraft route optimizations are used offline. Criteria for the route optimization are given ahead of the mission and the route is pre-computed. An algorithm that can re-optimize the route in-flight within available computational time limits is developed. A receding horizon control problem can be formulated and addressed in [370]. At every sample, an online optimization algorithm is used to update the route taking into account situation changes and disturbances. Consider the path point $u = k$. The planning of the path for $u \geq k$ is done as

$$x_{k+1}(u) = \sum_{n=0}^{N_{\max}} a_n \tilde{\psi}(u - n - k) + x_k(u) + \Delta x_k(u) \tag{2.160}$$

$$y_{k+1}(u) = \sum_{n=0}^{N_{\max}} b_n \tilde{\psi}(u - n - k) + y_k(u) + \Delta y_k(u) \tag{2.161}$$

$$z_{k+1}(u) = \sum_{n=0}^{N_{\max}} c_n \tilde{\psi}(u - n - k) + z_k(u) + \Delta z_k(u) \tag{2.162}$$

where $x_{k+1}(u), y_{k+1}(u), z_{k+1}(u)$ is the trajectory as computed before the update and $\Delta x_k(u), \Delta y_k(u), \Delta z_k(u)$ are corrections to deal with disturbances. The expansion weights a_n, b_n, c_n are recomputed at each step by solving an optimization problem similar to equation (2.153). The corrections $\Delta x_k(u), \Delta y_k(u), \Delta z_k(u)$ are introduced in equations (2.160)–(2.162) because disturbances such as the wind cause the position and velocity of the aircraft at time $u = k$ to be different from those given by the nominal trajectory $x_k(k), y_k(k), z_k(k)$. The corrections allow the route optimizer to generate a new route from the actual position of the aircraft. Assume that at the path point coordinate $u = k$, the guidance and navigation system of the aircraft determines a position deviation $\Delta x_k(u), \Delta y_k(u), \Delta z_k(u)$ and a velocity deviation $\Delta V_x(k), \Delta V_y(k), \Delta V_z(k)$ from the previously planned route. The route is then adjusted around the aircraft position. Since the B-spline approximation equations (2.154)–(2.157) gives a trajectory that is piece-wise second-order polynomial, the most natural way of computing the correction is also a piece-wise second-order polynomial spline. By matching the trajectory coordinates and derivatives at $u = k$, the correction can be computed as

$$\Delta x_k(u) = \Delta x(k)\tilde{\alpha}(u - k) + \Delta V_x(k)\tilde{\beta}(u - k) \tag{2.163}$$

$$\Delta y_k(u) = \Delta y(k)\tilde{\alpha}(u - k) + \Delta V_y(k)\tilde{\beta}(u - k) \tag{2.164}$$

$$\Delta z_k(u) = \Delta z(k)\tilde{\alpha}(u - k) + \Delta V_z(k)\tilde{\beta}(u - k) \tag{2.165}$$

where

$$\tilde{\alpha}(w) = \left\{ \begin{array}{ll} 1 - 0.5w^2 & 0 < w \leq 1 \\ 0.5(w - 2)^2 & 1 < w \leq 2 \end{array} \right\} \tag{2.166}$$

$$\tilde{\beta}(w) = \left\{ \begin{array}{ll} w - 0.75w^2 & 0 < w \leq 1 \\ 0.25(w - 2)^2 & 1 < w \leq 2 \end{array} \right\} \tag{2.167}$$

The receding horizon update at point $u = k$ computes a trajectory of the form equation (2.160) to (2.162) where the expansion weights $a_n, b_n, c_n (n = 1, \ldots, N_{\max})$ are such as to minimize a modified optimality index of the form equation (2.153). The modified optimality index takes into account only the "future" part of the trajectory for $u \geq k$.

2.4.3.2.3 Multi-Scale Route Representation

The route has to be dynamically updated in-flight to compensate for both disturbances and changes in the overall mission structure such as the emergence or disappearance of threats and targets. Route planning in the vicinity of the current location can usually be done in detail because the current local information can be expected to be reliable. It must also be done quickly before the aircraft moves too far. The longer term planning is also less critical at the current time. An optimal trade-off between different optimization horizons can be effected with limited computational resources using a multiscale representation. A technique similar to wavelet expansion is used as it provides an efficient way to represent a signal, such as a trajectory over time at multiple temporal and frequency scales.

2.4.4   FUZZY PLANNING

Many traditional tools for formal modeling, reasoning and computing are crisp, deterministic and precise character. However, most practical problems involve data that contain uncertainties. There have been a great amount of research in **probability theory, fuzzy set theory, rough set theory, vague set theory, gray set theory, intuitionistic fuzzy set theory** and **interval math**. Soft set and its various extensions have been applied with dealing with decision-making problems. They involve the evaluation of all the objects which are decision alternatives.

The imposition of constraints, such as aircraft dynamic constraints and risk limits, correspond to skills level decision-making [32]. The evaluation function used to calculate path costs is rule based, reflecting the rules level. Finally, the selection of an evaluation function and scheduling of planning activities, such as in an anytime framework, mimics the knowledge of level.

Heuristics primary role is to reduce the search space and thus guide the decision-maker onto a satisfying or possibly optimal solution in a short space of time. It is applicable to mission planning as flight plans tend to follow the standard flight profile. Additionally, due to the time pressure of online replanning, the satisfying heuristic could be used to quickly find a negotiable path rather than deliver an optimal solution that is late.

The planner can consider as **decision criteria:** obstacles, roads, ground slope, wind and rainfall. **Decision variables** are aggregated into a single cost value which is used in a heuristic search algorithm. The rain criterion, for example, is represented by the membership function: light rain, moderate rain and heavy rain. IF-THEN rules are then used to implicate the degree of membership to the output membership function on the mobility (i.e., difficulty of traversal) universe. A vector neighborhood can be identified as a suitable method for overcoming the limited track angle resolution to ensure path optimality. Uncertainty is often mitigated with online replanning, multi-objective decision-making and incorporation of tolerance into the planned path.

In the context of mission planning, each decision criterion has associated with it a constraint and ideal value. For example, there is an ideal cruise velocity and a maximum and minimum airspeed limit. Constraints can be enforced with simple boundary checking and correspond to the skills level of cognition. Optimization of the selected paths, however, requires evaluation of multiple decision rules as per the rules level. The safety objective may be decomposed into a risk criterion, no-fly zones criterion, wind and maximum velocity constraints. Additionally, an approximation of the dynamic constraints of the aircraft is required to ensure that the final problem is traversable. Each membership function objective can be decomposed into individual decision criteria [371,375].

2.4.4.1   Fuzzy Decision Tree Cloning of Flight Trajectories

A graph-based mission design approach is developed in [393], involving maneuvers between a large set of trim trajectories. In this approach, the graph is not an approximation of the dynamics in general, but rather each node consists of one of the available maneuvers and a connection exists if there is a low-cost transition between trim trajectories. The method can be divided into two primary sections: the offline mission planning phase and the onboard phases. The mission planning steps are relatively expensive computationally and involve creating, cataloging and efficiently storing the maneuvers and boundary conditions, as well as describing the relationship between the discretized dynamics and fuel estimates. On the other hand, the onboard process quickly leverages the work done on the ground and stored in memory to quickly generate a transfer using search and correction methods. The onboard and offline portions are complimentary images of each other: the mission planning stages translate and compress the continuous system dynamics into a discrete representation, whereas the onboard portion searches and selects discrete information and reconstructs continuous trajectories [59,73,115].

Trajectory optimization aims at defining optimal flight procedures that lead to time/energy efficient flights. A **decision tree** algorithm is used to infer a set of **linguistic decision rule** from a set of 2D obstacle avoidance trajectories optimized using MILP in [394]. A method to predict a discontinuous function with **fuzzy decision tree** is proposed and shown to make a good approximation to the optimization behavior with significantly reduced computational expense. Decision trees are shown to generalize to new scenarios of greater complexity than those represented in the training data and to make decisions on a time scale that would enable implementation in a real-time system. The transparency of the rule-based approach is useful in understanding the behavior exhibited by the controller. Therefore, the decision trees are shown to have the potential to be effective online controllers for obstacle avoidance when trained on data generated by a suitable optimization technique such as MILP.

Adaptive dynamic programming (ADP) and **Reinforcement learning** (RL) are two methods for solving decision-making problems where a performance index must be optimized over time. They are able to deal with complex problems, including features such as uncertainty, stochastic effects, and nonlinearity. Adaptive dynamic programming tackles these challenges by developing optimal control methods that adapt to uncertain systems over time. Reinforcement learning takes the perspective of an agent that optimizes its behavior by interacting with an initially unknown environment and learning from the feedback received.

The problem is to control an autonomous aircraft to reach a target obscured by one or more threats by following a near optimal trajectory, minimized with respect to path length and subject to constraints representing the aircraft dynamics. The path must remain outside known threat regions at all times. The **threat size** is assumed to be constant, and the threats and target are assumed to be stationary. Furthermore, it is assumed that the aircraft flies at constant altitude and velocity and is equipped with an autopilot to follow a reference heading input. The output from the controller shall be a change in demanded heading $\Delta\chi$, and the inputs are of the form $\{R_{\text{target}}, \theta_{\text{target}}, R_{\text{threat}}, \theta_{\text{threat}}\}$ where $R_{\text{target}}, \theta_{\text{target}}$ are range and angle to target, and $R_{\text{threat}}, \theta_{\text{threat}}$ are the ranges and angles to any threats present. All angles are relative to the autonomous aircraft current position and heading.

The system has two modes: learn and run. When in the learning mode, the MILP is used to generate heading deviation decisions. The heading deviations are summed with the current heading and used as a reference input to the aircraft and recorded with the optimization inputs in a training set. Once sufficient training runs ($\sim$100 in [394]) have been performed, the run mode is engaged where the decision tree is used to generate the heading deviation commands and performance evaluated.

Predicting a heading deviation and choosing to fix the frame of reference relative to the aircraft heading results in a data representation that is invariant under global translation and rotation. The independence of these basic transformations reduces the problem space and improves generalization by allowing many different scenarios to be mapped on to a single representation.

The optimization solves for the minimum time path to the target using a linear approximation to the aircraft dynamics. Variables are time to target N and acceleration $a(k)$ for predicted steps $k = 0, \ldots, N-1$.

Problem 80. *The optimization problem can be formulated as follows:*

$$min_{N,a(k)} \left[N + \tilde{\gamma} \sum_{k=0}^{N} \|a(k)\|^2 \right] \tag{2.168}$$

subject to $\forall k \in \{0, \ldots, (N-1)\}$

$$R(0) = R_0 \quad V(0) = V_0 \tag{2.169}$$

$$V(k+1) = V(k) + a(k)\delta z$$
$$R(k+1) = R(k) + V(k)\delta z + \tfrac{1}{2}a(k)\delta z^2 \tag{2.170}$$

$$\|a(k)\|_2 \leq a_{\max} \quad \|V(k)\|_2 \leq a_{\max} \tag{2.171}$$

$$\left\|R(N) - R_{\text{target}}\right\|_\infty \leq D_T \quad \|R(k) - R_{\text{threat}}\|_2 \geq R_0 \tag{2.172}$$

The cost equation (2.168) primarily minimizes time to target N in steps of δz, with a small weight $\tilde{\gamma}$ on acceleration magnitudes to ensure a unique solution. Equations (2.169) are initial conditions constraints defining position, heading and velocity. Equations (2.170) are a forward Euler approximations to the aircraft dynamics and kinematics. Equations (2.171) are constraints on maximal accelerations and velocities. Equation (2.172) ensures that the aircraft is within a given tolerance of the target at time N and that at all times the distance from the aircraft to the obstacle, r_{threat}, is outside the obstacle radius R_0. These are non-convex constraints, feature that makes the problem resolution difficult. The 2-norm constraint of equation (2.172) is implemented by approximating it with one of a choice of linear constraints and using binary variables to selectively relax all but one of these. The resulting optimization is a mixed linear integer programming.

The heading deviation command $\Delta\chi$ is found by taking the difference of the current heading and the heading predicted by the optimization. The heading predicted by the optimization is found from the velocity vectors $V(k+1)$ that are part of its output. Heading deviation is calculated in place of the required heading to obtain the translation and rotation invariance. The optimization is integrated with the model using the Algorithm 21 shown as follows.

Algorithm 21 Receding Horizon MILP Controller

---

1. Convert $\{R_{\text{target}}, \theta_{\text{target}}, R_{\text{threat}}, \theta_{\text{threat}}\}$ to $\{r_{\text{target}}, r_{\text{threat}}\}$
2. Solve the optimization problem
3. Derive initial heading deviation $\Delta\chi$ from optimization output
4. Set new desired heading χ_{k+1}
5. Run simulation for δz, i.e. one time step of the optimization
6. Return to step 1

A linguistic decision tree is a tree structured set of linguistic rules formed from a set of attributes that describe a system. The leaf nodes of the tree represent a **conditional probability distribution** on a set, $\mathbb{L}_t$, of descriptive labels covering a target variable given the branch is true. The algorithm constructs a decision tree on the basis of minimizing the **entropy** in a training database across random set partitions on each variable represented in the training data. The rules are formulated from sets of labels, LA, describing the universe of each attribute that provides information about the current state. The **degree of appropriateness** of a particular label, L, for a given instance quantifies the belief that L can be appropriately used to describe X. An automatic algorithm is used to define the appropriateness measures that best partition the data. The set of focal elements $\mathbb{F}$ for the same universe are defined as all the sets of labels that might simultaneously be used to describe a given X.

Definition 81. *Entropy is a scalar value measure of the compactness of a distribution. When a probability distribution is used to represent the knowledge, the smaller the entropy of the distribution, the more probability mass is assigned to a smaller area of the state space and thus the more informative the distribution is about the state. The entropy $h(X)$ of a multivariate Gaussian probability distribution over the variable X can be calculated from its covariance matrix as follows:*

$$h(X) = \frac{1}{2}\log\left((2\pi e)^n |\mathbf{P}|\right) \tag{2.173}$$

In order to try to mimic the data most readily available to an aircraft and to achieve the rotational and translational invariance, ranges and bearings are used to relate threat locations to the aircraft position. Each attribute that describes the current state, i.e., $\{R_{\text{target}}, \theta_{\text{target}}, R_{\text{threat}}, \theta_{\text{threat}}\}$, is covered by a set of labels or focal elements.

In practice, up to nearly 20 focal elements can be used to capture final details, the target attribute in this application is the aircraft heading deviation, $\Delta\chi$ which is covered by a similar set of labels. The **membership** of the **focal** elements can be thought of as the belief that a particular label is an appropriate description for a given value of the attribute. The membership is used to enable the probability that a particular branch is a good description of the current simulation state.

Remark 82. *The characteristic of bearing data is that it is circular. This can be achieved by merging the two boundary fuzzy sets from a linear domain to give a coverage in a polar plot where angle denotes bearing and radius denotes attribute membership of each focal element.*

2.4.4.2 Fuzzy Logic for Fire-Fighting Aircraft

The focus of this section is a single fire-fighting aircraft that plans paths using fuzzy logic. In this scenario, the system uses basic heuristics to travel to a continuously updated target while avoiding various, stationary or moving obstacles [368]. In a fire-fighting scenario, it is important that the aircraft is capable of performing evasive maneuvering in real time. Effective Collaboration between all the relevant agents can be difficult in a situation where communication can be minimal, and therefore, obstacle avoidance is crucial.

This method performs a multi-solution analysis supervised by a fuzzy decision function that incorporates the knowledge of the fire-fighting problem in the algorithm. This research explores the use of fuzzy logic in obtaining a rule base that can represent a fire-fighting heuristics. In the fire-fighting scenario, using information from ground troops and the incident commander on the target drop location $(x_t, y_t, z_t)^T$ and the system's current location $(x_0, y_0, z_0)^T$, the system drives the difference between its heading angle and the angle to the target to zero. The system has a sensing range that is considered able to sense obstacles of $\pm\frac{\pi}{2} rad$ within a certain radius. When obstacles are detected within this area, the system alters its velocity and heading angle using information about obstacles distances and angles and the location of the target to avoid and then recover from the obstruction. Similarly, when the system reaches its target location, the target alters its velocity to slow down and apply the fire retardant.

For this setup, four inputs can be used for the fuzzification interface and two outputs are given after defuzzification. Inputs into the system are distance from obstacle, angle from the obstacles, heading angle error and distance to the target. With these inputs and a rule base, the control input is obtained, that is the percentage of the maximum velocity and the heading angle is outputted from the fuzzy inference system and used as inputs into the system. The main objective of the controller when there is no obstacles within its sensing range is to plan a direct path to the target. With an inertial frame as a reference, the heading angle θ is measured as the angle of the agent's current heading from the horizontal and the target angle χ is measured as the angle to the target from the horizontal. Therefore with information about the agent location (x_0, y_0, z_0) and the target location (x_t, y_t, z_t), the corresponding angles are determined by

$$\chi = \arctan\left(\frac{y_0}{x_0}\right) \qquad \phi = \arctan\left(\frac{y_t}{x_t}\right) \qquad e = \phi - \chi \qquad (2.174)$$

The agent tries to drive the heading angle error e to zero by making small heading angle adjustments according to simple IF-THEN rules. Since the target location is considered to be known, the target distance D_t is easily determined within the distance formula and given information. When no obstacles are within the sensing range, the control objective is simple. Once the agent reaches the target, it will slow down to apply its fire retardant to the location. Once this is done, the new target location is the aircraft base to allow the agent to refuel and reload. If the agent senses an obstacle along its path to the target, it must slow down and change its heading to avoid collision. The obstacle distance D_0 and angle β are obtained from information from the agent's sensors.

Rules relating the inputs and outputs for the fuzzy logic controller are set up in the form of IF-THEN statements and are based on heuristics and human experience. There is a total of 40 rules in [368] for this setup that can be broken up into two situations: if there is an obstacle within the sensing range or not. If there is no obstacle detected, the default is set to "very far" away and moving toward the target is the agent's primary objective. However, when an obstacle is within range, the agent must slow down and change course to avoid it. Again, once it is clear of this obstacle, it can continue its path toward the target. The obstacle angle can be described by NB (Negative Big), NM (Negative Medium), NS (Negative Small), PS (Positive Small), PM (Positive Medium) and PB (Positive Big). The distance to the obstacle is described by either Close, Medium, Far or Very far away (Out of Range). Similarly, the target angle is NB (Negative Big), NM (Negative Medium), NS (Negative Small), ZE (zero), PS (Positive Small), PM (Positive Medium) or PB (Positive Big). The output velocity is Slow, Fast or Very Fast and the output angle change is parallel to the target angle. The sensing radius used is considered the safe distance from an obstacle. The first two sets of rules are for an input of obstacle distance very far, while the target distance and heading angle vary. The second three sets describe the change in heading speed and angle when an obstacle is detected. The last set case is for when the obstacles are at extreme angles and pose no threat of collision.

2.4.5 COVERAGE PROBLEM

The **coverage** of an unknown environment is also known as the **sweeping problem**, or mapping of an unknown environment. Basically, the problem can either be solved by providing ability for localization and map building first, or by directly deriving an algorithm that performs sweeping without explicit mapping of the area, an average event detection time, instead of a **measure of coverage**, can be used for evaluating the algorithm.

2.4.5.1 Patrolling Problem

In [200,201], the following base perimeter **patrol problem** is addressed: an UAV and a remotely located operator cooperatively perform the task of perimeter patrol. **Alert stations** consisting of **unattended ground sensors** (UGSs) are located at key locations along the perimeter. Upon detection of an incursion in its sector, an alert is flagged by the UGSs. The statistics of the alerts' arrival process are assumed known. A camera-equipped UAV is on continuous patrol along the perimeter and is tasked with inspecting UGSs with alerts. Once the UAV reaches a triggered UGS, it captures imagery of the vicinity until the controller dictates it to move on. The objective is to maximize the information gained and at the same time reduce the expected response time to alerts elsewhere [334]. The problem is simplified by considering discrete-time evolution equations for a finite fixed number m of UGSs locations. It is assumed that the UAV has access to real-time information about the status of alerts at each alert station. Because the UAV is constantly on patrol and is servicing a triggered UGS, the problem is a **cyclic polling system** in the domain of discrete-time-controlled queuing system. The patrolled perimeter is a simple closed curve with $N \geq m$ nodes that are spatially uniformly separated of which m are the alert stations (UGSs locations).

The objective is to find a suitable policy that simultaneously minimizes the service delay and maximizes the information gained upon loitering. A stochastic optimal control problem is thus considered [376]. A Markov decision process (MDP) is solved in order to determine the optimal control policy [67,182]. However, its large size renders exact dynamic programming methods intractable. Therefore, a state aggregation-based approximate linear programming method is used instead, to construct provably good sub-optimal patrol policies [382]. The state space is partitioned, and the optimal cost-to-go or value function is restricted to be a constant over each partition. The resulting restricted system of linear inequalities embeds a family of MCs of lower dimension, one of which can be used to construct a lower bound on the optimal value function. The perimeter patrol problem exhibits a special structure that enables tractable linear programming formulation for the lower bound [27].

Definition 83. *Markov Chain A discrete-time MC is a tuple $\mathcal{M} = \langle S, P, s_{\text{init}}, \Pi, L \rangle$ where S is a countable set of states, $P : S \times S \to [0,1]$ is the transition probability function such that for any state*

$s \in S, \sum_{s' \in S} P(s,s') = 1$, $s_{\text{init}} \in S$ is the initial state, Π is a set of atomic propositions and $L : S \to 2^{\Pi}$ is a labeling function

An observable first-order discrete MC is encoded as the matrix of state transition properties. Its rows sum to one, but the columns do not necessarily do so. A state S_i in a **MC** is said to be absorbing if $a_{ii} = 1$. Otherwise, such a state is said to be transient.

Definition 84. *Hidden Markov Model* *A hidden Markov model (HMM) is a Markov model in which the states are hidden. Rather than having access to the internal structure of the HMM, all that is available are observations that are described by the underlying Markov model. The hidden Markov model λ is described in terms of*

1. *N, the number of states in the model.*
2. *M, the number of distinct observation symbols per state. The individual symbol is $V = \{V_1, ..., V_M\}$*
3. *A, the state transition probability distribution. As for MCs $A = \{a_{ij}\} = Prob$ $(q_t = S_j | q_{t-1} = S_i)$*
4. *B, the observation symbol probability distribution for state $B = \{b_{ij}\} = Prob\,(V_k(t) | q_j = S_i)$, B is known as the emission matrix*
5. *π, the initial state distribution $\pi = \{\pi_i\} = Prob(q_1 | S_i)$*

Definition 85. *Markov Decision Process* *A Markov decision process (MDP) is defined in terms of the tuple $\langle S, A, T, R \rangle$, where*

- *S is a finite set of environmental states.*
- *A is a finite set of actions.*
- *$T : S \times A \to S$ is the state transition function. Each transition is associated with a transition probability $T(s, a, s')$, the probability of ending in state s', given that the agent starts in s and executes action a.*
- *$R : S \times A \to S$ is the immediate reward function, received after taking a specific action from a specific state.*

Definition 86. *Finite Markov Decision Process* *A particular finite MDP is defined by its state and action sets and by the one-step dynamics of the environment. Given any state s and action a, the probability of each possible nest state, s', is*

$$P_{ss'}^a = Prob\left\{s_{t+1} = s' | s_t = s, a_t = a\right\} \tag{2.175}$$

where $P_{ss'}^a$ represents transition probabilities, and t denotes a finite time step.

In the MDP, the value of $P_{ss'}^a$ does not depend on the past state transition history. The agent receives a reward r every time it carries out the one-step action. Given any current state s and action a, together with any next state s', the expected value of the next reward is

$$R_{ss'}^a = E\left[r_{t+1} | s_r = s, a_t = a, s_{t+1} = s'\right] \tag{2.176}$$

where $P_{ss'}^a$ and $R_{ss'}^a$ completely specify the dynamics of the finite MDP. In the finite MDP, the agent follows the policy Π. The policy Π is a mapping from each state s and action a to the probability $\Pi(s,a)$ of taking action a when in state s. In the stochastic planning calculation, based on the MDP, the policy Π is decided so as to maximize the value function $V^{\Pi}(s)$. The $V^{\Pi}(s)$ denotes the expected return when starting in S and following Π thereafter. The definition of $V^{\Pi}(s)$ is

$$V^{\Pi}(s) = E_{\Pi}\left[\sum_{k=0}^{\infty} \gamma^k r_{t+k+1} | s_t = s\right] \tag{2.177}$$

where E_Π denotes the expected value given when the agent follows the policy Π and γ is the discount rate $0 < \gamma < 1$. If the values of $P^a_{ss'}$ and $R^a_{ss'}$ are known, dynamic programming is used to calculate the best policy Π that maximizes the value function $V^\Pi(s)$. When the values of $P^a_{ss'}$ and $R^a_{ss'}$ are unknown, a method such as online reinforcement learning is useful in obtaining the best policy Π in the learning environment. After the planning calculation has finished, a greedy policy that selects action value a that maximizes $V^\Pi(s)$ is optimal.

Definition 87. *Path A path through a MDP is a sequence of states, i.e.,*

$$\omega = q_0 \xrightarrow{(a_0, \sigma^{q_0}_{a_0})(q_1)} q_1 \longrightarrow \ldots q_i \xrightarrow{(a_i, \sigma^{q_i}_{a_i})(q_{i+1})} \longrightarrow \ldots \quad (2.178)$$

where each transition is induced by a choice of action at the current step $i \geq 0$. The ith state of a path ω is denoted by $\omega(i)$ and the set of all finite and infinite paths by $Path^{fin}$ and $Path$, respectively.

A control policy defines a choice of actions at each state of a MDP. Control policies are also known as **schedules** or **adversaries** and are formally defined as follows:

Definition 88. *Control Policy: A control policy μ of an MDP model $\mathbb{M}$ is a function mapping a finite path, i.e., $\omega^{fin} = q_0, q_1, \ldots, q_n$ of $\mathbb{M}$ onto an action in $\mathbb{A}(q_n)$. A policy is a function: $\mu : Path^{fin} \to Act$ that specifies for every finite path, the next action to be applied. If a control policy depends only on the last state of ω^{fin}, it is called a **stationary policy**.*

For each policy μ, a probability measure $Prob_\mu$ over the set of all paths under μ, $Path_\mu$ is induced. It is constructed through an infinite state MC as follows: under a policy μ, a MDP becomes an MC that is denoted D_μ whose states are the finite paths of the MDP. There is a one-to-one correspondence between the paths of D_μ and the set of paths $Path_\mu$ in the MDP. Hence, a probability measure $Prob_\mu$ over the set of paths $Path_{\mu^{fin}}$ can be defined by setting the probability of $\omega^{fin} \in Path^{fin}$ equal to the product of the corresponding transition probabilities in D_μ [396].

With fuzzy logic, the cost elements are expressed as fuzzy membership functions reflecting the inherent uncertainty associated with the planned trajectory, the obstacles along the path and the maneuvers the aircraft is required to perform as it navigates through the terrain. If employed, the algorithm A* can use heuristic knowledge about the closeness of the goal state from the current state to guide the search. The cost of every searched cell, n, is composed of two components: the cost of the least-cost route (found in the search so far) from the start cell to cell n, and the heuristic (i.e., estimated) cost of the minimum-cost route from cell n to the destination cell. Given a search state space, an initial state (start node) and final state (goal node), the algorithm A* will find the optimal (least cost) path from the start node to the goal node, if such a path exists. The generated cell route is further optimized and smoothed by a filtering algorithm.

The filtered route is a series of consecutive way-points that the autonomous aircraft can navigate through. The supervisory module reads the objectives and the status of the mission and based on that it configures the search engine and assigns weights to the route's three cost components. Furthermore, the supervisory module chooses the start and the destination cells for the search engine depending on the current status of the aircraft, i.e., whether it is stationary or already navigating toward a destination and needs to be redirected to another destination. The learning-support module acquires route cost data from the search engine at certain map landmarks and updates a cost database that is used later to provide better heuristics to guide the search engine. Thus, a **two-point boundary value problem** (TPBVP) has to be solved for creating a reference path to be followed by the tracking system.

2.4.5.2 Routing Problem

The **UAV sensor selection and routing problem** is a generalization of the **orienteering problem**. In this problem, a single aircraft begins at a starting location and must reach a designated destination prior to time T. Along with the starting and ending points, a set of locations exists with an associated benefit that may be collected by the aircraft. Mission planning can be viewed as a complex version of path planning where the objective is to visit a sequence of targets to achieve the objectives of the mission [99,100]. The integrated sensor selection and routing model can be defined as an MILP formulation [339].

In [171], a path planning method for sensing a group of closely spaced targets is developed that utilizes the planning flexibility provided by the sensor footprint, while operating within dynamic constraints of the aircraft. The path planning objective is to minimize the path length required to view all of the targets. In addressing problems of this nature, three technical challenges must be addressed: coupling between path segments, utilization of the sensor footprint and determination of the viewing order of the targets. A successful path planning algorithm should produce a path that is not constrained by end points or heading that utilizes the full capability of the aircraft's sensors and that satisfies the dynamic constraints on the aircraft. These capabilities can be provided by discrete-time paths which are built by assembling primitive turn and straight segments to form a flyable path. For this work, each primitive segment in a discrete step path is of specified length and is either a turn or a straight line. Assembling the left turn, right turn and straight primitives creates a tree of flyable paths. Thus, the objective for the path planner is to search the path tree for the branch that accomplishes the desired objectives in the shortest distance. The learning real-time A* algorithm can be used to learn which branch of a defined path tree best accomplishes the desired path planning objectives.

Field operations should be done in a manner that minimizes time and travels over the field surface. A coverage path planning in 3D has a great potential to further optimize field operations and provide more precise navigation [160].

Another example is given as follows: given a set of stationary ground targets in a terrain (natural, urban or mixed), the objective is to compute a path for the reconnaissance aircraft so that it can photograph all targets in minimum time, because terrain features can occlude visibility. As a result, in order for a target to be photographed, the aircraft must be located where both the target is in close enough range to satisfy the photograph's resolution and the target is not blocked by terrain. For a given target, the set of all such aircraft positions is called the target's visibility region. The aircraft path planning can be complicated by wind, airspace constraints, aircraft dynamic constraints and the aircraft body itself occluding visibility. However, under simplifying assumptions, if the aircraft is modeled as a Dubins vehicle, the targets visibility regions can be approximated by polygons and the path is a closed tour [344]. Then, the 2D reconnaissance path planning can be reduced to the following: for a Dubins vehicle, find a shortest planar closed tour that visits at least one point in each of a set of polygons. This is referred to as the **polygon-visiting Dubins traveling salesman problem** (PVDTSP). Sampling-based roadmap methods operate by sampling finite discrete sets of poses (positions and configurations) in the target visibility regions in order to approximate a PVDTSP instance by a **finite-one in set traveling salesman problem** (FOTSP). The finite one in set traveling salesman problem is the problem of finding a minimum-cost closed path that passes through at least one vertex in each of a finite collection of clusters, the clusters being mutually exclusive finite vertex sets. Once a roadmap has been constructed, the algorithm converts the FOTSP instance into an ATSP instance so that a standard solver can be applied.

Another example is the case of a UAV that has to track, protect or provide surveillance of a ground based target. If the target trajectory is known, a deterministic optimization or control problem can be solved to give a feasible UAV trajectory. The goal in [18] is to develop a feedback control policy that allows a UAV to optimally maintain a nominal standoff distance from the target without full knowledge of the current target position or its future trajectory. The target motion is assumed to be random and described by a 2D stochastic process. An optimal feedback control is developed for fixed speed and fixed altitude UAV to maintain a nominal distance from a ground target in a way that it anticipates its unknown future trajectory. Stochasticity is introduced in the problem by assuming that the target motion can be modeled as Brownian motion, which accounts for possible realizations of the unknown target kinematics. The tracking aircraft should achieve and maintain a nominal distance d to the target. To this end, the expectation of an infinite horizon cost function is minimized, with a discounting factor and with penalty for control. Moreover, the possibility for the interruption of observations is included by assuming that the duration of observation times of the target is exponentially distributed, giving rise to two discrete states of operation. A Bellman equation based on an approximating MC that is consistent with the stochastic kinematics is used to compute an optimal control policy that minimizes the expected value of a cost function based on a nominal UAV target distance.

2.4.5.3 Discrete Stochastic Process for Aircraft Networks

This section considers a network of autonomous aircraft. Each one is equipped with a certain kind of onboard sensors, for example, a camera or a different sensor, taking snapshots of the ground area [406]. The general aim of this network is to explore a given area, i.e., to cover this area using several applications: target or even detection and tracking in an unknown area, monitoring geographically inaccessible or dangerous areas (e.g., wildfire or volcano), or assisting emergency personnel in case of disasters.

The objective is to provide a simple analytical method to evaluate the performance of different mobility patterns in terms of their coverage distribution. To this end, a stochastic model can be proposed using a MC. The states are the location of the aircraft, and the transitions are determined by the mobility model of interest. Such a model can be created for independent mobility models such as the random walk and random direction.

However, for a cooperative network, in which each autonomous aircraft decides where to move based on the information received from other aircraft in its communication range, creating a simple Markov model is not straightforward. Therefore, in addition to providing the necessary transition probabilities for random walk and random direction, an approximation to these probabilities is also proposed for a cooperative network. While intuitive rules for the movement paths are chosen when one or more autonomous aircraft meet each other, the proposed model can be extended such that other rules can be incorporated.

Several **mobility models** for autonomous agents have been proposed recently. Some of these are synthetic like the random walk and random direction, others are realistic and all of them are used mainly to describe the movement of the users in a given environment. In the autonomous aircraft domain, such models are good for comparison of different approaches but can give incorrect results when the autonomous aircraft is performing cooperative tasks. Mobility can increase throughput energy efficiency, coverage and other network parameters. Therefore, the analysis of mobility models has become a highlight to design the mobility of the nodes in a way to improve the network importance.

2.4.5.3.1 Markov Chain

A discrete-time discrete value stochastic process is introduced that can be used to analyze the coverage performance of autonomous aircraft network. Nodes can operate independently or in a cooperative manner. The system area is modeled as a 2D lattice where autonomous aircraft move from one grid point to another in each time step. It is assumed that an autonomous aircraft can only move to the four nearest-neighboring grid points: the **von Neumann neighborhood** of radius 1. The probability of moving to a neighboring grid point is determined by the mobility model of interest. Following two main components of the proposed MC are presented: state probabilities and transition probabilities.

In this model, the states are defined as [current location, previous location]. Depending on the location, the number of associated states is different. If the current location is at a corner, boundary or middle grid point, there are 2, 3 and 4 associated states, respectively: P_f, P_b, P_l and P_r are, respectively, the probabilities to move forward, backward, left and right. The steady-state probabilities of this MC are denoted by $\pi = [\pi(i,j,k,l)]$ and the transition probability matrix by T, where the entities of the matrix are the transition, probabilities between the states $[(i,j);(k,l)]$. Accordingly, the transient-state probabilities are denoted by $\pi^{(n)} = \left[\pi^{(n)}_{i,j,k,l}\right]$ at time step n. The following relations for the steady-state and transient-state probabilities can thus be written as follows:

$$\begin{aligned}
\vec{\pi} &= \pi.T \qquad \text{For steady state} \\
\pi^{(n)} &= \pi^{(0)} T^n \qquad \text{For transient state} \\
\lim_{n\to\infty} \pi^{(n)} &= \vec{\pi}
\end{aligned} \tag{2.179}$$

where $\sum \pi_{i,j,k,l} = 1$. The initial state $\pi^{(0)}$ can be chosen to be $[1,0,\ldots,0]$, since the solution for $\vec{\pi}$ is independent of the initial condition. From these linear equations, the steady- and transient-state probabilities can be obtained. This is used to determine the coverage of a given mobility pattern.

2.4.5.3.2 Coverage Metrics

The steady-state coverage probability distribution for an $a \times a$ area is denoted by $\mathbf{P} = [P(i,j)], 1 \le i \le a, 1 \le j \le a$. The probability matrix represents the percentage of time a given location (i,j) is occupied and can be computed by adding the corresponding steady-state probabilities obtained from equation (2.179):

$$P(i,j) = \sum_{k,l} \pi(i,j;k,l) \tag{2.180}$$

where $(k,l) = \{(i-1,j),(i,j-1),(i+1,j),(i,j+1)\}$ for the non-boundary states. The (k,l) pairs for boundary states can be determined in a straightforward manner. The transient coverage probability distribution $\mathbf{P}^{(n)} = [P(i,j)], 1 \le i \le a, 1 \le j \le a$:

$$P^{(n)} = \sum_{k,l} \pi^{(n)}(i,j;k,l) \tag{2.181}$$

Using the obtained $P^{(n)}$, the probability that location (i,j) is covered by time step can be computed as follows:

$$C^{(n)}(i,j) = 1 - \prod_{v=0}^{n} \left(1 - P^{(v)}(i,j)\right) \tag{2.182}$$

In the case of multiple autonomous aircraft, the state probabilities can be computed. Given the steady-state coverage distribution matrix of the autonomous aircraft k is P_k entities obtained using relation (2.180) and assuming independent/decoupled mobility, the steady-state coverage distribution of an $m-$autonomous aircraft network can be obtained as

$$p^{\text{multi}}(i,j) = 1 - \prod_{k=1}^{m} (1 - P_k(i,j)) \tag{2.183}$$

The transient behavior of the $m-$aircraft network can be computed similarly, by substituting the (i,j) entry of the transient coverage probability matrix $\mathbf{P}^{(k)}_n$ from relations (2.181)–(2.183). Some potential **metrics** of interest are now defined besides the coverage distribution of a mobility model in a grid, at time step n for a grid of size $a \times a$:

1. **Average coverage:**

$$E\left[C^{(n)}\right] = \frac{1}{a^2} \sum_{i,j} C^{(n)}(i,j) \tag{2.184}$$

2. **Full coverage:**

$$\varepsilon^{(n)} = \Pr\left(C^{(n)} = \vec{1}_{a\times a}\right) = \prod_{i,j} C^{(n)}(i,j) \tag{2.185}$$

where $\vec{1}_{a\times a}$ is an $a \times a$ matrix of ones. These metrics carry some valuable information regarding the coverage performance, e.g., how well a given point is covered, how well the whole area is covered and how much time would be necessary to cover the whole area.

2.4.5.3.3 Transition Probabilities: Independent Mobility

The state transition probabilities for the random walk and direction mobility models are first discussed where the transition probabilities are very intuitive. For random walk, the knowledge of the previous location is not necessary. Therefore, the states of the analytical tool (i,j,k,l) can be further simplified to (i,j). For random walk, it is assumed that at each time step, the autonomous aircraft can go to any one of the neighboring grid points with equal probability. Clearly, the number of neighboring points changes depending on the location. On the other hand, for random direction model, the direction is changed only when the autonomous aircraft reaches the boundary of the grid. Therefore, the previous location, which is also equivalent to direction for the lattice, needs to be taken into account. For both of these schemes, as well as for the cooperative scheme at the boundaries and corners, the next location is chosen randomly among the available neighboring points with equal probability.

2.4.5.3.4 Transition Probabilities: Cooperative Mobility

A method to approximate the coverage performance of a cooperative mobile network is proposed in this section. In such a network, the nodes interact with each other: exchange information, whenever they meet. The objective is to come up with an appropriate transition probability matrix that can be used by the proposed stochastic tool. For independent mobility, the proposed MC can be easily extended to multiple autonomous aircraft. However, for cooperative mobility, this MC is not sufficient to model the interactions. The states of an MC that exactly models all the interactions will grow exponentially with the number of autonomous aircraft [359]. Therefore, an approximation method can be proposed to model the behavior of the aircraft in a way that would allow to treat the cooperative mobility as independent mobility [397].

To decouple the actions for the aircraft from each other, for an $m-$aircraft network the following probabilities are defined:

$$P_X = \sum_{k=0}^{m-1} P_{X|k} \Pr\left(k+1\,\text{nodes meet}\right), X \in \{B,F,L,R\} \tag{2.186}$$

where the backward, forward, left-turn and right-turn probabilities are given by the decision metric $P_{X|k}$ of the cooperative mobility as well as the number of aircraft that will meet. With the assumption that any node can be anywhere in the grid with equal probability, probability that exactly k aircraft out of a total of m aircraft will also be at (i,j) is given by the binomial distribution:

$$\Pr\left(k+1\,\text{nodes meet}\right) = \binom{m-1}{k}\left(\frac{1}{a^2}\right)^k\left(1-\frac{1}{a^2}\right)^{m-1-k} \tag{2.187}$$

The entries of the corresponding transition probability matrix can be computed using relations (2.186)–(2.187), given the decision metric $P_{X|k}$.

2.4.5.4 Sensor Tasking in Multi-Target Search and Tracking Applications

The problem of managing uncertainty and complexity of planning and executing an **Intelligence Surveillance Reconnaissance** (ISR) mission is addressed in this section. This Intelligence

Surveillance Reconnaissance mission uses a network of UAV sensor resources. In such applications, it is important to design uniform coverage dynamics such that there is little overlap of the sensor footprints and little space left between the sensor footprints. The **sensor footprints** must be uniformly distributed so that it becomes difficult for a target to evade detection. For the search of a stationary target, the uncertainty in the position of the target can be specified in terms of a fixed probability distribution. The spectral multi-scale coverage algorithm makes the sensors move so that points on the sensor trajectories uniformly sample this stationary probability distribution. Uniform coverage dynamics coupled with sensor observations helps to reduce the uncertainty in the position of the target.

Coverage path planning determines the path that ensures a complete coverage in free workspace. Since the aircraft has to fly over all points in the free workspace, the coverage problem is related to the covering salesman problem. Coverage can be a static concept; that is, it is a measure of how a static configuration of agents covers a domain or samples a probability distribution. Coverage can also be a dynamic concept; that is, it is a measure of how well the points on the trajectories of the sensor cover a domain. Coverage gets better and better as every point in the domain is visited or is close to being visited by an agent. Uniform coverage uses metric inspired by the ergodic theory of dynamical system. The behavior of an algorithm that attempts to achieve uniform coverage is going to be inherently multi-scale. Features of large size are guaranteed to be detected first, followed by features of smaller and smaller size [388,390,392].

Definition 89. *Ergodic dynamics: A system is said to exhibit **ergodic dynamics** if it visits every subset of the phase space with a probability equal to the measure of that subset. For a good coverage of a stationary target, this translates to requiring that the amount of time spent by the mobile sensors in an arbitrary set be proportional to the probability of finding the target in that set. For good coverage of a moving target, this translates to requiring that the amount of time spent in certain tube sets be proportional to the probability of finding the target in the tube sets.*

A model is assured for the motion of the targets to construct these tube sets and define appropriate metrics for coverage. The model for the target motion can be approximate and the dynamics of targets for which precise knowledge is not available can be captured using stochastic models. Using these metrics for uniform coverage, centralized feedback control laws are derived for the motion of the mobile sensors.

For applications in environmental monitoring with a mobile sensor network, it is often important to generate accurate spatio-temporal maps of scalar fields such as temperature or pollutant concentration. Sometimes, it is important to map the boundary of a region. In [275], a multi-vehicle sampling algorithm generates trajectories for non-uniform coverage of a nonstationary spatio-temporal field characterized by spatial and temporal decorrelation scales that vary in space and time, respectively. The sampling algorithm uses a nonlinear coordinate transformation that renders the field locally stationary so that existing multi-vehicle control algorithm can be used to provide uniform coverage. When transformed back to original coordinates, the sampling trajectories are concentrated in regions of short spatial and temporal decorrelation scales.

For applications of multiagent persistent monitoring, the goal can be to patrol the whole mission domain while driving the uncertainty of all targets in the mission domain to zero [381]. The uncertainty at each target point is assumed to evolve nonlinearly in time. Given a closed path, multiagent persistent monitoring with the minimum patrol period can be achieved by optimizing the agent's moving speed and initial locations on the path [126,401].

Remark 90. *The main difference between multiagent persistent monitoring and dynamic coverage lies in that dynamic coverage task is completed when all points attain satisfactory coverage level while the persistent monitoring would last forever.*

2.4.5.4.1 Coverage Dynamics for Stationary Targets

There are N mobile agents assumed to move by either the first-order or second-order dynamics. An appropriate metric is needed to quantify how well the trajectories are sampling a given probability distribution μ. It is assumed that μ is zero outside a rectangular domain $\mathbb{U} \in \mathbb{R}^n$ and that the agent trajectories are confined to the domain $\mathbb{U}$. For a dynamical system to be ergodic, the fraction of the time spent by a trajectory must be equal to the measure of the set. Let $B(X,R) = \{R : \|Y - X\| \leq R\}$ be a spherical set and $\chi(X,R)$ be the indicator function corresponding to the set $B(X,R)$. Given trajectory $X_j : [0,t] \longrightarrow \mathbb{R}^n, j = 1,\ldots N$, the fraction of the time spent by the agents in the set $B(X,R)$ is given as

$$d^t(X,R) = \frac{1}{Nt} \sum_{j=1}^{N} \int_0^t \chi(X,R)(X_j)(\tau)d\tau \tag{2.188}$$

The measure of the set $B(X,R)$ is given as

$$\bar{\mu}(X,R) = \int_{\mathbb{U}} \mu(Y)\chi(X,R)(Y)dY \tag{2.189}$$

For ergodic dynamics, the following relation must be verified:

$$\lim_{t \to \infty} d^t(X,R) = \bar{\mu}(X,R) \tag{2.190}$$

Since the equation above must be true for almost all points X and all radii R, this motivates defining the metric:

$$E^2(t) = \int_0^R \int_{\mathbb{U}} \left(d^t(X,R) - \bar{\mu}(X,R)\right)^2 dXdR \tag{2.191}$$

$E(t)$ is a metric that quantifies how far the fraction of the time spent by the agents in the spherical sets is from being equal to the measure of the spherical sets. Let the distribution C^t be defined as

$$C^t(X) = \frac{1}{Nt} \sum_{j=1}^{N} \int_0^t \delta(X - X_j(\tau))d\tau \tag{2.192}$$

Let $\phi(t)$ be the distance between C^t and μ as given by the Sobolev space norm of negative index H^{-1} for $s = \frac{n+1}{2}$, i.e.,

$$\phi^2(t) = \left\|C^t - \mu\right\|_{H^{-s}}^2 = \sum_K \Lambda_k |s_k(t)|^2 \tag{2.193}$$

where

$$s_k(t) = C_k(t) - \mu_k \qquad \Lambda_k = \frac{1}{\left(1 + \|k\|^2\right)^s} \tag{2.194}$$

$$C_k(t) = \langle C^t, f_k \rangle \qquad \mu_k = \langle \mu, f_k \rangle \tag{2.195}$$

Here, f_k are Fourier basis functions with wave number vector k. The metric $\phi^2(t)$ quantifies how much the time averages of the Fourier basis functions deviate from their spatial averages, but with more importance given to large-scale modes than small-scale modes. The case is considered where the sensors are moving by the first-order dynamics described by

$$\dot{X}_j(t) = U_j(t) \tag{2.196}$$

The objective is to design feedback laws:

$$U_j(t) = F_j(X) \tag{2.197}$$

so that the agents have ergodic dynamics. A model predictive control problem is formulated to maximize the rate of decay of the coverage metric $\phi^2(t)$ at the end of a short time horizon, and the feedback law is derived in the limit as the size of the receding horizon goes to zero. The cost function is taken to be the first time-derivative of the $\phi^2(t)$ at the end of the horizon $[t, t + \Delta t]$, i.e.,

$$C(t, \Delta t) = \sum_K \Lambda_k s_k(t + \Delta t) \dot{s}_k(t + \Delta t) \tag{2.198}$$

The feedback law in the limit as $\Delta t \to 0$ is given as

$$U_j(t) = -U_{\max} \frac{B_j}{\|B_j(t)\|_2} \tag{2.199}$$

where:

$$B_j(t) = \sum_K \Lambda_k s_k(t) \nabla f_k(X_j(t)) \tag{2.200}$$

and $\nabla f_k(t)$ is the gradient field of the Fourier basis functions f_k.

2.4.5.4.2 Coverage Dynamics for Moving Targets

Let the target motion be described by a deterministic set of ordinary differential equations:

$$\dot{Z}(t) = V(Z(t), t) \tag{2.201}$$

where $Z(t) \in \mathbb{U}, \mathbb{U} \subset \mathbb{R}^3$ is the zone in which the target motion is confined over a period $[0, T_f]$. Let T be the corresponding mapping that describes the evolution of the target position; that is, if the target is at point $Z(t_0)$ at time $t = t_0$, its position $Z(t_f) = T(Z(t_0), t_0, t_f)$.

Given a set $\mathbb{A} \subset \mathbb{U}$, its inverse image under the transformation $T(., t_0, t_f)$ is given as:

$$T^{-1}(., t_0, t_f)(\mathbb{A}) = \{Y : T(Y, t_0, t_f) \in \mathbb{A}\} \tag{2.202}$$

The initial uncertainty in the position of the target is specified by the probability distribution $\mu(0, X) = \mu_0(X)$.

Let $[P^{t_0, t_f}]$ be the family of Perron–Frobenius operators corresponding to the transformations $T(., t_0, t_f)$, i.e.,

$$\int_{\mathbb{A}} [P^{t_0, t_f}] \mu(t_0, Y) dY = \int_{\mathbb{A}} \mu(t_f, Y) dY = \int_{T^{-1}(., t_0, t_f)(\mathbb{A})} \mu(t_0, Y) dY \tag{2.203}$$

At time t, the spherical set $B(X, R)$ with radius R and center X as well as the corresponding tube set is considered:

$$H^t(B(X, R)) = \{(Y, \tau) : \tau \in [0, t] \text{ and } T(Y, \tau, t) \in B(X, R)\} \tag{2.204}$$

The tube set $H^t(B(X, R))$ is a subset of the extended space-time domain and is the union of the sets

$$T^{-1}(., \tau, t)(B(X, R)) \times \{\tau\}, \forall \tau \in [0, t] \tag{2.205}$$

This tube set can be thought of as the set of all points in the extended space–time domain that end up in the spherical set $B(X, R)$ at time t when evolved forward in time according to the target dynamics.

The probability of finding a target within any time slice of the tube set is the same, i.e.,

$$\mu\left(\tau_1, T^{-1}(., \tau_1, t)(B(X, R))\right) = \mu\left(\tau_2, T^{-1}(., \tau_2, t)(B(X, R))\right) = \mu(t, B(X, R)) \tag{2.206}$$

$\forall \tau_1, \tau_2 \leq t$.

This is because none of the possible target trajectories either leave or enter the tube set $H^t(B(X,R))$. Let the sensor trajectories be $X_j : [0,t] \to \mathbb{R}^2, \forall j = 1, \ldots, N$. The fraction of the time spent by the sensor trajectories $(X_j(t),t)$ in the tube set $H^t(B(X,R))$ is given as:

$$d^t(X,R) = \frac{1}{Nt} \sum_{j=1}^{N} \int_0^t \chi T^{-1}(.,\tau,t)(B(X,R))(X_j(\tau)) d\tau \tag{2.207}$$

or

$$d^t(X,R) = \frac{1}{Nt} \sum_{j=1}^{N} \int_0^t \chi B(X,R)(T(X_j(\tau),\tau,t)) d\tau \tag{2.208}$$

$\chi B(X,R)$ is the indicator function on the set $B(X,R)$. $d^t(X,R)$ can be computed as the spherical integral

$$d^t(X,R) = \int_{B(X,R)} C^t(Y) dY \tag{2.209}$$

of a distribution

$$C^t(X) = \frac{1}{Nt} \sum_{j=1}^{N} \int_0^t P^{\tau,t} \delta X_j(\tau)(x) d\tau \tag{2.210}$$

referred to as the coverage distribution. $\delta X_j(\tau)$ is the delta distribution with mass at the point $X_j(\tau)$. The coverage distribution C^t can be thought of as the distribution of points visited by the mobile sensors when evolved forward in time according to the target dynamics.

For uniform sampling of the target trajectories, it is desirable that the fraction of time spent by the sensor trajectories in the tube must be close to the probability of finding a target trajectory in the tube which is given as

$$\mu(t,B(X,R)) = \int_{B(X,R)} \mu(t,Y) dY = \int_{T^{-1}(.,0,t)(B(X,R))} \mu_0(Y) dY \tag{2.211}$$

This motivates defining the metric

$$\Psi^2(t) = \left\| C^t - \mu(t,.) \right\|_{H^{-s}}^2 \tag{2.212}$$

Using the same receding horizon approach as described before for stationary targets, feedback laws similar to that in equation (2.199) are obtained.

During search missions, efficient use for flight time requires flight path that maximize the probability of finding the desired subject. The probability of detecting the desired subject based on UAV sensor information can vary in different search areas due to environment elements like varying vegetation density or lighting conditions, making it likely that the UAV can only partially detect the subject. In [325], an algorithm that accounts for partial detection in the form of a difficulty map is presented. It produces paths that approximate the payoff of optimal solutions, the path planning being considered as a discrete optimization problem. It uses the mode goodness ratio heuristic that uses a Gaussian mixture model to prioritize search sub-regions. The algorithm searches for effective paths through the parameter space at different levels of resolution. The task difficulty map is a spatial representation of sensor detection probability and defines areas of different levels of difficulty.

2.4.6 RESOURCE MANAGER FOR A TEAM OF AUTONOMOUS AIRCRAFT

Knowledge of meteorological properties is fundamental to many decision processes. It is useful if related measurement processes can be conducted in a fully automated fashion. The first type of cooperation that the autonomous aircraft may exhibit is to support each other if there is evidence that an interesting physical phenomenon has been discovered. The second type of cooperation that the aircraft can exhibit through their control algorithm is when an aircraft is malfunctioning or may be malfunctioning. If an aircraft internal diagnostic indicates a possible malfunction, then it will send out an omnidirectional request to the other aircraft for help to complete its task. Each autonomous aircraft will calculate its priority for providing help. The autonomous aircraft send their priority for providing help message back to the requesting aircraft. The requester subsequently sends out a message informing the group of the ID of the highest priority aircraft. The high-priority aircraft then proceeds to aid the requester. The support provided by the helping aircraft can take on different forms. If the requester suspects a malfunction in its sensors, the helper may measure some of the same points originally measured by the autonomous aircraft in doubt. This will help establish the condition of the requester's sensors. If additional sampling indicates the requester is malfunctioning and represents a liability to the group, it will return to base. In this case, the supporter may take over the mission of the requester [379].

Whether or not the supporter samples all the remaining sample points of the requester, subsequently abandoning its original points depends on the sample points priorities. If it is established that the requester is not malfunctioning or the requester can still contribute to the mission's success, it may remain in the field to complete its current mission [14].

2.4.6.1 Routing with Refueling Depots for a Single Aircraft

A single autonomous aircraft routing problem with multiple depots is considered. The aircraft is allowed to refuel at any depot. The objective of the problem is to find a path for the UAV such that each target is visited at least once by the aircraft, the fuel constraint is never violated along the path for the UAV and the total fuel required by the UAV is a minimum. An approximation algorithm for the problem is developed and a fast construction and improvement heuristic is proposed for the solution [386].

As small autonomous aircraft typically have fuel constraints, it may not be possible for them to complete a surveillance mission before refueling at one of the depots. For example, in a typical surveillance mission, an aircraft starts at a depot and is required to visit a set of targets. It is assumed that the fuel required to travel a given path for the UAV is directly proportional to the length of the path. To complete this mission, the aircraft might have to start at one depot, visit a subset of targets and then reach one of the depots for refueling before starting a new path. If the goal is to visit each of the given targets once, the UAV may have to repeatedly visit some depots for refueling before visiting all the targets. In this scenario, the following problem arises:

Problem 91. *Routing Problem: Given a set of targets and depots and a UAV where the aircraft is initially stationed at one of the depots, find a path for the UAV such that each target is visited at least once, the fuel constraint is never violated along the path for the UAV and the travel cost for the aircraft is a minimum. The travel cost is defined as the total fuel consumed by the aircraft as it traverses its path.*

The main difficulty in this problem is mainly combinatorial [319]. As long as a path of minimum length can be efficiently computed from an origin to a destination of the autonomous aircraft, the motion constraints of the autonomous aircraft does not complicate the problem. The UAV can be modeled as a Dubins aircraft. If the optimal heading angle is specified at each target, the problem of finding the optimal sequence of targets to be visited reduces to a generalization of the TSP [356].

The autonomous aircraft must visit each target at a specified heading angle. As a result, the travel costs for the autonomous aircraft may be asymmetric. Symmetry means that the cost of traveling from target A with the heading χ_A and arriving at target B with heading χ_B may not equal the cost of traveling from target B with heading χ_B and arriving at target A with the heading χ_A.

Definition 92. *An $\alpha-$approximation algorithm for an optimization problem is an algorithm that runs in polynomial time and finds a feasible solution whose cost is at most $\alpha-$times the optimal cost for every instance of the problem.*

This guarantee α is also referred to as the approximation factor of the algorithm. This approximation factor provides a theoretical upper bound on the quality of the solution produced by the algorithm for any instance of the problem. These upper bounds are known a priori. The bound provided by the approximation factor is generally conservative.

Let $\mathbb{T}$ denote the set of targets and $\mathbb{D}$ represent the set of depots. Let $s \in \mathbb{D}$ be the depot where the UAV is initially located. The problem is formulated on the complete directed graph $G = (\mathbb{V}, E)$ with $\mathbb{V} = \mathbb{T} \cup \mathbb{D}$. Let f_{ij} represent the amount of fuel required by the aircraft to travel from vertex $i \in \mathbb{V}$ to vertex $j \in \mathbb{V}$. It is assumed that the fuel costs satisfy the triangle inequality, i.e., for all distinct $i, j, k \in \mathbb{V}$, $f_{ij} + f_{jk} \geq f_{ik}$. Let L denote the maximum fuel capacity of the aircraft. For any given target $i \in \mathbb{I}$, it is assumed that there are depots d_1 and d_2 such that $f_{d_1 i} + f_{i d_2} \leq aL$, where $0 < a < 1$ is a fixed constant. It is also assumed that it is always possible to travel from one depot to another depot (either directly or indirectly, by passing through some intermediate depots, without violating the fuel constraints). Given two distinct depots d_1 and d_2, let $\ell'_{d_1 d_2}$ denote the minimum fuel required to travel from d_1 to d_2. Then, let β be a constant such that $\ell'_{d_1 d_2} \leq \beta \ell'_{d_1 d_2}$ for distinct d_1 and d_2. A tour for the aircraft is denoted by a sequence of vertices $T = (s, v_1, v_2, \ldots, v_p, s)$ visited by the aircraft, where $v_i \in \mathbb{V}$ for $i = 1, \ldots, p$. A tour visiting all the targets can be transformed to a tour visiting all the targets and the initial depot and vice-versa.

Problem 93. *The objective of the problem is to find a tour $T = (s, v_1, v_2, \ldots, v_p, s)$ such that $\mathbb{T} \subseteq \{v_1, v_2, \ldots, v_p\}$, the fuel required to travel any sub-sequence of vertices of the tour $(d_1, t_1, t_2, \ldots, t_k, d_2)$ starting at a depot d_1 and ending at the next visit to a depot d_2, while visiting a sequence of targets $(t_1, t_2, \ldots, t_k) \in \mathbb{T}$ must be at most equal to L, i.e.,*

$$f_{d_1 t_1} + \sum_{i=1}^{k-1} f_{t_i t_{i+1}} + f_{t_k d_2} \leq L \tag{2.213}$$

The travel cost $f_{sv_1} + \sum_{i=1}^{p-1} f_{v_i v_{i+1}} + f_{v_p s}$ is a minimum. Let x_{ij} denote an integer decision variable which determines the number of directed edges from vertex i to vertex j in the network, $x_{ij} \in \{0, 1\}$ if either i or vertex j is a target.

The collection of edges chosen by the formulation must reflect the fact that there must be a path from the depot to every target. Flow constraints are used to formulate this connectivity constraints. In these flow constraints, the aircraft collects $|T|$ units of a commodity at the depot and delivers one unit of commodity at each target as it travels along its path. p_{ij} denotes the amount of commodity flowing from vertex i to vertex j, and r_i represents the fuel left in the aircraft when the ith target is visited.

Problem 94. *The problem can be formulated as an MILP as follows:*

$$\min \sum_{(i,j) \in \mathbb{E}} f_{ij} x_{ij} \tag{2.214}$$

Subject to degree constraints

$$\sum_{i \in \mathbb{V}/\{k\}} x_{ik} = \sum_{i \in V/\{k\}} x_{ki} \qquad \forall k \in \mathbb{V} \tag{2.215}$$

$$\sum_{i \in \mathbb{V}/\{k\}} x_{ik} = 1, \forall k \in \mathbb{T} \tag{2.216}$$

Capacity and flow constraints

$$\sum_{i\in\mathbb{V}/\{s\}}(p_{si}-p_{is})=|T| \tag{2.217}$$

$$\sum_{j\in\mathbb{V}/\{i\}}(p_{ji}-p_{ij})=1 \quad \forall i\in\mathbb{T} \tag{2.218}$$

$$\sum_{j\in\mathbb{V}/\{i\}}(p_{ji}-p_{ij})=0 \quad \forall i\in\mathbb{D}/\{s\} \tag{2.219}$$

$$0\leq p_{ij}\leq |T|x_{ij} \quad \forall i,j\in\mathbb{V} \tag{2.220}$$

Fuel constraints

$$-M(1-x_{ij})\leq r_j-r_i+f_{ij}\leq M(1-x_{ij}) \quad \forall i,j\in\mathbb{T} \tag{2.221}$$

$$-M(1-x_{ij})\leq r_j-L+f_{ij}\leq M(1-x_{ij}) \quad \forall i\in\mathbb{D},\forall j\in\mathbb{T} \tag{2.222}$$

$$-M(1-x_{ij})\leq r_i-f_{ij} \quad \forall i\in\mathbb{T},\forall j\in\mathbb{D} \tag{2.223}$$

$$x_{ij}\in\{0,1\},\forall i,j\in\mathbb{V};either\ i\ or\ j\ is\ a\ target \tag{2.224}$$

$$x_{ij}\in\{0,1,2,\ldots,|T|\},\forall i,j\in\mathbb{D}; \tag{2.225}$$

Equation (2.215) states that the in-degree and out-degree of each vertex must be the same, and equation (2.216) ensures that each target is visited once by the aircraft. These equations allow for the aircraft to visit a depot any number of times for refueling. The constraints equations (2.217)–(2.220) ensure that there are $|T|$ units of commodity shipped from one depot and the aircraft delivers exactly one unit of commodity at each target. In equations (2.221)–(2.225), M denotes a large constant and can be chosen to be equal to $L+\max_{i,j\in\mathbb{V}}f_{ij}$. If the UAV is traveling from target i to target j, equations (2.221) ensure that the fuel left in the aircraft after reaching target j is $r_j=r_i-f_{ij}$. If the UAV is traveling from depot i to target j, equation (2.222) ensures that the fuel left in the aircraft after reaching target j is $r_j=L-f_{ij}$. If the UAV is directly traveling from any target to a depot constraint (2.224) must be at least equal to the amount required to reach the depot.

An approach to trajectory generation for autonomous aircraft using **MILP** and a modification of the A* algorithm to optimize paths in dynamic environments particularly having pop-ups with a known future probability of appearance. Each pop-up leads to one or several possible evasion maneuvers, characterized with a set of values used as decision-making parameters in an integer linear programming model that optimizes the final route by choosing the most suitable alternative trajectory, according to the imposed constraints such as fuel consumption and spent time. The decision variables are the UAV state variables, i.e., position and speed. The constraints are derived from a simplified model of the UAV and the environment where it has to fly [367].

2.4.6.2 Routing with Refueling Depots for Multiple Aircraft

The multiple autonomous aircraft routing problem can also be considered. Given a set of heterogeneous aircraft, find an assignment of targets to be visited by each UAV along with the sequence in which it should be visited so that each target is visited at least once by an autonomous aircraft, all the UAV return to their respective depots after visiting the targets and the total distance traveled by the collection of autonomous aircraft is minimized.

Problem 95. *Let there be n targets and m aircraft located at distinct depots, and let $\mathbb{V}(T)$ be the set of vertices that correspond to the initial locations of the aircraft (targets) with the m vertices, $\mathbb{V}=\{V_1,\ldots,V_m\}$ representing the aircraft (i.e. the vertex i corresponds to the ith aircraft) and $T=\{T_1,\ldots,T_n\}$ representing the targets. Let $\mathbb{V}^i=\mathbb{V}_i\cup T$ be the set of all the vertices corresponding to the ith aircraft, and let $C^i:\mathbb{E}^i\to\mathbb{R}^+$ denote the cost function with $C^i(a,b)$ representing the cost of traveling from vertex a to vertex b for aircraft i. The cost functions are considered to be asymmetric,* *i.e., $C^i(a,b) \neq C^i(b,a)$. An aircraft either does not visit any targets or visits a subset of targets in T. If the ith aircraft does not visit any target, then its tour $TOUR_i = \emptyset$ and its corresponding tour $C(TOUR_i) = 0$. If the ith aircraft visits at least one target, then its tour may be represented by an ordered set $\left\{V_i, T_{i_1}, \ldots, T_{i_{r_i}}, V_i\right\}$ where $T_{i_\ell}, \ell = 1, \ldots, r_i$ corresponds to r_i distinct targets being visited in that sequence by the ith aircraft. There is a cost $C(TOUR_i)$ associated with a tour for the ith aircraft visiting at least one target defined as:*

$$C(TOUR_i) = C^i(V_i, T_{i_1}) + \sum_{k=1}^{r_i-1} C^i(T_{i_k}, T_{i_{k+1}}) + C^i(T_{i_{r_1}}, V_i) \qquad (2.226)$$

Find tours for the aircraft so that each target is visited exactly once by some aircraft and the overall cost defined by $\sum_{i \in V} C(TOUR_i)$ is minimized.

The approach is to transform the routing problem into a single ATSP and use the algorithms available for the ATSP to address the routing problem [343].

In the generalized TSP, a major issue is that its mathematical formulation involves both continuous and integer decision variables [143]. To solve the problem, it is necessary to determine the following topics which minimize the mission completion time:

1. the order of the visit of the points.
2. the number of take-offs and the number of points that have to be visited between each take-off and landing.
3. the continuous path that the aircraft has to follow.

Multi-Criteria Decision Analysis (MCDA) technique is a process that allows to make decisions in the presence of multiple potentially conflicting criteria. Common elements in the decision analysis process are a set of design alternatives, multiple decision criteria and preference information representing the attitude of a decision-maker in favor of one criterion over another, usually in terms of weighting factors. Because of different preferences and incomplete information, uncertainty always exists in the decision analysis process [385].

To effectively select the most appropriate decision analysis method for a given decision problem, an approach consisting of the following steps can be proposed:

1. Define the problem
2. define evaluation criteria
3. calculate appropriateness index
4. evaluate decision analysis method
5. choose the most suitable method
6. conduct sensitivity analysis.

An integrated approach based on graph theory for solving the deadlock problem in the cooperative decision-making and control is presented in [104]. The vehicle team can contain a group of fixed-wing UAV with different operational capabilities and kinematic constraints. Because heterogeneity, one task cannot be performed by arbitrary vehicles in the heterogeneous group. The task assignment problem is described as a combinatorial optimization problem. Each assignment that allocates multiple vehicles to perform multiple tasks on multiple targets is a candidate solution. The execution time that the mission takes to be accomplished is chosen as the objective function to be minimized. A vehicle that performs multiple tasks on targets needs to change its path, waiting for other if another target needs to change its path, waiting for another vehicle that executes a former or simultaneous task has not finished or arrived. This creates risks of deadlocks. Two or more vehicles may fall into a situation of infinite waiting due to shared resources and precedence constraints among various tasks. A task-precedence graph of solutions is constructed and analyzed for detecting deadlocks. In addition, the topological sort of tasks is used for the path elongation of vehicles. Thus, deadlock-free solutions are obtained and the path coordinate is done.

The focus is to find a method to manage the non-deadlock condition and the time constraint. Consequently, the combinatorial optimization problem could be processed. All initial assignments are encoded according to a scheme that makes candidate solutions satisfy the constraints. Each feasible assignment of tasks is a feasible solution of the combinatorial optimization problem. After an initial assignment is generated, it must be checked whether it encodes deadlocks because the non-deadlock condition is a prerequisite for the subsequent process. An initial assignment is first processed into two groups according to two types of task relation. The first subgraph, the task executing subgraph is derived by the vehicle-based task group. The second subgraph, the task constraint subgraph is derived by the target tasks.

2.5   CONCLUSION

In the first part of this chapter, path and trajectory planning is presented. Trim is concerned with the ability to maintain flight equilibrium with controls fixed. Then, an algorithm for 2D and 3D open-loop path planning is derived for the system presented in the previous section. Then, the Zermelo navigation problem is considered in the sequel. Parametric paths are investigated, depending on the initial and final configurations. Smoother paths can be obtained by asking for the continuity of the derivatives of the path curvature and torsion. Maneuvers should be kept only to join two trim flight paths. Finally, some parametric curves such as polynomials, Pythagorean hodograph and η^3 splines are presented.

In the second part of the chapter, guidance and collision/obstacle avoidance topics are investigated into static and dynamic environments. In the author's opinion, there is no algorithm better than another. Depending on the mission, some have a better performance than the others. Only practitioners can choose the algorithm suitable for their case study.

Mission planning is presented in the last part of this chapter: route optimization, fuzzy planning, coverage problem and resource manager are this topics of the important subject.

3 Orienteering and Coverage

3.1 INTRODUCTION

In this chapter, operations concern generic aerial robotic problems, such as **orienteering** and **coverage**, in applications such as surveillance, search and rescue, geo-location, exploration, monitoring and mapping. The use of Unmanned Aerial Vehicles (UAVs), instead of ground robots, provides several advantages. The capacity to fly allows avoiding obstacles and having a bird's-eye view. A Minkowski sum of an obstacle expands the obstacle according to UAV's size, while simultaneously, the UAV shrinks to a reference point. In large outdoors, the Minkowski sum represents the inaccessible area for the UAV. If the Minkowski sums of different obstacles intersect, then they will be merged into one to form a closed inaccessible area for the UAV. The space outside the union region is regarded as the free space for the UAV, and consequently, the UAV can follow paths in it [225]. If they have a small size, they can also navigate in narrow outdoor and indoor environments and they represent only a limited invasive impact. Before proceeding further on orienteering and coverage, some basic features of operational research involved in UAV missions are presented. These are the vehicle routing problem (VRP), traveling salesperson problem (TSP), postperson problem and knapsack problem. The material in this section is not intended to be exhaustive, but rather to provide a sufficient introduction to those who may be unfamiliar with such topics.

3.2 OPERATIONAL RESEARCH PRELIMINARIES

This section presents some fundamental algorithms in UAV mission planning such as the VRP, the TSP and its variants, the postperson problem be it Chinese or rural and finally the knapsack method.

3.2.1 GENERAL VEHICLE ROUTING PROBLEM

The **general vehicle routing problem** (GVRP) is the problem of finding distinct feasible tours maximizing the profit determined by the accumulated revenue of all orders served by a vehicle reduced by the cost for operating the tours [151]. In the GVRP, a transportation request is specified by a non-empty set of a pickup, delivery and/or service location which have to be visited in a particular sequence by the same vehicle, the time windows in which these locations have to be visited and the revenue gained when the transportation request is served. The GVRP is a combined load acceptance and routing problem, which generalizes the VRP and **pickup and delivery problem** (PDP) [297].

Definition 96. *A **tour** of a vehicle is a journey starting at the vehicle's start location and ending at its final location, passing all other locations the vehicle has to visit in the correct sequence and passing all locations belonging to each transportation request assigned to the vehicle in the correct respective sequence.*

Definition 97. *A tour is **feasible** if and only if for all orders assigned to the tour compatibility constraints hold at each point in the tour time window and capacity restrictions hold.*

The objective is to find distinct feasible tours maximizing the profit, which is determined by the accumulated revenue of all served transportation requests, reduced by the accumulated costs for operating these tours.

The most widely studied VRPs are the **capacitated VRP** and the **VRP with time windows** (VRPTW). **Variable neighborhood search** (VNS) is a meta-heuristic based on the idea of systematically changing the neighborhood structure during the search. **VNS** systematically exploits the following observation:

1. A **local optimum** with respect to one neighborhood structure is not necessary as for another.
2. A **global optimum** is a local optimum with respect to all possible neighborhood structures.
3. For many problems, **local optima** with respect to one or several neighborhoods are relatively close to each other.

3.2.2 TRAVELING SALESPERSON PROBLEM

The most basic strategic problem is often how to choose the order of UAV way-points in a set of possible locations.

Problem 98. *Traveling salesperson problem (TSP): Given a set of n points in the plane, TSP asks if there exists a tour for the UAV that visits all these points exactly once.*

In general, the TSP includes two different kinds:

1. The **symmetric traveling salesperson problem** (STSP): There is only one way between two adjacent cities; i.e., the distance between cities A and B is equal to the distance between cities B and A.
2. The **asymmetric traveling salesperson problem** (ATSP): There is no such symmetry, and it is possible to have two different costs or distances between two cities.

TSP is a representative of a large class of problems known as combinatorial problems. Among them, TSP is one of the most important, since it is easy to describe, but difficult to solve. There are many variations to the TSP where a tour of minimum length has to be found that passes through every target location precisely once. Some are listed below:

1. The **Euclidean traveling salesperson problem** (ETSP) is a TSP where the distances between the vertices are precisely the Euclidean distances of the target locations in the plane.
2. The **Dubins traveling salesperson problem** (DTSP) refers to a kinematically constrained vehicle, such as an airplane at a constant altitude.
3. The **TSP with neighborhoods** (TSPN) extends the TSP to the case where each vertex of the tour is allowed to move in a given region. This approach takes into account the communication range or the sensor footprint of the UAV.
4. The **Euclidean traveling salesperson problem with neighborhoods** (ETSPN) seeks for a shortest Euclidean path passing through the regions.
5. The **TSP**, also known as delivery man problem or minimum latency problem, aims to find a tour, or a Hamiltonian cycle, on a given graph that minimizes the total traveling time plus the waiting time sum across every customer (relatively to some fixed node).
6. The **dynamic traveling repairman problem** (DTRP) in which the UAV is required to visit a dynamically changing set of targets [46] is an example of distributed task allocation problems.
7. The **k-traveling repairman problem** (k-TRP) is with differentiated waiting times, which is a variation of the TRP and TSP with cumulative costs. The waiting time, or latency, of customer k is the total time involved in the path 1 to k in the tour and can also be thought of as delay for service. The traveling time is sometimes viewed as latency of the repairman. Given an undirected graph $G = (V, E)$ and a source vertex $s \in V$, the k-TRP, also known as the minimum latency problem, asks for k tours, each starting at s and together covering all the vertices such that the sum of the latencies experienced by the customers is minimum. The latency of a customer p is defined to be the distance traveled before visiting p for the first time.

At the higher decision-making level, the dynamics of the UAV are usually not taken into account and the mission planner might typically choose to solve the ETSP. The first step determines the order in which the way-points should be visited by the UAV. At the lower level, a path planner takes as an input this way-point ordering and designs feasible trajectory between the way-points respecting the dynamics of the UAV. Even if each problem is solved optimally, however, the separation into two successive steps can be inefficient since the sequence of points chosen by the TSP algorithm is often hard to follow for the physical system. In order to improve the performance of the UAV system, mission planning and path planning steps should be integrated [134].

3.2.2.1 Deterministic Traveling Salesperson

A salesperson has to visit several cities (or road junctions). Starting at a certain city, the salesperson wants to find a route of minimum length which traverses each of the destination cities exactly once leading back to his starting point. Modeling the problem as a complete graph with n vertices, the salesperson wishes to make a tour or **Hamiltonian cycle**, visiting each cycle exactly once and finishing at the initial city. A TSP instance is given by a complete graph G on a node set $V = \{1,2,\ldots,m\}$ for some integer m and by a cost function assigning a cost c_{ij} to the arc (i,j) for any i,j in V. The salesperson wishes to make the tour whose the total cost is minimum where the total cost is the sum of the individual costs along the edges of the tour.

Remark 99. *There are different approaches for solving the TSP. Classical methods consist of heuristic and exact methods. Heuristic methods like **cutting planes** and **branch and bound** can only optimally solve small problems, whereas **Markov chains** and **tabu search** are good for large problems. Besides, some algorithms based on greedy principles such as **nearest neighbor** and **spanning tree** can be introduced as efficient solving methods. New methods such as **nature-based optimization** algorithms, **evolutionary computation**, **neural networks**, **ant systems**, **particle swarm optimization**, **simulated annealing**, bee colony optimization, intelligent water drops algorithms and artificial immune systems are among the solving techniques inspired by observing nature [93].*

G has a traveling salesperson tour with cost at most k. Given a TSP instance with m nodes, any tour passing once through any city is a feasible solution and its cost leads to an upper bound to the least possible cost. Algorithms that construct in polynomial time with respect to m feasible solutions, and thus upper bounds for the optimum value, are called heuristics. In general, these algorithms produce solutions but without any quality guarantee as to how far is their cost from the least possible one. If it can be shown that the cost of the returned solution is always less than k times the least possible cost, for some real number $k > 1$, the heuristic is called a $k-$approximation algorithm. The data consist of weights assigned to the edges of a finite complete graph, and the objective is to find a Hamiltonian cycle, a cycle passing through all the vertices, of the graph while having the minimum total weight. $c(A)$ denotes the total cost of the edges in the subset $A \subseteq E$:

$$c(A) = \sum_{(u,v)\in A} c(u,v) \tag{3.1}$$

In many practical situations, the least costly way to go from a place u to a place w is to go directly, with no intermediate steps. The cost function c satisfies the triangle inequality if for all the vertices, $u,v,w \in V$

$$c(u,w) \leq c(u,v) + c(v,w) \tag{3.2}$$

This triangle inequality is satisfied in many applications; not all, it depends on the chosen cost. In this case, the minimum spanning tree can be used to create a tour whose cost is no more than twice that of the minimum tree weight, as long as the cost function satisfies the triangle inequality. The pseudocode of the TSP approach can be presented in Algorithm 22.

Algorithm 22 TSP with Triangle Inequality

---

1. Select a vertex $r \in G.V$ to be a root vertex
2. Compute a minimum spanning tree T for G from root r using MST-PRIM(G, c, r)
3. Let H be a list of vertices, ordered according to when they are first visited in a pre-order tree walk of T
4. A state transition rule is applied to incrementally build a solution
5. Return the Hamiltonian cycle H

Minimum Spanning Trees: Procedure MST-PRIM(G, c, r)

1. For each $u \in G.V$
2. $u \cdot \text{key} = \infty$
3. $u \cdot \pi = \text{NULL}$
4. $r \cdot \text{key} = 0$
5. $Q = G.V$
6. While $Q \neq 0$
7. $u = \text{EXTRACT-MIN}(Q)$
8. for each $v \in G \cdot \text{Adj}[u]$
9. if $v \in Q$ and $w(u, v) < v \cdot \text{key}$
10. $v \cdot \pi = u$
11. $v \cdot \text{key} = w(u, v)$

Lines 1–5 in Procedure MST-PRIM of Algorithm 22 set the key of each vertex to ∞ (except for the root r, whose key is set to 0 so that it is the first vertex-processed), set the parent of each vertex to NULL, and initialize the min-priority queue Q to contain all the vertices. The algorithm maintains the following three-part loop invariant. Prior to each iteration of the while loop of lines 6–11 [56],

1. $A = \{(v, v.\pi) : v \in V - \{r\} - Q\}$.
2. The vertices already placed into the minimum spanning tree are those in $V - Q$.
3. For all vertices $v \in Q$, if $v \cdot \pi \neq \text{NULL}$, then $v \cdot \text{key} < \infty$ and $v.key$ is the weight of a light edge $(v, v.\pi)$ connecting v to some vertex, already placed into the minimum spanning tree.

Line 7 in the procedure identifies a vertex $u \in Q$ incident on a light edge that crosses the cut $(V - Q, Q)$. Removing u from the set Q adds it to the set $V - Q$ of vertices in the tree, thus adding $(u, u.\pi)$ to A. The for loop of lines 8–11 updates the key and π attributes of every vertex v adjacent to u, but not in the tree, thereby maintaining the third part of the loop invariant.

A generalization of the TSP is the DTRP for dynamic systems [181]. In the DTRP approach, customers are arising dynamically and randomly in a bounded region R, and when customers arrive, they wait for the repairman to visit their location and offer a service that will take a certain random amount of time. The repairman is modeled as a dynamic system whose output space contains R and the objective is the average time a customer has to wait to be serviced. Such problems appear in search and rescue or surveillance missions [44,47]. In [118], a scenario is considered in which the target points are generated dynamically, with only prior statistics on their location and a policy is designed to minimize the expected time a target waits to be visited. This formulation is a variation of the DTRP, with the addition of differential constraints on the UAV motion. The analysis can be broken into two limiting cases:

1. **Light load:** Targets are generated sporadically. The challenge is to design loitering policies with the property that when a target appears, the expected wait for the closest UAV to arrive is minimized. It reduces to a choice of waiting locations, and the solutions are known from

the locational optimization literature. These results are applicable to coverage problems in which the UAVs spread out uniformly to comb the environment efficiently.

2. **Heavy load:** Targets are generated rapidly. The optimal policy for nonholonomic vehicles relies on Euclidean traveling salesperson tours through large sets of targets. It is strongly related to works concerned with the generation of efficient cooperative strategies of several UAVs moving through given target points and possibly avoiding obstacles or threats.

3.2.2.2 Stochastic Traveling Salesperson

In UAV applications, the TSPs are often not deterministic, and some of the parameters are not known with certainty at the decision-making moment. The stochastic model has been used to represent this uncertainty, including the consideration of probability in the presence of customers, the demand level and the service time at customer's site usually assuming a known distribution governs some of the problem's parameters. In the case of a reconnaissance mission planning, before assigning the UAV to fly over dangerous targets, the flight time, fuel usage and forbidden areas on the flight path are often uncertain; only the belief degree of these quantities can be obtained [308]. For solving VRPs exactly, algorithms based on **column generation** and **Lagrangian relaxation** are the state of the art. Typically, the master program is a set-partitioning and the sub-problem a variant of the shortest-path problem (SPP) with resource constraints [295].

The TSP and the SPP, with deterministic arc costs but uncertain topologies, assume that any arc exists with a certain probability, and there are only two possible scenarios for a route connected by a set of arcs: success or failure. The nonexistence or failure of any single arc within the selected route a priori would result in the failure of the entire route due to the lack of recourse. To measure the risk of failure and to find the lowest cost route with an acceptable level of reliability, the **Bernoulli trial** of each arc's existence is assumed independent of each other for convenience. In risk-constrained stochastic network flow problems, these models are useful in situations where both the travel cost and the reliability are important to the decision makers. For example, in the case that a city is struck by an earthquake, a quick emergency response is required to send the rescue workers and humanitarian supplies to the damage zones via the most efficient route. The consequence of not considering the possibility that the roads may be broken due to uncertain factors such as aftershocks, heavy traffic and weather conditions, however, could be crucial as significant resources would be trapped and not be delivered in time. Therefore, a more reliable route needs to be planned in advance to prevent the route failure with a desired level of confidence. In addition, there is always a trade-off between minimizing the total cost and maximizing the reliability. Solving the proposed models would benefit decision makers to find a good balance in between. By setting different confidence levels, the corresponding optimal costs can be obtained, and therefore, the trade-off between the total cost and the risk level can be achieved. The cost could be significantly saved by slightly decreasing the route reliability, and on the other hand, the risk could be largely reduced by selecting another route with only a little higher total cost. While considering independent failures of all arcs, the number of scenarios of a reliable routing problem would exponentially increase as the number of arcs in the input network increases, and the problem of a large-sized or even moderate-sized network would be very difficult to be solved [173].

A **stochastic TSP** is studied where the n targets are randomly and independently sampled from a uniform distribution. The dynamic version of the TSP is stated as follows:

Problem 100. *Dynamic stochastic traveling salesperson problem: Given a stochastic process that generates target points, is there a policy that guarantees that the number of unvisited points does not diverge over time? If such stable policies exist, what is the minimum expected time that a newly generated target waits before being visited by the UAV?*

Algorithm 23 Stochastic TSP Algorithm

1. In the first phase of the algorithm, a tour is constructed with the following properties:
   a. It visits all non-empty beads once.
   b. It visits all rows in sequence top to down, alternating between left to right and right to left passes, and visiting all non-empty beads in a row.
   c. When visiting a non-empty bead, it visits at least one target in it.
2. Instead of considering single beads, meta-beads composed of two beads each and proceed in a way similar to the first phase are now considered with the following properties:
   a. The tour visits all non-empty meta-beads once.
   b. It visits all meta-beads rows in sequence top to down, alternating between left to right and right to left passes, and visiting all non-empty meta-beads in a row.
   c. When visiting a non-empty meta-bead, it visits at least one target in it.

The **stochastic TSP** can consider the scenario when n target points are stochastically generated according to a uniform probability distribution function (PDF). The proposed recursive **bead-tiling algorithm** to visit these points consists of a sequence of phases; during each phase, a closed path is constructed [259]. The methodology is presented in Algorithm 23. This process is iterated, and at each phase, meta-beads composed of two neighboring meta-beads from the previous phase are considered. After the last recursive phase, the leftover targets are visited using the alternating algorithm.

The **stochastic routing problem** is the generalization of the arc routing methods. Consider a complete graph $G = (V, E)$ on n nodes on which a routing problem is defined. If every possible subset of the node set V may or may not be present on any given instance of the optimization problem, then there are 2^n possible instances of the problem, all the possible subsets of V. Suppose instance S has probability $\text{prob}(S)$ of occurring. Given a method $\mathbb{U}$ for updating the a priori solution f to the full-scale optimization problem on the original graph $G = (V, E)$, $\mathbb{U}$ will then produce for problem instance S a feasible solution $t_f(S)$ with value $L_f(S)$. In the case of the TSP, $t_f(S)$ would be a tour through a subset S of nodes and $L_f(S)$ the length of that tour. Then, given that the updating method $\mathbb{U}$ has already been selected, the natural choice for the a priori solution f is to minimize the expected cost:

$$E[L_f] = \sum_{S \subseteq V} \text{prob}(S) L_f(S) \qquad (3.3)$$

with the summation being over all the subsets of V. The weighted average over all problem instances of the values $L_f(S)$ obtained is minimized by applying the updating method $\mathbb{U}$ to the a priori solution f.

The **probabilistic TSP** is essentially a TSP in which the number of points to be visited in each problem instance is a random variable. Considering a problem of routing through a set of n known points, on any given instance of the problem, only a subset S consisting of $|S| = k$ out of n points $(0 \leq k \leq n)$ must be visited. Ideally, the tour should be re-optimized for every instance by re-optimization in many cases. Unfortunately, re-optimization might turn out to be too time-consuming. Instead, an a priori tour should be found through all n points. On any given instance of the problem, the k points present will then be visited in the same order. The updating method for the probabilistic TSP is therefore to visit the point on every problem instance in the same order as in the a priori tour, skipping those points that are not present in that problem instance. The **probabilistic TSP** is to design an a priori route for each UAV, in which the route is followed exactly, simply skipping locations not requiring a visit. The goal is to find an a priori tour with the least expected cost. The probabilistic TSP represents a strategic planning model in which stochastic factors are considered explicitly. If n nodes are assumed spread over a bounded area, each node has a given probability of requiring a visit. The probability of requiring a visit is the **coverage problem**. Service requirements are assumed independent across nodes. The a priori tour is one in which the nodes are visited with the least expected length. The a priori tour is one in which the nodes are visited in the order given by the tour and nodes requiring a visit are simply skipped [226].

3.2.3 POSTPERSON PROBLEM

The postperson problem is divided into two topics: the Chinese postperson problem (CPP) and the rural postperson problem (RPP).

3.2.3.1 Chinese Postperson Problem

This section focuses on the **CPP** and its variations, which involves constructing a tour of the road network traveling along each road with the shortest distance. Starting at a given point, the postperson tries to find a route of minimum length to traverse each street at least once and lead back to the post office. The CPP seeks an optimal path that visits a predefined subset of graph edges at least once. An optimal path is defined as the lowest cost coverage path given the current graph information. Typically, the road network is mapped to an undirected graph $G = (V, E)$ and edge weights $w : E \rightarrow \mathbb{R}^+$, where the roads are represented by the edge set E and road crossings are represented by the node set V. Each edge is weighted with the length of the road or the amount of time needed to pass it. The CPP algorithm involves first constructing an even graph from the road network graph. This even graph has a set of vertices with an even number of edges attached to them. This is required as any traverse of the junction by approaching on one road and leaving on another, which means that only an even number of edges will produce an entry and exit pair for the tour. As the road network graph roads may have junctions with an odd number of edges, some roads are chosen for duplication in the graph. The technique chooses a set of roads with the shortest combined length to minimize duplication. The tour of the even graph is calculated by determining the Euler tour of the graph, which visits every edge exactly once or twice for duplicated edge. The CPP works well for applications where it is necessary to traverse every part of the space.

The environment is initially known in the form of a prior map. This prior map is converted into a graph structure with goal locations as nodes in the graph and paths between goals as edges in the graph. The first step in solving the coverage problem is to assume the prior map is accurate and generate a tour that covers all the edges in the graph. This CPP pseudocode can be represented as Algorithm 24. Its optimal tour consists of traversing all the edges in the graph at least once and starting and ending at the same node.

Algorithm 24 Chinese Postperson Problem Algorithm

---

1. **Input:** s (start vertex), G (connected graph where each edge has a cost value)
2. **Output:** P (value if tour found or empty is no tour found)
3. For $i \in 2, \ldots, n - 1$ do
4. if sEven(G), then $P = $ FindEulerCycle(G, s)
5. else
   $$O = \text{FindOddVertices}(G)$$
   $$O' = \text{FindAllPairsShortestPath}(O)$$
6.   Mate $= $ FindMinMatching(O')
   $$G' = (G, \text{Mate})$$
   $$P = \text{FindEulerCycle}(G', s)$$
7. end
8. Return P

The first step in Algorithm 24 is to calculate the degree of each vertex. If all the vertices have even degree, then the algorithm finds an Euler tour using the end-pairing technique. If any vertices have odd degree, a minimum weighted matching among the odd degree vertices is found using all pairs' shortest-path graph of the odd vertices. Because the matching algorithm requires a complete graph, all pairs' shortest-path algorithm is a way to optimally connect all the odd vertices. The matching finds the minimum cost set of edges that connect the odd nodes. Finally, the algorithm finds a tour on the new Eulerian graph. The end-pairing technique is used to generate the Eulerian cycle from the graph, consisting of two steps:

1. First, it builds cycles that intersect at least one vertex.
2. Next, the cycles are merged together two at a time by adding one cycle onto another at the intersecting vertex.

The cycle building step is shown in Algorithm 25. During each step of the algorithm, edges are added to a path sequence and removed from the graph until the starting node of the path is encountered. In the original end-pairing algorithm, the heuristic for choosing the next edge to add to the sequence consisted of picking a random edge incident to the current node. To maintain a small coverage environment, edges are chosen in such a way that the path travels away from the start and then travels back always visiting the farthest unvisited edges until it reaches the start. Essentially, the coverage path should be always walking along the boundary of the coverage subgraph. This will allow the edges around the start to be as connected as possible while separating the coverage and travel subgraphs.

Algorithm 25 Cycle Building Algorithm

---

1. **Input:** s (start vertex), G (graph)
2. **Output:** C (cycle found)
3. Begin
4. $C = s$
5. $i = s$
6. $e = \text{NextEdgeHeuristic}(G, s, i)$
7. $i = \text{OtherEndPoint}(e, i)$
8. While $i \neq s$ do
9. $e = \text{NextEdgeHeuristic}(G, s, i)$
10. $i = \text{OtherEndPoint}(e, i)$
11. $C = [C; i];$
12. RemoveEdge (G, e)
13. End
14. Return C

When environmental changes are found online, replanning is done on the updated graph with different starting and ending vertices. To remedy this disparity, an artificial edge is added from the current UAV location c to the ending vertex s in the graph. This edge (c, s) is assigned a large cost value to prevent it from being doubled in the solution. Using this modified graph, a tour from s to s is found. The edge (c, s) is then deleted from the graph and the tour. The algorithm adjusts the coverage path to start at the current location and travel to the end location.

Algorithm 26 Online Coverage Algorithm

1. **Input:** s (start vertex), c (current vertex), $G = (C, T)$ (graph where each edge has a label and a cost value), C is the subset of coverage edges, T is a subset of travel edges, and OTP is a subset of optimal travel paths (OTPs).
2. **Output:** tour found P
3. Begin
4. $G' = G$
5. If $c \neq s$, then $G' = [G, (c, s, \text{INF})]$
6. If sConnected(C), then $P = \text{CPP}(s, G')$
7. else $P = \text{RPP}(s, G', \text{OTP})$
8. If $c \neq s$ and $P \neq [\,]$, then
9. RemoveEdge $[P, (c, s, \text{INF})]$
10. end
11. Return P

Remark 101. *As shown in Algorithm 26, if the unvisited edges in the new problem are connected, the CPP algorithm is run; otherwise, the RPP algorithm is run.*

The **Reeb graph** can be used as input to the CPP to calculate a Eulerian circuit, which consists of a closed path traversing every cell at least once. The Reeb graph is a construction that originated in Morse theory to study a real-valued function defined on a topological space. The structure of a Morse function can be made explicit by plotting the evolution of the component of the level set. The Reeb graph is a fundamental data structure that encodes the topology of a shape. It is obtained by contracting to a point the connected components of the level-sets (also called contours) of a function defined on a mesh. Reeb graphs can determine whether a surface has been reconstructed correctly, can indicate problem areas and can be used to encode and animate a model. The Reeb graph has been used in various applications to study noisy data, which creates a desire to define a measure of similarity between these structures. A Eulerian circuit can be achieved by doubling selected edges of the Reeb graph, although no edge needs to be duplicated more than once. The Eulerian circuit is the solution of the linear programming problem:

$$\text{Minimize } z = \sum_{e \in E} c_e \cdot x_e \tag{3.4}$$

subject to:

$$\sum_{e \in E} a_{ne} \cdot x_e - 2w_n = k_n; \forall n \in V, x_e \in \mathbb{N}, \forall e \in E; w_n \in \mathbb{N} \tag{3.5}$$

where $\sum_{e \in E} a_{ne} \cdot x_e$ is the number of added edges to node $n \in V$. To be Eulerian, an odd number of edges must be added to nodes with odd degree and an even number of edges must be added to nodes with even degree; a_{ne} is equal to 1 if node n meets edge e and 0 otherwise; x_e is the number of added copies of edge e in the solution; w_n is an integer variable that will force $\sum_{e \in E} a_{ne} \cdot x_e$ to be odd for odd nodes and even for even nodes; k_n is 1 for nodes with odd degree and 0 otherwise; and c_e is a real number representing the cost of edge e. To prevent repeat coverage, cells corresponding to doubled Reeb graph edges are split into non-overlapping top and bottom sub-cells. In [310], a cell-splitting scheme guarantees that sub-cells share the same critical points as their parent, by interpolating between the bounding critical points and between the top and bottom boundaries of the original cell. At the end of the analysis phase, the resulting Eulerian circuit outlines a cyclic path through all connected cells in the environment.

For the road network search using multiple UAVs, a variation of the typical CPP algorithm is required, so that it can consider the operational and physical characteristics of the UAV in the search problem. Since the fixed-wing UAV cannot change its heading angle instantaneously due to the physical constraint, the trajectory has to meet the speed and turn limits of the UAV.

Different metrics can be used, such as the Euclidean distance between the vertices, the number of connected coverage components in the graph, the number of OTPs in the partition set, the number of branches in the search tree, the percentage of replanning calls that are CPP rather than RPP calls and the computation time in seconds [17]. Computational requirements are often a limiting factor in the capability of real-time trajectory optimization algorithm. As UAVs perform more complex missions with higher levels of autonomy, trajectory optimization algorithms need to be efficient and have the flexibility to adapt to different missions.

The CPP has a lot of variations such as the **capacitated CCP** (CCPP), which capacitates the total edge cost, or the **rural Chinese postperson problem**, which visits certain roads, but not necessarily all of them, and the **windy Chinese postperson problem**, which has different values for the same edge according to the direction. The **k-CPP** deals with the deployment of several postmen.

Problem 102. *The k-CPP can be formulated as follows: given a connected edge-weighted graph G and integers p and k, decide whether there are at least k closed walks such that every edge of G is contained in at least one of them and the total weight of the edges in the walks is at most p.*

Min-Max k-CPP (MM-k-CPP) algorithms are described for multiagent road network search. **MM k-CPP** is a variation of k-CPP, which considers the route of the similar length. This objective can be required if the UAV should finish road search mission with the minimum mission completion time.

3.2.3.2 Rural Postperson Problem

In many practical applications, it is not necessary to traverse every part of the space. The RPP seeks a tour that traverses a required subset of the graph edges using the extra graph edges as travel links between the required edges. Optimal solutions exist that formulate the RPP as an integer linear program and solve it using **branch and bound**. Another approach introduces new dominance relations, such as computing the minimum spanning tree on the connected components in a graph to solve large problem instances. Additionally, many TSP heuristics have been extended to the RPP [307]. There are two sets of graph edges: required and optional. The required edges are defined as coverage edges, and the optional as travel edges. Any solution would include all coverage edges and some combinations of travel edges.

Definition 103. *A **coverage or travel vertex** is a vertex in the graph incident to only coverage or travel edges, respectively. A border vertex is a vertex in the graph incident to at least one coverage edge and at least one travel edge. A travel path is a sequence of travel segments and travel vertices connecting a pair of border vertices.*

Branch and bound is a method of iterating through a set of solutions until an optimal solution is found:

1. In the **branching step**, the algorithm forms n branches of sub-problems where each sub-problem is a node in the branch-and-bound tree. The solution to any sub-problem could lead to a solution to the original problem.
2. In the **bounding step**, the algorithm computes a lower bound for the sub-problem.

These lower bounds enable branch and bound to guarantee a solution is optimal without having to search through all possible sub-problems. Branch and bound is a general method for handling hard problems that are slight deviations for low-complexity problems. Finally, an **OTP** is a travel path connecting a pair of border vertices such that it is the lowest cost path between the vertices, and the vertices are not in the same cluster. OTPs are the shortest paths between clusters of coverage segments that do not cut through any part of a coverage cluster. All the OTPs are computed by finding the lowest cost path p_{ij} between each pair of border vertices v_i and v_j in different clusters. If p_{ij} is a travel path, it is saved as an OTP. If it is not a travel path, then v_i and v_j do not have an OTP between them (i.e., $p_{ij} = $ NULL). The OTP becomes the partition set. The iterations are set within the branch-and-bound framework; at each branch step, the algorithm generates a new sub-problem by either including or excluding an OTP.

Algorithm 27 Rural Postperson Problem

1. **Input:** s (start vertex), $G = (C, T)$ (graph where each edge has a label and a cost value), C is the subset of coverage edges, T is a subset of travel edges, and OTP is a subset of optimal travel paths.
2. **Output:** P (tour found)
3. Begin
4. PQ = []
5. $G' = [G, \text{OTP}]$ where $\forall \text{OTP}, \text{cost}(p_{ij}) = 0$
6. $P = \text{CPP}(s, G')$, add to $PQ(PQ, [G', P])$
7. While ! is Empty (PQ) do
8. $[G', P] = \text{PopLowestCost(PQ)}$;
9. $P_{ij} = \text{FindMaxOTP}(G')$
10. If $P_{ij} = []$, then return P;
11. $G'' = \text{IncludeEdge}(G', P_{ij})$
12. $P1 = \text{CPP}(s, G')$
13. AddToPQ($pq, [G'', P]$
14. RemoveEdge $e(G', P_{ij})$
15. $P2 = \text{CPP}(s, G'')$
16. AddToPQ($pq, [G'', P2]$
17. end
18. Return P

Pseudocode of the RPP is presented in Algorithm 27. At the beginning, cost 0 is assigned to the unlabeled **OTPs** and solves the problem with all the OTPs as required edges using the CPP are shown in Algorithm 24. This problem and the CPP costs are pushed onto a priority queue. The CPP cost is the lower bound on the problem since all the OTPs have zero cost. While the queue is not empty, the lowest cost sub-problem is selected from the queue. For the sub-problem, the algorithm selects an unlabeled OTP P_{ij} with the best path cost. By employing this strategy of closing the OTP with the highest path cost, the aim is to increase the lower bound with the highest amount, which may help focus the search to the correct branch and prevent extraneous explorations. Once an OTP P_{ij} is selected, two branches are generated, which are as follows:

1. The first branch includes P_{ij} in the solution, this time with real path cost assigned.
2. The second branch omits P_{ij} from the solution.

A solution to each branch is found using the CPP algorithm. Because each solution is generated with a cost of 0 assigned to the unlabeled OTP in the sub-problem, the costs of the inclusion and exclusion of CPP solutions lower bounds on the cost of the RPP problem with and without using P_{ij} for travel, respectively. These new sub-problems are added to the priority queue and the algorithm iterates until the lowest cost problem in the queue contains no OTP. The solution to this problem is the optimal solution to the RPP since it has either included or excluded every single OTP in the solution and has a path cost that is equal to or lower than the lower bounds of the other branches. The branch-and-bound algorithm for the RPP is an exponential algorithm with a complexity $O\left(|V|^3 2^t\right)$, where t is the number of OTP and $|V|$ is the number of vertices in the graph.

While most research on arc routing problems focuses on the static environment, there has been work that addresses dynamic graphs such as the **dynamic rural postperson problem** (DRPP). Dynamic changes occur when the environment differs from the original map. There are two categories of planners that handle these differences [244], which are as follows:

1. **Contingency planners** model the uncertainty in the environment and plan for all possible scenarios.
2. **Assumptive planners** presume the perceived world state is correct and plan based on this assumption.

If disparities arise, the perceived state is corrected and replanning occurs. The lower-complexity assumptive planning is used in order to generate solutions quickly. In the presented planner, an initial plan is found based on the graph of the environment. As the UAV uncovers differences between the map and the environment during traversal, the algorithm propagates them into the graph structure. This may require a simple graph modification such as adding, removing or changing the cost of an edge. But it can also result in more significant graph restructuring. These changes may convert the initial planning problem into an entirely different problem. For the coverage problems, most changes in the environment are discovered when the UAV is actively traversing the space. These online changes are typically detected when the UAV is not at the starting location, but at a middle location along the coverage path. At this point, some of the edges have already been visited. Because it is not necessary to revisit the edges that are already traversed, the visited edges in the previous plan are converted to travel edges.

3.2.3.2.1 Cluster Algorithm

The algorithm is based on the cluster first, route second. In the first step, the edge set E is divided into k clusters, and then, a tour for each cluster is computed. This algorithm pseudocode can be represented as a constructive heuristic method and described by Algorithm 28.

Algorithm 28 Cluster Algorithm

---

1. Determine the set of k representative edges $f_1, \ldots, f_k$ of cluster F_i for each vehicle. Let f_1 be the edge having the maximum distance from the depot and f_2 be the edge having the maximum distance from f_1. The rest of successive edges are successively determined by maximizing the minimum distance to the already-existing representatives. Then, the remaining edges are assigned to the cluster according to the weighted distance between e and f_i. Consider the distance between representative edges and depot, number of assigned edge to the cluster F_i and cost of the cluster.
2. Include edges for connectivity. Add edges between every vertex and depot, and determine minimum spanning tree that includes original edges in each cluster for connection between the edges.
3. The rural Chinese postperson problem: Compute CPP route of required subset of edges out of total edges by the conventional CPP.

3.2.3.2.2 Rural Chinese Postperson Problem

Unlike the cluster algorithm, the first route algorithm follows a route first, cluster second. In a first step, postperson tour that covers all edges is computed, and then, this tour is divided by k tour segments that have the similar length. This method is described in Algorithm 29.

Algorithm 29 Rural Chinese Postperson Algorithm

---

1. Compute an optimal postperson route C^* using the conventional CPP.
2. Compute splitting nodes: $(k-1)$ splitting nodes $v_{p_1}, \ldots, v_{p_{k-1}}$ on C^* are determined in such a way that they mark tour segments of C^* approximately having the same length. Approximated tour segment length L_j is computed by using the shortest-path tour lower-bounded $s_{\max}$,

$$s_{\max} = \frac{1}{2} \max_{e=u,v \in E} w(SP(v_1, u)) + w(e) + w(SP(v_1, v_1)) \tag{3.6}$$

$$L_j = \frac{j}{k} \left(w(C^* - 2s_{\max}) \right) + s_{\max}, 1 \leq k \leq N - 1 \tag{3.7}$$

where N denotes the number of UAVs, $w(\alpha)$ represents the distance of the sub-tour α, and SP represents the shortest path between nodes considering road network. Then, the splitting node v_{p_j} is determined as being the last node such that $w(C^*_{v_{p_j}}) \leq L_j, C^*_{v_n}$ is the sub-tour of C^* starting at the depot node and ending at v_n.
3. k-postmen tours: Construct k tours $C = (C_1, \ldots, C_k)$ by connecting tour segments with shortest paths to the depot node.

3.2.4 KNAPSACK PROBLEM

Differently from ground vehicles, the UAV has to fly along the road only to cover a certain edge that is not connected. This modified search problem can be formulated as a **multi-choice multi-dimensional knapsack problem**, which is to find an optimal solution minimizing flight time. Classical multi-dimensional knapsack problem is to pick up items for knapsacks for maximum total values so that the total resource required does not exceed the resource constraint of knapsacks. For applying multi-dimensional knapsack problem to the road network search, UAVs are assumed as the knapsacks, the roads to be searched are resources and limited flight time or energy of each UAV is the capacity of knapsacks. Multi-dimensional knapsack problem formulation allows to consider the limitations of each UAV flight time and different types of roads, vehicles and minimum turning radius and get the sub-optimal solution of the coordinated road search assignment. Moreover, for fixed-wing UAVs, the Dubins path planning produces the shortest and flyable paths taking into consideration their dynamical constraints; thus, the Dubins path is used to calculate the cost function of the modified search problem [13].

The classical knapsack problem is defined as follows:

Problem 104. *Knapsack problem: Given an knapsack capacity $C > 0$ and a set $I = \{1, \ldots, n\}$ of items, with profits $P_i \geq 0$ and weights $w_i \geq 0$, the knapsack problem asks for a maximum profit subset of items whose total weight does not exceed the capacity. The problem can be formulated using the following mixed-integer linear program (MILP):*

$$\max \left\{ \sum_{i \in I} p_i x_i : \sum_{i \in I} w_i x_i \leq C, x_i \in \{0, 1\}, i \in I \right\} \tag{3.8}$$

where each variable x_i takes the value 1 if and only if item i is inserted in the knapsack.

The two classic approaches for solving this problem exactly are branch and bound and dynamic programming [138]. These algorithms work in two phases: in the forward phase, the optimal value of the profit function is calculated, and in the backtracking phase, an actual solution is determined using this optimal profit. Similar to the TSP, the knapsack problem has online and offline versions. Assuming the weights are identical among items, the offline version of the knapsack problem becomes trivial: a greedy algorithm can be used that selects the items with the maximal value one by one until no more items can be added. However, when the values of the items are not fully known ahead of time, the knapsack problem with identical weights is an interesting problem with many applications. The profitability of an object type is the ratio of its profit to weight. In the UAV domain, many objective functions may be incorporated, such as time, energy and flight time. Different and more complex models can be discussed, including the binary multiple-criteria knapsack problem, problems with many constraints and multi-period as well as time-dependent models.

3.3 ORIENTEERING

The problem of routing one UAV over **points of interest** (POI) can be formulated as an **orienteering problem**. The **orienteering problem** is a generalization of the TSP. It is defined on a graph in which the vertices represent geographical locations at which a reward can be collected. The orienteering problem includes the TSP and the knapsack problem [304].

Definition 105. *The **orienteering problem** is defined on a network where each node represents a POI and each arc represents travel between two nodes. Each node can be associated with a weight, and each arc with a travel time. The goal is to find a maximum prize path with a travel time. The orienteering problem can be generalized to cases with time window constraints.*

Definition 106. *A **time window** is a time interval that starts with the earliest time a task can start, and ends with the latest time the task can end. If the earliest time is not given, the latest time is referred to as the deadline. A time window is said to be closed if both start and end times are given.*

3.3.1 ORIENTEERING PROBLEM FORMULATION

The orienteering problems represent a family of VRPs which take into account practical situations where giving service to customers is optional and produces a profit if it is done within a time limit. Representing customers by vertices of a graph, the orienteering problem aims at selecting a subset of vertices and designing a route not longer than a pre-specified time distance limitation, to maximize the total collected profit. A team of uncapacitated vehicles located at a depot must give service to a set of regular customers, while another set of optional customers is available to be potentially serviced. Each optional customer generates a profit that is collected if it is serviced. The problem aims at designing vehicle routes with a time duration not longer than a pre-specified limit, serving all regular customers and some of the optional customers to maximize the total profit collected [253].

3.3.1.1 Nominal Orienteering Problem

First, the nominal orienteering problem is considered in which all input parameters are assumed to be deterministic. If N represents the set of targets and $|N|$ its cardinality, the depot location representing the UAV recovery point is denoted by the vertex $0 \notin N$ and $N^+ = N \cup \{0\}$; each target $i \in N$ is associated with a value p_i; and the orienteering problem is formulated on a complete graph $G = (N^+, A)$ with $|N| + 1$ vertices. A weight f_{ij}, representing the expected fuel consumption between targets i, j plus the expected fuel required to record target j, is associated with each arc $(i, j) \in A$. The fuel capacity of the UAV is denoted by F. A binary decision variable x_{ij} is introduced for every arc $(i, j) \in A$, $x_{ij} = 1$ if arc(i, j) is used in the tour. An auxiliary variable u_i is introduced to denote the position of vertex i in the tour. The goal is to find a tour of maximum profit, feasible with respect to the fuel constraint, which starts and ends at the recovery point. Based on these definitions, the formulation of the **nominal orienteering problem** is the following:

Problem 107. *Orienteering problem (OP)*

$$\max \sum_{i \in N} p_i \sum_{j \in N^+\{i\}} x_{ij} \qquad (3.9)$$

subject to

$$\sum_{(i,j)\in A} f_{ij}x_{ij} \leq F \quad \text{Capacity constraints} \tag{3.10}$$

$$\sum_{i\in N} x_{0i} = \sum_{i\in N} x_{i0} = 1 \quad \text{The tour starts and ends at the depot} \tag{3.11}$$

$$\sum_{i\in N^+\{j\}} x_{ji} = \sum_{i\in N^+\{j\}} x_{ij} \leq 1, \forall j \in N \tag{3.12}$$

Flow conservation ensures that a vertex is visited at most once.

$$u_i - u_j + 1 \leq (1 - x_{ij})|N|, \forall i,j \in N \quad \text{Prevents the construction of subtours} \tag{3.13}$$

$$1 \leq u_i \leq |N|, \forall i \in N \quad \text{Boundary constraints} \tag{3.14}$$

$$x_{ij} \in \{0,1\}, \forall i,j \in N \quad \text{Integrality constraints} \tag{3.15}$$

Remark 108. *The orienteering problem can be considered as a variant of the TSP. In contrast to the TSP, in which the goal is to minimize the tour length to visit all the targets, the orienteering problem objective is to maximize the total sum of the collected rewards, while the reward-collecting tour does not exceed the specified travel budget. Thus, the orienteering problem is a more suitable formulation for cases where visiting all the targets is unfeasible with the given travel budget.*

In operations research, the environmental coverage solution uses the representation of the environment as a graph, and algorithms such as the TSP or postperson problems are used to generate optimal solutions. In the graph representation, nodes in the graph are locations in the environment and edges in the graph are the paths between the locations. Each edge has a cost assigned to it where the cost can represent measurements such as Euclidean distance between locations, terrain traversability, travel time or a combination of several metrics. Additionally, each edge is undirected. One possible approach of resolution is to seek a path that visits all the edges or a designated edge subset of the graph [234].

In order to improve the effectiveness of a reconnaissance mission, it is important to visit the largest number of interesting target locations possible, taking into account operational constraints related to fuel usage, weather conditions and endurance of the UAV. Given the uncertainty in the operational environment, robust planning solutions are required [120]. Some locations can be more relevant than others in terms of **information gathering**, and therefore, priorities are usually assigned to the locations. In order to optimize data collection, the UAV should fly a tour, including target locations of higher priority, starting and ending at the recovery point. The tour planning can be modeled as the orienteering problem, wherein the target locations correspond to the nodes, profits are associated with the nodes to model the target location priorities, the arcs represent the flight path from one target location to the other and fuel consumption for such a flight path is modeled by the weight on the associated arc. The depot represents the recovery point of the UAV. Often there are obstacles present in the area being covered. A common technique is to consider obstacles in the environment as static. This allows the coverage area to be decomposed into unobstructed cells. Each cell is then treated individually and covered independently. The major disadvantage of this approach is the requirement of an a priori map of the environment and the obstacles within. The decomposed cells may well be too small for under-actuated vehicles to move around in. Other limitations include excessive overlap in the coverage and wasted time traveling from one cell to the other.

Approaches that deal with dynamic obstacles detect that an obstacle is present which should not have been there. In the cases of large obstacles, it might be better to split up the coverage area into cells. Another example is a **Voronoi** diagram where the paths are edges in the graph and the path intersections are nodes. This is one way to generate optimal paths for some of the problems in continuous space coverage [157]. In general, the routing approach consists in reducing a general routing problem to the SSP. The specification of routes is a problem in its own right. If the route network is modeled as a directed graph, then the routing problem is the discrete problem of finding paths in a graph, which must be solved before the speed profile is sought [258].

3.3.1.2   Robust Orienteering Problem

In reality, the input parameters of a UAV planning problem may be uncertain. Since a UAV operates in a dynamic but uncertain environment, effective mission planning should be able to deal with environment changes and changing expectations. In fact, weather circumstances like wind have a great impact on the fuel consumption of the UAV. Since replanning costs can be significant, it is important to generate UAV tour plans that are robust. So the **robust orienteering problem** (ROP) is introduced. In UAV mission planning, **sustainability** (robustness) of an initial plan is highly valued. More specifically, the flight plan that is constructed before the actual start of the UAV flight should be designed in such a way that the probability of visiting all planned targets is sufficiently high. The ROP is suitable for designing the initial plan of the UAV, since it provides a tool to balance the probability of feasibility of the initial plan against the profit value of the planned tour. The actual fuel realizations are not known yet at the planning stage. It will depend on these fuel realizations whether or not all planned targets can be visited during the actual flight of the UAV. In this paragraph, a formal description of the **ROP** is given. The ROP explicitly considers uncertainty in the weight of the arcs f_{ij}, whose realizations are assumed to lie in the interval $\left[\bar{f}_{ij} - \sigma_{i,j}, \bar{f}_{ij} + \sigma_{i,j}\right]$, where $\bar{f}_{ij}$ is the expected fuel consumption on the arc from target i to j. When modeling real-life instances of the UAV problem, the expected fuel consumption $\bar{f}_{ij}$ is based on the current weather circumstances: the speed and the direction of the wind. Consequently, the possible correlations between the weights of the arcs are already captured in these expected values, and thus, the deviations from the expected values are assumed to be uncorrelated noise. The formulation of the ROP that incorporates robustness against fuel uncertainty is the following:

Problem 109. *Robust orienteering problem (ROP)*

$$\max \sum_{i \in N} p_i \sum_{j \in N^+\{i\}} x_{ij} \tag{3.16}$$

subject to

$$\sum_{(i,j) \in A} \bar{f}_{ij} x_{ij} + \sum_{s \in S} \rho_s \|y^s\|_s^* \leq F \quad \textit{Capacity constraints} \tag{3.17}$$

$$\sum_{s \in S} y_{ij}^s = \sigma_{ij} x_{ij}, \forall (i,j) \in A \tag{3.18}$$

$$\sum_{i \in N} x_{0i} = \sum_{i \in N} x_{i0} = 1 \quad \textit{The tour starts and ends at the depot} \tag{3.19}$$

$$\sum_{i \in N^+\{j\}} x_{ji} = \sum_{i \in N^+\{j\}} x_{ij} \leq 1, \forall j \in N \tag{3.20}$$

Flow conservation ensures that a vertex is visited at most once.

$$u_i - u_j + 1 \leq (1 - x_{ij})|N|, \forall i, j \in N \quad \textit{Prevents the construction of sub-tours} \tag{3.21}$$

$$1 \leq u_i \leq |N|, \forall i \in N \quad \textit{Boundary constraints} \tag{3.22}$$

$$x_{ij} \in \{0,1\}, \forall i, j \in N \quad \textit{Integrality constraints} \tag{3.23}$$

$$y_{ij}^s \in \mathbb{R}, \forall s \in S, (i,j) \in A \tag{3.24}$$

Constraint equation (3.17) makes the problem in general nonlinear. However, for the uncertainty sets such as the L^∞, L^2, L^1 ball and the intersections defined by these balls, the problem remains tractable. A planning approach that complements the robust UAV tour with agility principles can be introduced. Three different policies can be used, which are as follows:

1. The first approach selects the **most profitable target** out of the set of available targets.
2. The second one selects the **target with the highest ratio** between the profit value and the total expected fuel required, to fly to and record the target and fly back to the recovery point.
3. The third policy is a **re-optimization policy** where the expected fuel consumption is used to find the remaining part of the tour that would be optimal for the deterministic case with nominal fuel consumption. At each target, only the first target of the optimal nominal tour is selected as the target to visit next.

Additional details on this implementation can be found in [120]. The ROP can be extended with **agility principles** used to make decisions during the flight of the UAV in which all fuel realizations that have been revealed so far are taken into account. First, given the fuel realizations, the initial tour will be followed as long as possible. This might imply that the UAV has to return to its recovery point before all planned targets are reached. In case of beneficial fuel realization, the resulting extra fuel capacity can be exploited to increase the total profit value obtained during the entire mission. Since visiting the planned targets is of primary interest, additional targets are considered only after reaching the final target of the planned tour. The uncertain sets are designed to find a solution by taking a certain part of the uncertain parameters into consideration. Consequently, robust optimization allows to tune the level of conservatism applied by the choice of the uncertainty set.

3.3.1.3   UAV Team Orienteering Problem

In the team orienteering problem, a team of m UAVs are scheduled to serve a set of nodes, and each node is associated with a reward. The objective is to maximize the total received reward, while the travel time of each route must be not more than a time limit. The team orienteering problem is defined on a complete graph $G = (V, E)$, where $V = \{0, 1, \ldots, n+1\}$ is the set of nodes and $E = \{(iV, j) | i, j \in \}$ is the set of edges, c_{ij} is the travel time of edge $(i, j) \in E$, and r_i is the reward of node i; a feasible path must start at node 0 and finish at node $n+1$; and its travel time cannot exceed a time limit T_{max}.

Problem 110. *Team orienteering problem: Let $R(x_k)$ be the total received reward of a path $x_k \in \Omega$. The problem is to maximize the total received reward:*

$$\max \sum_{x_k \in \Omega} R(x_k) y_k \tag{3.25}$$

subject to:

$$\sum_{x_k \in \Omega} a_{ik} y_k \leq 1; \quad i = 1, \ldots, n \tag{3.26}$$

$$\sum_{x_k \in \Omega} y_k \leq m; y_k \in \{0, 1\} \tag{3.27}$$

where Ω is the set of all feasible paths, $a_{ik} = 1$ if path x_k visits node i; otherwise, $a_{ik} = 0$, $y_k = 1$ if path $x_k \in \Omega$ is traveled; otherwise, $y_k = 0$.

The first constraint means that each node only can be visited at most once, and the second constraint ensures that there are at most m paths in a feasible solution. Several exact algorithms have been developed; however, there is no algorithm currently able to find an optimal solution. An alternative approach is meta-heuristic, which aims to yield a satisfactory solution within reasonable time [203].

An extension of this problem is the multi-constraint team orienteering problem with multiple time windows. Often, soft time windows are used for which a penalty is incurred in the objective function in case of time window violations. However, in the case of UAV mission planning, there are hard time windows. In a dynamic situation, a UAV tour should be designed by taking into account the possibility that during the flight emergency recordings, new targets become known and should take place: time-sensitive targets, which have priority over the foreseen targets. They are only worthwhile visiting within a predefined time limit; the emergency reaction time is defined by the mission type. Therefore, the UAV will be sent to such a target as soon as it appears if remotely possible, considering the uncertainty in the time required to travel to the target. In this dynamic situation, at each moment of the flight, the UAV is either on its way to a foreseen target, on its way to a new target that has just appeared, recording or waiting to start recording at a target, or on its way back to the depot. Profits are obtained by recording foreseen targets that are reached in time and for which the recording time could be completed, without interruption. The available mission time will be partly devoted to new targets, and partly to foreseen targets. In case a planned tour to foreseen targets is purposely located in the proximity of the locations where new targets are expected to appear, the UAV will likely be in time to record a new target if it appears. On the other hand, when the planned tour does not take the possible locations of new targets into account beforehand, the expected profit to be obtained from foreseen targets might be higher. It is therefore worthwhile to consider in advance how much emphasis should be put on the possible appearance of these new, time-sensitive targets when designing the planned tour: the online stochastic UAV mission planning problem with time windows and time-sensitive targets.

A different class of fast algorithms based on a decomposition of the orienteering problem into a knapsack problem (assignment of tasks with values to a UAV with limited task budget) and a subsequent TSP (selection of the most efficient route for assigned tasks) is developed in [109]. Fast knapsack algorithms are based on selecting tasks in terms of their marginal value per additional resource cost. For orienteering problems, the resource cost is hard to evaluate because it requires the solution of the subsequent TSP. An approach based on spanning trees that allow for the estimation of increased resource costs, leading to a fast algorithm for selecting tasks to be performed by each UAV, is thus developed in Algorithm 30.

Algorithm 30 Orienteering Knapsack Algorithm

---

1. Initialization: The attention should be restricted to all tasks that are within round-trip distance of the source node. The domain is partitioned circularly and equally, and the total reward of tasks present in each sector is calculated. The goal is to start the trees with some geographic diversity, with anticipation of directions where the higher-value tasks exist.
2. Tree growing: Starting from source vertex, a tree using a greedy knapsack approach is grown. A reward-to-connection cost ratio is used to select the appropriate next vertex. After no tree can add additional vertices without exceeding its budget estimate, the cost of a tour is refined for each tree, estimating the cost of a tour using the topological order imposed from a depth-first traversal of the tree.
3. Tour construction: The resulting tree from the first stage, tree growing, identifies a set of vertices S with tasks that should be performed.
4. Tour improvement: The reward-to-incremental cost ratio is used to unselect each vertex with respect to the tours, and insert vertex in order of the biggest cost ratio as long as the cost of the tour meets the budget constraint.
5. Tour refinement: After more vertices are inserted into the tour, the obtained tour may not be an optimal TSP tour on this enlarged vertex subset. In this case, a new tour must be found for the current vertex subset and repeat the tour improvement step until no further improvements can be made. This step is not needed in most cases, but it is a simple step.

The **Dubins orienteering problem** is a generalization of the orienteering problem for fixed-wing UAVs [245]. Its objective is to maximize the total collected rewards by visiting a subset of the given target locations by a fixed-wing UAV, while the length of the collecting tour does not exceed a given travel budget. Therefore, a solution of the Dubins orienteering problem requires determining particular heading angles at the target locations to minimize the length of Dubins maneuvers between the targets. Regarding computational complexity, the Dubins orienteering problem can be considered as more challenging than the Euclidean orienteering problem, as changing only one heading angle or target location in the reward-collecting path usually enforces the change of all heading angles of nearby connected target locations.

The proposed solution in [121] starts by constructing a tour using the available information, and hence determining the first target to be visited. After completing recording at this target and if no time-sensitive target has appeared, the next foreseen target to visit is determined by replanning the tour based on past travel and recording time realizations. This replanning is based on the **maximum coverage stochastic orienteering problem with time windows** (MCS-OPTW). The MCS-OPTW planning approach provides a path from the current location to the depot, containing only foreseen targets. The next target planned to be visited is the first target in this path. In executing this replanning procedure, the MCS-OPTW balances two objectives: maximizing the expected profit obtained by recording foreseen targets and maximizing the expected **weighted location coverage** (WLC) of the path. The WLC relates the distance of the arcs to locations where new targets are expected to appear. Hence, by this second objective, the MCS-OPTW selects foreseen targets such that the UAV will be sent in the direction of areas where new targets are expected to appear. In both objectives, the expected values are determined based on predefined probability distributions of the travel and recording times.

3.3.2   UAV SENSOR SELECTION

The **UAV sensor selection and routing problem** is a generalization of the **orienteering problem**.

Problem 111. *UAV sensor selection and routing problem: A single UAV begins at a starting location and must reach a designated destination prior to time T. Along with the starting and ending points, a set of locations exists with an associated benefit that may be collected by the UAV. Find the optimal path for a UAV orienteering with a given sensor.*

Mission planning can be viewed as a complex version of path planning where the objective is to visit a sequence of targets to achieve the objectives of the mission. The integrated sensor selection and routing model can be defined as a **MILP formulation** [234]. A successful path planning algorithm should produce a path that is not constrained by end points or heading that utilizes the full capability of the aircraft's sensors and that satisfies the dynamic constraints on the UAV.

In [171], a path planning method for sensing a group of closely spaced targets is developed that utilizes the planning flexibility provided by the sensor footprint, while operating within dynamic constraints of the UAV. The path planning objective is to minimize the path length required to view all of the targets. In addressing problems of this nature, three technical challenges must be addressed:

1. Coupling between path segments,
2. Utilization of the sensor footprint,
3. Determination of the viewing order of the targets.

For Dubins vehicles, capabilities can be provided by discrete time paths that are built by assembling primitive turn and straight segments to form a flyable path. For this work, each primitive segment in a discrete step path is of specified length and is either a turn or a straight line. Assembling the left turn, right turn and straight primitives creates a tree of flyable paths. Thus, the objective for the path planner is to search the path tree for the branch that accomplishes the desired objectives in the shortest distance. Other parametric curves can be used, such as Cartesian polynomials, different kinds of splines and Pythagorean hodograph. [56]. The learning real-time A* algorithm can be used to learn which branch of a defined path tree best accomplishes the desired path planning objectives.

Another example follows.

Problem 112. *Given a set of stationary ground targets in a terrain (natural, urban or mixed), the objective is to compute a path for the reconnaissance UAV so that it can photograph all targets in minimum time, because terrain features can occlude visibility.*

As a result, in order for a target to be photographed, the UAV must be located where both the target is in close enough range to satisfy the camera's resolution and the target is not blocked by terrain. For a given target, the set of all such UAV positions is called the **target's visibility region**. The UAV path planning can be complicated by wind, airspace constraints, dynamic constraints and the UAV body itself occluding visibility. However, under simplifying assumptions, if the UAV is modeled as a Dubins vehicle, the target's visibility regions can be approximated by polygons and the path is a closed tour [239]. Then, the 2D reconnaissance path planning for a fixed-wing UAV, also called Dubins vehicle because of its limited curvature, can be reduced to the following:

Problem 113. *For a Dubins vehicle, find a shortest planar closed tour that visits at least one point in each of a set of polygons. This is referenced to as the **polygon visiting Dubins traveling salesperson problem** (PVDTSP).*

Sampling-based roadmap methods operate by sampling finite discrete sets of poses (positions and configurations) in the target visibility regions in order to approximate a PVDTSP instance by a **finite-one in set traveling salesperson problem** (FOTSP). The FOTSP is the problem of finding a minimum cost closed path that passes through at least one vertex in each of a finite collection of clusters, the clusters being mutually exclusive finite vertex sets. Once a roadmap has been constructed, the algorithm that converts the FOTSP instance into an **ATSP** instance to solve a standard solver can be applied.

The desired behavior for a fixed-wing UAV is to maximize sensor coverage of the target. The objective function that drives this behavior is a weighted sum of separate objectives:

$$J = \int_{t_0}^{t_f} \left(W_1 u_1^2 + W_2 u_2^2 + W_3 u_3^2 + W_4 u_4^2 \right) + \int_{t_0}^{t_f} W_5 \left[(r_x - r_x^d)^2 + (r_y - r_y^d)^2 + (r_z - r_z^d)^2 \right] \quad (3.28)$$

where W_{1-5} are the given weights, u_{1-4} are the control inputs, (r_x, r_y, r_z) are the actual 3D positions and (r_x^d, r_y^d, r_z^d) represent the square of the distance to the target. The first four terms penalize control effort, and the fifth term weighs the square of the distance to the target. Constraints on the problem include the equations of motion and dynamic limitations. The target position and velocity vectors are continuously provided to the algorithm. Between path planning computations, the winds can be assumed as constant. A standard **VRP** can be considered but with demands, which are probabilistic in nature rather than deterministic. The problem is then to determine a fixed set of routes of minimal expected total length, which corresponds to the expected total length of the fixed set of routes plus the expected value of extra travel distances that might be required. The extra distance will be due to the possibility of demand on one or more routes that may occasionally exceed the capacity of a vehicle and force it to go back to the depot before continuing on its route.

The following two solution-updating methods can be defined:

1. Under method $\mathbb{U}_a$, the UAV visits all the points in the same fixed order as under the a priori tour but serves only location points requiring service during that particular problem instance. The total expected distance traveled corresponds to the fixed length of the a priori tour plus the expected value of the additional distance that must be covered whenever the demand on the route exceeds vehicle capacity.

2. Method $\mathbb{U}_b$ is defined similarly to $\mathbb{U}_a$, with the difference that location points with no demand on a particular instance of the UAV tour are simply skipped.

If each point x_i requiring a visit with a probability p_i independently of the others has a unit demand and the UAV has a capacity q, then an a priori tour through n points has to be found. On any given instance, the subsets of points present will then be visited in the same order as they appear in the a priori tour. Moreover, if the demand on the route exceeds the capacity of the vehicle, the UAV has to go back to the recovery point before continuing on its route. The problem of finding such a tour of minimum expected total length is defined as a **capacitated probabilistic TSP**.

3.4   COVERAGE

A coverage algorithm in robotics is a strategy on how to cover all points in terms of a given area using a set of sensors and a robot. It describes how the UAV should move to cover an area completely, ideally in a safe, energy-consuming and time-effective manner taking into account the wind effect. Initially, the focus was coverage of structured and semi-structured indoor areas [252,298]. With the introduction of GPS, the focus has turned to outdoor coverage; however, due to nearby buildings and trees, GPS signals are often corrupted. The unstructured nature of an outdoor environment makes covering an outdoor area with all its obstacles and simultaneously performing reliable localization a difficult task [237].

Remark 114. *The **coverage** of an unknown environment is also known as the **sweeping problem**, or mapping of an unknown environment. Basically, the problem can be solved either by providing ability for localization and map building first, or by directly deriving an algorithm that performs sweeping without explicit mapping of the area. Instead of a **measure of coverage**, an average event detection time can be used for evaluating the algorithm.*

It is a challenge to develop a system that can handle any coverage tasks in indoor and outdoor environments with various UAV platforms with different kinematic and dynamic constraints. Assessing each task and finding the most suitable algorithm for that task are necessary. Yet, some of these tasks can be grouped together in the same category allowing the usage of the same coverage approach. There are many challenges when working in the outdoor environment, such as weather conditions, lighting conditions and unstructured surroundings. In many real-world applications, there exists the added complication of obstacles within the coverage area. For safety reasons, it is important that the obstacles are avoided. Furthermore, large obstacles in the vicinity of operations tend to corrupt GPS signals, which is most commonly used for outdoor localization. Hence, it is desirable to employ alternative localization techniques.

Remark 115. *In the literature, the term **coverage** has also been applied to the problem of distributing a group of mobile sensor units within an environment such that their positioning achieves maximum coverage of an area of interest. Most of these approaches assume that the mobile units do not move after reaching their desired positions, unless the configuration of the environment is altered during run-time.*

The TSP aims to identify a least cost Hamiltonian tour on a given network. All nodes of the network must be visited exactly once in this TSP. Identifying a tour over a subset of nodes so that the others are within a reasonable distance of some tour stop can be more desirable. The aim of the **time-constrained maximal coverage salesperson problem** is to find a tour visiting a subset of the demand points, so as to maximize the demand covered subject to a time constraint. The demands of the vertices that are on the tour are assumed to be fully covered, while only a percentage of the demand of a vertex is covered if it is not visited but is within a specified distance of some tour stop. Routing of UAVs for information gathering against intruders is a perfect application for this problem [242]. The first problem incorporating the coverage concept into a routing scheme is the covering salesperson problem. It is the problem of identifying a minimum length Hamiltonian tour over a subset of vertices in a way that every vertex not on the tour lies within a certain distance of some visited vertex. A generalization is to have an additional cost incurred for every node visited by the tour, while each node is associated with a weighted demand representing the minimum number of times it has to be covered. It can be classified into three categories: maximizing profit under a distance constraint, minimizing distance under a profit constraint or a combination of distance minimization and profit maximization.

The coverage planning problem is related to the covering salesperson problem where an agent must visit a neighborhood of each city. Coverage algorithms can be classified as heuristic or complete depending on whether or not they provably guarantee complete coverage of the free space. The following classifications can be proposed [139]:

1. **Classical exact cellular decomposition methods** break the free space down into simple, nonoverlapping regions called cells. The union of all the cells exactly fills the free space. These regions, which contain no obstacles, are easy to cover and can be swept by the UAV using simple motions.

2. **Morse-based cellular decomposition approach** is based on critical points of Morse functions. A Morse function is one whose critical points are nondegenerate. A critical point is a value where either the function is not differentiable or all its partial derivatives are zero. By choosing different Morse functions, different cell shapes are obtained.

3. **Landmark-based topological coverage approach** uses simpler landmarks to determine an exact cellular decomposition termed slice decomposition. Due to the use of simpler landmarks, slice decomposition can handle a larger variety of environments.

4. **Grid-based methods** use a representation of the environment decomposed into a collection of uniform grid cells. Most grid-based methods are resolution-complete; i.e., their completeness depends on the resolution of the grid map.

5. **Graph-based coverage** is interesting for environments that can be represented as a graph. In particular, it can take into account that the prior map information provided as a graph might be incomplete, accounting for environmental constraints in the environment, such as restrictions in certain directions in the graph and providing strategies for online replanning when changes in the graph are detected by the UAV's sensors when performing coverage scenarios.

6. **Coverage under uncertainty scenarios** is useful when the lack of a global localization system such as GPS makes the UAV accumulate drift, and hence a growing uncertainty about its pose. Although the topological representations, such as the adjacency graph, are tolerant to localization error, the performance of coverage algorithms, even if using such representations, is still affected.

7. **Maximum weighted coverage problem** is useful when a limited number of bases should be open such that the combined coverage is maximized [204]. It is also known as the maximal covering location problem.

The **swap local search method** iteratively improves an initial feasible solution by closing a subset of opened bases and opening a different subset of bases with strictly increased coverage. The two subsets are chosen such that they have the same cardinality. The simplification steps, and the resulting instance transformations, can be characterized as follows:

1. Remove unnecessary bases and demand points,
2. Create a bijection between bases and demand points,
3. Use the inherent symmetry of the coverage problem,
4. Simplify the structure of the instance.

Another classification for coverage can be proposed into three classes: **barrier coverage**, **perimeter coverage** and **area coverage**. Using UAVs needs a coverage path planning algorithm and a coordinated patrolling plan [4]. **Perimeter surveillance** missions can be approached as a patrolling mission along a defined path: the perimeter. On the other hand, **area surveillance** missions could be divided in two different problems: an area coverage path planning problem and a patrolling mission along that path. A frequency-based patrolling approach involves the frequency of visits as the parameter to optimize, if the area has to be covered again and again. Coverage path planning tries to build an efficient path for the UAVs, which ensures that every point in the area can be monitored from at least one position along the path. The challenge is to maximize the area to cover with a given amount of sensors, assuming a defined coverage range for each sensor.

3.4.1 BARRIER COVERAGE

Among intruder surveillance applications, barrier coverage is a widely known coverage model to detect intruders. A barrier is a line of sensors across the entire field of interest. The sensing ranges of two neighbor sensors in the barrier are overlapped, and thus, the intruders are guaranteed to be detected. Barrier coverage minimizes the probability of undetected intruder passing through the arrangement in a static arrangement of UAVs forming a barrier [294].

Definition 116. *A path P is said to be 1-covered if it intercepts at least one distinct sensor.*

Definition 117. *A sensor network is said to be strongly 1-barrier covered if:*

$$P(any\ crossing\ path\ is\ 1\text{-}covered) = 1 \qquad (3.29)$$

Barrier coverage of sensor networks provides sensor barriers that guard boundaries of critical infrastructures [84]. Many considerations can be taken into account. Some approaches examined **weak/strong barrier** and **k−barrier** in terms of barrier detection capability, whereas others focused on the network lifetime of barriers and attempted to extend a barrier's network lifetime by reducing the number of barrier members. Others were concerned about the cost of barrier construction and took advantage of the mobility of mobile camera sensors to reduce the number of sensors required for barrier construction. The k-barrier detects intruders by constructing k weak/strong barriers. The relationship between the number of sensors required and the success rate of the barrier construction can be analyzed to find the optimal number of sensors to scatter in a random deployment context. Collaboration and information fusion between neighboring sensors reduce the number of barrier members and prolong the network lifetime of barriers using a **sleep-wake-up schedule**.

3.4.1.1 Barrier Coverage Approach

The problem of barrier coverage is classified on the basis of the type of region [309]:

1. Barrier coverage along a landmark can be formulated along a line or a point on it. The problem of barrier coverage along line W with direction $\bar{\theta}$ is formulated as follows:

$$W = \left\{ p \in \mathbb{R}^2 : u^T p = d_1 \right\}, \bar{\theta} = \beta + \pi/2 \qquad (3.30)$$

where $u = [\cos\beta, \sin\beta]$ is a unit vector with a given $\beta \in [-\pi/2, \pi/2]$ measured with respect to x-axis and d_1 is a given scalar associated with W. The UAVs are supposed to make a barrier of length L from W and are evenly deployed by maximizing the length L.

2. Barrier coverage between two landmarks: In this problem, a barrier of sensors is supposed to ensure coverage between two landmarks. The problem of barrier coverage between two landmarks L_i and L_j is formulated as follows.

Problem 118. *Let u be a unit vector associated with L_i and L_j with $u = \frac{L_i - L_j}{\|L_i - L_j\|}$. The unit vector $u = [\cos\beta, \sin\beta]^T$ for some $\beta \in [-\pi/2, \pi/2]$ characterizes the bearing of L_j*

relative to L_i. An associated scalar is also defined as $\bar{\theta} = \beta + \pi/2$. A line L using L_i and L_j is defined as $L = \{p \in \mathbb{R}^2 : (L_j - L_i)^T u^{\perp} = 0\}$. n points h_i are defined on L as: Let there be n UAVs and two distinct landmarks $L1$ and $L2$. Then, a decentralized control law for barrier coverage between the landmarks is formulated if for almost all initial sensor positions, there exists a permutation $\{z_1, z_2, \ldots, z_n\}$ of the set $\{1, 2, \ldots, n\}$ such that the following condition holds: $\lim_{k \to \infty} \|p_{z_i}(kT) - h_i\|$.

The problem of barrier coverage can be categorized on the basis of approaches used:

1. **Nearest neighbor rule:** In the basic method, one simply stores each training instance in memory. The power of the method comes from the retrieval process. Given a new test instance, one finds the stored training case that is nearest according to some distance measure, notes the class of the retrieved case and predicts the new instance will have the same class. Many variants exist on this basic algorithm. Another version retrieves the k-closest instances and base predictions on a weighted vote, incorporating the distance of each stored instance from the test case; such techniques are often referred to as k-nearest neighbor algorithms. However, the simplicity of this method does not mean it lacks power.

2. **Artificial potential field:** The main feature of this approach is its scalar potential field, which represents both a repulsive force from obstacles and an attractive force to the goal. Therefore, a UAV's path from its starting position to the goal is found by threading the valleys of the potential field. The principle is quite simple and gives good results in many cases. However, since local minima of the potential are sometimes produced, the UAV is trapped before it can reach the goal in such cases. In addition, the generation of a potential field may require a large computation time in a complicated environment that includes concave objects. However, many possibilities of hybridization exist allowing avoiding this local minima.

3. **Virtual force field:** This technique is predicated on the gravitational force field. It lies in the integration of two known concepts: certainty grids for obstacle representation and potential fields for navigation. It is popular because of its simplicity, online adaptive nature and real-time prompting.

4. **Generalized non-uniform coverage:** This approach may concern the problem of border patrol, or adaptive sampling in 2D or 3D environments where the non-uniformity in the sampled field is dominant in one dimension. This is also closely related to information gathering and sensor array optimization problems.

The set of n UAVs initially situated at arbitrary positions $x_1(0), \ldots, x_n(0)$ is located in the interval $[0, 1]$. The density of information at each point is measured by a function $\rho : [0, 1] \to (0, \infty)$, this function being bounded: $\rho_{\min} \leq \rho \leq \rho_{\max}$. The metric is defined by $d_\rho(a, b) = \int_a^b \rho(z) dz$. This metric expands regions where ρ is large and shrinks regions where ρ is small.

Definition 119. *Set of points coverage: The coverage of a set of points $x_1, \ldots, x_n$ relative to the density field ρ is defined as:*

$$\Phi(x_1, \ldots, x_n, \rho) = \max_{y \in [0,1]} \min_{i=1,;n} d_\rho(y, x_i) \tag{3.31}$$

The best (smallest) possible coverage is given by:

$$\Phi^* = \inf_{(x_1, \ldots, x_n) \in [0,1]^n} \Phi(x_1, \ldots, x_n, \rho) \tag{3.32}$$

One possible control law can be found in [224], using the fact that the non-uniform coverage problem can be made uniform by a transformation.

3.4.1.2   Sensor Deployment and Coverage

In coverage problem, sensors must be placed in a certain area to retrieve information about the environment. A model of the sensors needs to be elaborated since how the surroundings are perceived by the UAVs strongly affects the evolution of the algorithm. Most of the existing results on coverage consider sensors with symmetric, omnidirectional field of view, and only recently, agents with anisotropic and vision-based sensors have been considered. The major challenge in the real-world implementation of coverage algorithms lies in the communication and information exchange among the UAVs. Some definitions useful for the sequel are introduced.

Definition 120. *Field of view: The field of view of the optical sensor is defined as the extent of observable world that it can collect at any given time.*

In a planar application, the set $\mathbb{R}^2$ is partitioned as follows:

$$R_1 = \left\{ s \in \mathbb{R}^2 : s_x \leq 0 \right\}$$
$$R_2 = \left\{ s \in \mathbb{R}^2 : s_x > 0 \text{ and } \|s\| \leq 1 \right\}$$
$$R_3 = \left\{ s \in \mathbb{R}^2 : s_x > 0 \text{ and } \|s\| > 1 \right\} \tag{3.33}$$

Definition 121. *Visibility: The visibility is defined as:*

$$vis_{I_3}(s) = \left\{ \begin{array}{cc} 0 & s \in R_1 \\ s_x & s \in R_2 \\ \frac{s_x}{\|s\|} & s \in R_3 \end{array} \right\} \tag{3.34}$$

The visibility of all the points in R_1 is zero because the region R_1 lays on the back of the agent. The points in R_2 are on the front of the agent and close to it, so the visibility increases with the distance from the agent and with the centrality in the agent's field of view.

The problem of barrier coverage by stationary wireless sensors that are assisted by a UAV with the capacity to move sensors is considered in [84]. Assume that n sensors are initially arbitrarily distributed on a line segment barrier. Each sensor is said to cover the portion of the barrier that intersects with its sensing area. Owing to incorrect initial position, or the death of some of the sensors, the barrier is not completely covered by the sensors. Assume that n sensors $s_1, s_2, \ldots, s_n$ are distributed on the line segment $[0, L]$ of length L with endpoints 0 and L in locations $x_1 \leq x_2 \leq \ldots \leq x_n$. The range of all sensors is assumed to be identical and is equal to a positive real number $r > 0$. Thus, sensor s_i in position x_i defines a closed interval $[x_i - r, x_i + r]$ of length $2r$ centered at the current position x_i of the sensor, in which it can detect an intruding object or an event of interest. The total range of the sensors is sufficient to cover the entire line segment $[0, L]$, i.e., $2rn \geq L$. A gap is a closed sub-interval g of $[0, L]$ such that no point in g is within the range of a sensor. Clearly, an initial placement of the sensors may have gaps. The sensors provide complete coverage of $[0; L]$ if they leave no gaps [96].

3.4.2   PERIMETER COVERAGE

3.4.2.1   Coverage of a Circle

In the coverage of a circle using a network of UAVs considered as mobile sensors with nonidentical maximum velocities, the goal is to deploy the UAVs on the circle such that the largest arrival time from the mobile sensor network to any point on the circle is minimized. This problem is motivated by the facts that in practice the assumption of UAVs with identical moving speed often cannot be satisfied, and events taking place in the mission domain only last for a finite time period. When the sensing range of mobile sensors is negligible with respect to the length of a circle, reduction of the largest arrival time from a sensor network to the points on the circle will increase the possibility of capturing the events taking place on the circle before they fade away. To drive the sensors to the optimal locations such that the overall sensing performance of the sensor network is optimized, gradient descent coverage control laws based on **Voronoi** partition are developed for mobile sensors with limited sensing and communication capabilities to minimize a locational optimization function. A Voronoi diagram is a subdivision of a Euclidean space according to a given finite set of generating points such that each generating point is assigned a Voronoi cell containing the space which is closer to this generating point than to any other.

3.4.2.1.1 Voronoi Diagrams

Definition 122. *Planar ordinary Voronoi diagram:* *Given a set of finite number of distinct points in the Euclidean plane, all locations in that space are associated with the closest member with respect to the Euclidean distance. The result is a **tessellation** of that plane into a set of regions associated with members of the point set. This tessellation is called a planar ordinary Voronoi diagram generated by the point set, and the regions constituting the Voronoi diagram are Voronoi polygons. Voronoi edges are the boundaries of the polygons. An end point of a Voronoi edge is a Voronoi vertex.*

Remark 123. *Other metrics than the Euclidean distance can be used in that definition. These Voronoi diagrams can be generalized in a variety of ways: weighting of the points, considering regions associated with subsets of points rather than individual points, including obstacles in the space, considering regions associated with sets of geometric features other than points and examining Voronoi diagrams on networks and on moving points [240].*

3.4.2.1.1.1 Centroidal Voronoi diagrams

Definition 124. *Centroid:* *Given a region $U \subset \mathbb{R}^N$ and a density function ρ defined in U, the centroid z^* of U is defined by:*

$$z^* = \frac{\int_U y\rho(y)dy}{\int_U \rho(y)dy} \tag{3.35}$$

*The **Voronoi regions** $\{\hat{U}_i\}$, given k points $z_i, i = 1, \ldots, k$ in U, are defined as:*

$$\bar{U}_i = \left\{ s \in U; |x - z_i| < |x - z_j|; j = 1, \ldots, k, j \neq i \right\} \tag{3.36}$$

*The **centroidal Voronoi tessellation** corresponds to the situation where:*

$$z_i = z_i^*, i = 1, \ldots, k \tag{3.37}$$

3.4.2.1.1.2 Identification of bubbles

Definition 125. *Bubbles* *are the set of points of a mesh such that the value of a field ρ (e.g., the density) is below a predefined threshold for all points in the set and each point of the set can be reached starting from any point of the same set by moving along the directions of the model.*

Two bubbles never share points; otherwise, they are fused in a single bubble. The identification of such type of bubbles is like the problem of **cluster labeling**. A cluster-labeling procedure assigns unique labels to each distinct cluster of lattice points. Once a bubble has been identified with a unique label, its center of mass can be identified. Delaunay triangulation, involving only topological properties, can be computed starting from the circumcenters of the triangles. To differentiate nodes that lie at the boundary between two Voronoi cells from those that lie at the vertices, the number of gaps along the frame is counted. It is necessary to distinguish between nearest neighbors and second neighbors. Finally, the comparison of two different triangulations leverages a suitable isomorphism linking two triangulation graphs. Given the adjacency list, the differences can be checked and the location of topological changes can be extracted.

A benchmark problem of 1D coverage is the uniform coverage problem in which the distance between neighboring agents is required to reach a consensus. The sensing performance of a homogeneous sensor network is maximized when the sensors are uniformly deployed on a line or circle provided that the information density of all points on the line or circle is identical. In contrast, when the **density of information** is not uniform over the mission space, uniform coverage is generally undesirable and more sensors should be deployed in areas with high information density. Problems that are closely related to the coverage control problem on a circle include circular formation and multiagent consensus. In the circular formation problem, a team of mobile agents is required to form a formation on a circle and the desired distance between neighboring agents is generally prescribed a priori. In contrast, in the coverage control problem, the desired distance between sensors is unknown beforehand and depends on the coverage cost function to be optimized. In [269], a distributed coverage control scheme is developed for heterogeneous mobile sensor networks on a circle to minimize the coverage cost function while preserving the mobile sensors order on the circle. The difficulties caused by the heterogeneity of mobile sensors maximum velocities are as follows:

1. For a network of mobile sensors with identical maximum speed, the optimal configuration is uniform deployment of the sensors on the circle. However, it is still unclear under what conditions the coverage cost function is minimized when a network of heterogeneous mobile sensors is deployed.
2. Different constraints are imposed on the mobile sensors' control inputs due to the existence of nonidentical maximum velocity for each sensor. This complicates the proof of mobile sensors' order preservation and convergence analysis of the distributed coverage control scheme, by taking into consideration input constraints and order preservation of the mobile sensors.
3. Environmental conditions, such as wind and lightning conditions, should also be taken into account.

3.4.2.2 Dynamic Boundary Coverage

Bayesian search focuses on how to estimate the target's motion and position based on probability theory. The assumptions of Bayesian search are that the search area can be divided into finite cells/graphs and that each cell represents individual **probability of detection** (PD). The goal is to determine the optimal path to find the lost or moving target according to the **PDF**. The three steps of **Bayesian** search for a target are as follows:

1. Compute prior PDF according to motion information (e.g., flight dynamics and drift data).
2. Compute posterior PDF according to sensor information.
3. Move to the highest probability cell, scan this area and update the posterior PDF as the prior PDF.

The three steps are repeated until the target is found. Two optimization objectives are as follows: to maximize the PD or minimize the **expected time to detection** (ETTD). The advantages of Bayesian search are that the imperfect sensing detection and target's motion can be modeled as a probability distribution and the probability of each cell is updated according to real-time sensing data [299].

In **dynamic boundary coverage task**, a team of UAVs must allocate themselves around the boundary of a region or object according to a desired configuration or density. It is a dynamic boundary coverage, in which UAVs asynchronously join a boundary and later leave it to recharge or perform other tasks. In **stochastic coverage schemes** (SCSs), UAVs probabilistically choose positions on the boundary.

1. SCSs enable a probabilistic analysis of the graph for different classes of inputs identified by the joint PDF of UAV positions.
2. SCSs allow to model natural phenomena, such as **random sequential adsorption** (RSA), the clustering of ants around a food item and Renyi Parking, the process by which a team of cars parks without collisions in a parking lot.

Each UAV can locally sense its environment and communicate with other UAVs nearby. UAVs can distinguish between other robots and a boundary of interest, but they lack global localization: highly limited on-board power may preclude the use of GPS, or they may operate in GPS-denied environments. The UAVs also lack prior information about their environment. Each UAV exhibits random motion that may be programmed, for instance, to perform probabilistic search and tracking tasks, or that arises from inherent sensor and actuator noise. This random motion produces uncertainty in the locations of UAV encounters with a boundary. When a UAV attaches to the boundary, it selects an interval of the boundary of length R lying completely within the boundary.

Definition 126. *A SCS is the choice of multiple random points on the boundary. Formally, a SCS is a 1D point process (PP) realized on the boundary.*

A special case of a PP involves UAVs attaching to a boundary at predefined locations. In the **Poisson point process** (PPP), UAVs attach independently to the boundary, and one of its generalizations is the Markov Process. The independent attachments in PPP make them easy to analyze. On the other hand, interactions between UAVs are harder to handle and require generalizations of PPP [214]. The Poisson Voronoi diagram is an infinite Voronoi diagram referring to the situation in which points are located in space at random, according to the homogeneous PPP.

3.4.3 AREA COVERAGE

3.4.3.1 Preliminaries

The **area coverage** problem with UAVs can be solved from two different approaches:

1. In the online coverage algorithms, the area to cover is unknown a priori, and step-by-step, has to discover obstacles, compute their paths and avoid collisions. Both Voronoi spatial partitioning and coverage can be handled in a distributed manner, with minimal communication overhead.
2. In the offline algorithms, the UAVs have a map of the area and the obstacles, and can plan the path to cover the whole area.

The usefulness of a map not only depends on its quality but also on the application. In some domains, certain errors are negligible or not so important. That is why there is not one measurement for map quality. Different attributes of a map should be measured separately and weighed according to the needs of the application [261]. Those attributes can include the following:

1. **Coverage:** How much area was traversed/visited.
2. **Resolution quality:** To what level/detail are features visible.
3. **Global accuracy:** Correctness of positions of features in the global reference frame.
4. **Relative accuracy:** Correctness of feature positions after correcting (the initial error of) the map reference frame.
5. **Local consistencies:** Correctness of positions of different local groups of features relative to each other.
6. **Brokenness:** How often is the map broken. That is, how many partitions of it are misaligned with respect to each other by rotational offsets.

The objective of area coverage algorithm studies is as follows:

Problem 127. *Given a region in the plane, and given the shape of a cutter, find the shortest path for the cutter such that every point within the region is covered by the cutter at some position along the path.*

In a multi-UAV system, each UAV is equipped with communication, navigation and sensing capabilities allowing it to become a node of the mobile sensor network that can be steered to the region of interest. The coordination problem involves two challenges:

1. What information should the UAVs interchange?
2. How should each UAV read that information?

Coverage path planning is the determination of a path that a UAV must take in order to pass over each point in an environment.

Definition 128. *Coverage and k-coverage: Given a set of n UAVs*
$U = \{u_1, u_2, \ldots, u_n\}$ *in a 2D area X. Each UAV $u_i (i = 1, \ldots, n)$ is located at coordinate (x_i, y_i) inside X and has a sensing range of r_i called sensing radius. Any point X is said to be covered by u_i if it is within the sensing range of u_i, and any point in X is said to be $k-$covered if it is within at least k UAVs' sensing ranges.*

The goal of the directional $k-$coverage problem in camera sensor network is to have an object captured by k cameras.

Definition 129. *Connectivity: When two UAVs u_i and u_j are located inside X, u_i and u_j are connected if they can communicate with each other.*

Definition 130. *General sensibility: The general sensibility $S(u_i, P)$ of u_i at an arbitrary point P is defined as:*

$$S(u_i, P) = \frac{\lambda}{d(u_i, P)^K} \tag{3.38}$$

where $d(u_i, P)$ is the Euclidean distance between the UAV u_i and the point P, and positive constants λ and K are sensor technology-dependent parameters.

When a UAV has only to reach some point in an orbit around a location to cover a surrounding region, by placing and sequencing orbit centers minimizing their number and inter-orbit motion costs, efficient coverage may be obtained. At a high level, the strategy may be described as follows [107]:

1. Tessellate regions to be covered with sensing-footprint-sized circles.
2. Provide visitation points to graph tour generator, producing an orbit sequence.
3. Choose the nearest point on the tour for each vehicle and tour direction.
4. Follow the orbit sequence with each vehicle in parallel.
5. Return UAVs to their start locations.

A class of complex goals impose **temporal constraints** on the trajectories for a given system, referred to also as temporal goals. They can be described using a formal framework such as **linear temporal logic** (LTL), **computation tree logic** and $\mu-$**calculus**. The specification language, the discrete abstraction of the UAV model and the planning framework depend on the particular problem being solved and the kind of guarantees required. Unfortunately, only linear approximations of the UAV dynamics can be incorporated. Multilayered planning is used for safety analysis of hybrid systems with reachability specifications and motion planning involving complex models and environments.

3.4.3.1.1   Coverage Path Planning

Coverage path planning determines the path that ensures a complete coverage in free workspace. Since the UAV has to fly over all points in the free workspace, the coverage problem is related to the covering salesperson problem:

1. **Static:** It is a measure of how a static configuration of UAVs covers a domain or samples a probability distribution.
2. **Dynamic:** It is a measure of how well the points on the trajectories of the sensor cover a domain. Coverage gets better and better as every point in the domain is visited or is close to being visited by an agent.
3. **Uniform:** It uses metric inspired by the ergodic theory of dynamical system. The behavior of an algorithm that attempts to achieve uniform coverage is multi-scale.
4. In **persistent monitoring**, the goal can be to patrol the whole mission domain while driving the uncertainty of all targets in the mission domain to zero [268]. The uncertainty at each target point is assumed to evolve nonlinearly in time. Given a closed path, multiagent persistent monitoring with the minimum patrol period can be achieved by optimizing the agent's moving speed and initial locations on the path [280].

Features of large size are guaranteed to be detected first, followed by features of smaller and smaller size [290,292,297].

Definition 131. *A system is said to exhibit **ergodic dynamics** if it visits every subset of the phase space with a probability equal to the measure of that subset. For a good coverage of a stationary target, this translates to requiring that the amount of time spent by the mobile sensors in an arbitrary set be proportional to the probability of finding the target in that set. For good coverage of a moving target, this translates to requiring that the amount of time spent in certain tube sets be proportional to the probability of finding the target in the tube sets.*

A model is assured for the motion of the targets to construct these tube sets and define appropriate metrics for coverage. The model for the target motion can be approximate, and the dynamics of targets for which precise knowledge is not available can be captured using stochastic models. Using these metrics for uniform coverage, centralized feedback control laws are derived for the motion of the mobile sensors. For applications in environmental monitoring with a mobile sensor network, it is often important to generate accurate spatio-temporal maps of scalar fields, such as temperature or pollutant concentration. Sometimes, it is important to map the boundary of a region.

A different strategy toward coverage can be presented as the environment is partitioned with the objective of minimizing the amount of rotation the UAV has to perform, rather than minimizing the total traveled distance. An improved strategy is to reduce the total length of the coverage path. In another approach, a set of pre-calculated motion strategies are chosen in order to minimize repeating coverage. A grid-based approach can be used for planning a complete coverage path using a spanning tree formulation. The primary requirement for a solution to exist is that the environment must be decomposable into a grid with a pre-determined resolution. Coverage can also be achieved by partitioning the environment into a grid. The spanning tree techniques can be extended to achieve efficient grid-based coverage using a team of multiple UAVs [311]. In general, most of the techniques used for the distributed coverage of some region are based on cellular decomposition. The area to be covered is divided between the agents based on their relative locations. Two methods for cooperative coverage, one probabilistic and the other based on an exact cellular decomposition, can be discussed. Domains with non-uniform traversability can also be examined [251].

To facilitate the means for autonomous inspection, a UAV must be equipped with the necessary accurate control units, the appropriate sensor systems and the relevant global path planning intelligence. Such path planning algorithms should be able to lead to the quick computation of efficient paths that result in full coverage of the structure to be inspected, while respecting the on-board sensor limitations as well as the vehicle motion constraints that may apply. Due to the nature of the coverage problem, its inherent difficulties still pose hard limitations on the performance, efficiency and practical applicability of the proposed solutions especially when 3D structures are considered. The path planning algorithm employs a two-step optimization paradigm to compute good viewpoints that together provide full coverage while leading to a connecting path that has a low cost. In order to enable path planning for real 3D structures, advanced algorithms employ a two-step optimization scheme proved to be more versatile with respect to the inspection scenario.

1. Compute the minimal set of viewpoints that covers the whole structure: **art gallery problem** (AGP).
2. Compute the shortest connecting tour over all these viewpoints: TSP.

A recent application of these concepts allows some redundancy in the AGP such that it is able to improve the path in a post-processing step. This algorithm can deal with 3D scenarios. Close-to-optimal solutions are derived at the inherently large cost of computational efficiency. Within every iteration, the set of updated viewpoint configurations is selected such that combining all viewpoints full coverage is achieved, and the cost-to-travel between the corresponding vehicle configuration and the neighboring viewpoint configurations gets reduced. Subsequently, the optimally connecting and collision-free tour is re-computed. The proposed approach selects one admissible viewpoint for every triangle in the mesh of the structure to be inspected. In order to compute viewpoints that allow low-cost connections, an iterative re-sampling scheme is employed. Between each re-sampling, the best path connecting the current viewpoints is re-computed. The quality of the viewpoints is assessed by the cost to connect to their respective neighbors on the latest tour. This cost is minimized in the subsequent re-sampling, resulting in locally optimized paths. Initialization of the viewpoints is arbitrarily done such that full coverage is provided with, at this stage, non-optimized viewpoints [57].

Remark 132. *A large class of surveillance problems for passive targets in mapped regions can be converted to TSPs, which are solved offline to provide several alternative solutions stored offline for online use [189]. An end-to-end framework is constructed, integrating existing algorithms where available, and developing others where necessary, to solve this class of problems. In an instance of the TSP, the distances between any pair of n points are given. The problem is to find the shortest closed path (tour) visiting every point exactly once. This problem is often encountered in robotics and has been traditionally solved in two steps with the common layered controller architecture for UAVs [223,259].*

A strategic level regarding the choice of the way-points is made before the operational decisions regarding routing of UAVs between the different way-points. These two decisions are strongly related by factors such as UAV range and endurance, topography and communications, and the mission requirements. Different approaches have been developed such as a method that creates a spanning tree and generates the coverage paths as the boundary around it. Many different types of basic path patterns have been proposed for coverage algorithms. The most common patterns are parallel swath, also known as parallel milling, or zigzag patterns. The seed spreader algorithm describes an efficient, deterministic and complete coverage strategy for simple regions, by having the robot move in **back and forth**, or **lawnmower motion**, or **sweeping motions**. The following standard search patterns are as follows:

1. **Lawnmower Search** that consists in flying along straight lines with 180 degree turns at the end. Based on the sweep direction, there are two types of lawnmowers:
   a. **Parallel track search** if the search area is large and level, only the approximate location of the target is known and uniform coverage is desired.
   b. **Creeping line search** if the search area is narrow and long and the probable location of the target is thought to be on either side of the search track.

2. **Spiral search** and **Expanding square search** if the search area is small and the position of the target is known within close limits.
3. **Sector search** used similarly to the expanding square search; it offers several advantages: concentrated coverage near the center of the search area, easier to fly than the expanding square search and view of the search area from many angles;
4. **Contour search** used to patrol obstacles, often assumed to be polygonal.

The **Boustrophedon cellular decomposition** (BCD) algorithm is an extension to the seed spreader algorithm, which guaranteed complete coverage of bounded environments, with a variety of control Morse functions. Other typical patterns include inward and outward spirals, random walks and wall following or contour following. Unfortunately, these planners need absolute localization or not taking UAV kinematics into account and not dealing efficiently with obstacles. In terms of tasks, the UAV motion is not the primary one, but it is a necessity to perform the main coverage task. When multiple UAVs are used, a previous decomposition of the field to cover is required. Two approaches are commonly used: exact cell decompositions and approximate cell decomposition. After this task, the path for every vehicle to cover the area assigned is computed.

3.4.3.2 Boustrophedon Cellular Decomposition

The BCD allows to obtain a coverage path. This technique implies the area division in smaller sub-areas that can be covered with a simple back-and-forth method. This method can be extended to a multi-UAV system with limited communications [85]. The BCD is a generalization of the trapezoidal decomposition valid for non-polygonal obstacles. The exact cellular decomposition is the union of nonintersecting cells composing the target environment. Each cell is covered by simple back-and-forth motions. Once each cell is covered, the entire environment is covered. Therefore, coverage is reduced to finding an exhaustive path through a graph, which represents the adjacency relationships of the cells in the BCD. The approach is resolved by Algorithm 31 [306].

Algorithm 31 Boustrophedon Cellular Decomposition

1. Decompose the accessible area of the workspace into non-overlapping cells.
2. Construct an adjacent graph where each vertex is a cell and each edge connects two vertices corresponding to two adjacent cells.
3. Determine an exhaustive walk through the adjacent graph based on a depth-first-like graph search algorithm such that each vertex is visited at least once. Let V be the list that represents a consecutive sequence of vertices of a walk through the adjacent graph, as follows:

   a. Start with any cell resulting from the decomposition. Add it into V, and mark it as visited.
   b. Move to the first counterclockwise unvisited cell in the neighboring cells of the current cell. Add this cell into V and mark it as visited.
   c. Repeat the preceding step until reaching a cell whose neighboring cells have all been visited.
   d. Backtrack and add each visited cell into V until a cell with an unvisited neighboring cell is reached. Go to Step 3b.
   e. If no cell with an unvisited neighboring cell is found during the backtracking process, then the walk is the consecutive sequence of vertices in V.
4. As V is determined, drive the UAV to start at the first cell corresponding to the first vertex of V and move it to the next cell based on V. Perform the coverage task by using a Boustrophedon motion only when the UAV enters an unvisited cell. Repeat moving and covering until the cell corresponding to the final vertex in V is reached.

Remark 133. *One good strategy is to have long motions in parallel with one of the edges of the workspace assuming that there is a bounding rectangle with an edge collinear with one of the edges from the polygon that addresses the workspace. To have a pattern with straight lines that are as long as possible reduces the number of changes in direction so as to reduce exploration time. Another one is to take the direction of the dominant wind into consideration in order to avoid the lateral wind effect. An alternative method is to align the direction of coverage directly with the distribution of the free space, under the assumption that the length of sweep lines will be maximized along the dominant axis of the free space. Given the eigenspace decomposition for free space pixel coordinates, the direction of coverage is set orthogonally to the heading of the eigenvector with the largest eigenvalue.*

A crucial parameter required by this algorithm is the **coverage footprint**, which measures the spacing between consecutive parallel sweep lines in the back-and-forth motion path. Different factors contribute to the definition of the coverage footprint, depending on the intended application: the footprint width determines whether the UAV will arrive at the top or the bottom boundary of a completed cell. Another extra sweep line can be added to avoid this problem. The Boustrophedon family of algorithms ensures the complete coverage of an unknown environment, although none of these algorithms provide any guarantees on the optimality of the coverage path.

Remark 134. *A complete coverage strategy for unknown rectilinear environments using a square robot with contact sensing can perform an online decomposition of free space, where each resulting rectangular cell could be covered using back-and-forth seed spreader motions that are parallel to the walls of the environment.*

The stochastic trajectory optimization motion planning is used in [140] to reshape the nominal coverage path so that it adapts to the actual target structure perceived on-site via on-board sensors. This algorithm explores the space around an initial trajectory by generating noisy trajectories, which are then combined to produce an updated trajectory with lower cost in each iteration. It optimizes a cost function based on a combination of smoothness and application-specific costs, such as obstacles and constraints. General costs for which derivatives are not available can be included in the cost function as this algorithm does not use gradient information.

3.4.3.3 Spiral Path

An alternative to the Boustrophedon motion is the spiral motion pattern. These types of motion are applicable to multiple scenarios: regular or irregular areas, low or high density of obstacles. The spiral is defined by three parameters: initial point, initial direction and pattern direction. Different exploration areas can be studied in order to determine these parameters. The spiral algorithm has been designed to start inside the area and perform counterclockwise rotations to the perimeter of the area. Beginning within the area is useful for performing a search in a completely unknown terrain, but it can be inefficient if the UAV is initially deployed on the border of the field.

1. The initial point (C_x, C_y) is the node where the UAV starts the spiral path. The location of this node depends on the features of the area (shape or parity), and it has an influence on other features of the spiral (initial direction, pattern direction or final point).
2. The initial direction defines the first movement of the UAV. The algorithm uses only two values for this parameter: clockwise when the area is a square or a rectangle with more columns than rows, and counterclockwise when the area is a rectangle with more rows than columns. These initial selections are defined in this way in order to reduce the number of turns.
3. The pattern direction is defined as the sense of movement where the path gets closer to the area limits. This ensures that each loop of the spiral is closer to the boundary than the

previous one. If the motion is in other directions, the distance to the area limits will not change. The distance between the path and the boundary only decreases when the robot is moving in the pattern direction.

For example, in [11], the problem of locating a Unmanned Ground Vehicle (UGV) using a UAV is considered, maximizing the probability of finding the ground vehicle. The length of the spiral between the UAV and ground vehicle at time $T = 0$ is an important parameter. Knowing the description of the spiral $r = m\theta$, its length L is given by:

$$L = \frac{m}{2}\theta_{max}\sqrt{1 + \theta_{max}^2} + \frac{1}{2}ln\left|\theta_{max} + \sqrt{1 + \theta_{max}^2}\right| \tag{3.39}$$

With this equation and the UAV's average speed V_a, it is possible to determine the time for the UAV to navigate to the start point of the ground vehicle:

$$t = \frac{L}{V_a} \tag{3.40}$$

The location of the ground vehicle after the elapsed time needs to be predicted. One method for mapping the possible location includes describing circles with radius r:

$$r = V_g t \tag{3.41}$$

The focal length f can be determined as follows:

$$f = \frac{ccd_y/2}{\tan(\pi/7)} \tag{3.42}$$

The angle of view of the camera can be defined as:

$$aov_x = 2\arctan ccd_x/(2f)$$
$$aov_y = 2\arctan ccd_y/(2f) \tag{3.43}$$

Now, the field of view can be described as:

$$fov_x = h\tan aov_x$$
$$fov_y = h\tan aov_y \tag{3.44}$$

The maximizing factor to determine is the height of the UAV:

$$h = \max\left(\frac{fov_x}{\tan(aov_x)}, \frac{fov_y}{\tan(aov_y)}\right) \tag{3.45}$$

Maximizing the probability of capturing the ground vehicle implies that the field of view must be greater than or equal to the circular area describing the possible location of the ground vehicle.

3.4.3.3.1 3D Terrain Reconstruction

3D terrain reconstruction is a two-level problem: the first one considers the acquisition of the aerial images, and the second one considers the 3D reconstruction. The camera is assumed to be mounted on the UAV to obtain images fully covering the area of interest. As these images will be later used for terrain reconstruction, several considerations arise, which are as follows:

1. **Overlapping:** Consecutive pictures should have a given percentage of overlapping. The greater the overlap is, the higher the accuracy of the 3D model will be.

2. **Time contiguity:** The quality of the 3D texture will be higher when the pictures of contiguous areas of the terrain are taken at a similar time. Otherwise, uncorrelated shadows or visual differences may appear, leading to a more difficult reconstruction and a less-quality texture.
3. **Orientation:** It is desired to have the pictures taken in the same orientation because it leads to a simplification in the 3D reconstruction phase.

The UAV will be always heading to the same direction and will move sideways and backward when it is required. The pictures' orientation will be the same for all pictures taken, facilitating the reconstruction problem. The algorithm returns a path that will be traversed as a zigzag or back-and-forth motion composed by longitudinal (the rows), transverse and possibly slightly diagonal moves [296]. In order to generate back-and-forth motion, the distance between two rows is needed and its calculation depends on the defined vertical overlap and the camera footprint. Let v be the vertical overlap and let w be the width of the camera footprint. The distance d among rows is the vertical distance between both footprints. Taking into account the vertical overlapping, d is calculated as follows:

$$d = w \cdot (1 - v) \qquad (3.46)$$

The number of turns, n, to photograph a given polygon depends on the values of d, w and l_s where l_s is the length of the optimal line sweep direction given by Algorithm 32 for a single convex polygon coverage.

Algorithm 32 Optimal Line Sweep Direction

---

1. distance(e, v): Euclidean distance between edge e and vertex v
2. **For all** edges in the polygon **do**
   a. max-dist-edge =0.
   b. **for all** vertex in the polygon **do**
      i. **If** distance(edge, vertex) $\geq$ max-dist-edge **then**
      ii. max-dist-edge = distance (edge, vertex)
      iii. opposed-vertex = vertex
      iv. **end if**
   c. **end for.**
3. **If** (max-dist-edge $\leq$ optimal-dist) or (is first edge) **then**
   a. optimal-dist=max-dist-edge
   b. line-sweep=direction FROM edge TO opposed vertex
   c. **end if**
4. **End for**

For each segment, two turning points are needed, thus leading to a total number of turns given by:

$$n = \begin{cases} 2\lceil z/d \rceil & \text{if } z \bmod d \leq w/2 \\ 2(\lceil z/d \rceil + 1) & \text{if } z \bmod d \leq w/2 \end{cases} \qquad (3.47)$$

where $z = l_s - w/2$. The number of turns depends on z because d is fixed at the problem's formulation according to the resolution required for the images. A coverage alternative is defined according to two criteria: if the current direction of the line sweep or the opposite one is considered, and the way the coverage path is constructed: clockwise (the first turn is made to the right) or counterclockwise (the first turn is made to the left).

3.4.3.3.2 Signal-Searching Approach

The **signal-searching approach** is layered in two procedures: coverage area optimization and coverage path planning. The overall procedure for the task proposed is as follows:

1. Find a solution to the small area enclosed rectangle for the workspace shape.
2. Compute the obtained bounding rectangle orientation.
3. Perform a point sampling of the field according to sensor coverage range.
4. Compute the motion pattern with a determined orientation.

The pseudocode of this procedure is given in Algorithm 33 [144].

Algorithm 33 Signal Searching Algorithm

1. $r \leftarrow$ Effective area to cover (polyline)
2. $\Phi \leftarrow$ Extract orientation (r)
3. $x \leftarrow$ Way-point sampling $(r$, sensor range)
4. $p \leftarrow$ Motion pattern (x, Φ)
5. Return p

If an unshaped field is considered, the main direction used in the discretization process becomes highly relevant for reducing the number of cells to be visited by the UAV during the mission. Indeed, the area to cover might be optimized if this angle is considered as exploring orientation. Moreover, this angle can have an influence on the mission definition because it can be used together with the navigation information for guidance purposes. That is, it allows for smoother paths, and it ensures that the UAV goes from way-point to way-point in a straight line. The **small area enclosed rectangle** problem refers to finding the smallest rectangle (i.e., the minimum area rectangle) that encloses a polygon regardless of its shape. This approach can be solved in $O(n)$ with methods such as rotating calipers or minimum bounding rectangle. The UAV's heading during the searching is obtained from the small area enclosed rectangle main direction [302].

3.4.3.4 Distributed Coverage

Distributed coverage control is widely studied on continuous domains. A multi-UAV spanning tree coverage formulation can be applied over a general cell-based representation of free space. Others have applied genetic algorithms and visual landmarks to improve the speed of coverage. An information-theoretic path planner on a hexagonal grid representation of free space was also investigated, and a variant of the CPP solved the problem of boundary coverage, in which the objective was to cover the immediate area around the boundary of obstacles within the environment. Another possible approach is to employ potential fields to drive each UAV away from the nearby UAVs and obstacles. Alternatively, a prevailing approach is to model the underlying locational optimization problem as a continuous p-median problem and to employ Lloyd's algorithm as presented in Algorithm 34. As such, the UAVs are driven onto a local optimum, i.e., a centroidal Voronoi configuration, where each point in the space is assigned to the nearest agent, and each UAV is located at the center of mass of its own region. Later on, this method was extended for UAVs with distance-limited sensing and communications and limited power, as well as for heterogeneous UAVs covering nonconvex regions. Also, the requirement of sensing density functions was relaxed by incorporating methods from adaptive control and learning. Distributed coverage control was studied on discrete spaces represented as graphs. One possible approach is to achieve a centroidal Voronoi partition of the graph via pairwise gossip algorithms or via asynchronous greedy updates.

Definition 135. *The **Voronoi region** V_i^* of a sensor is defined by all points that are* closer *in the sense of the considered distance measure to that sensor than any other. For the Euclidean distance measure, the Voronoi region V_i associated with its generator p_i is as follows:*

$$V_i = \{q \in \mathbf{Q}, \|q - p_i\| \leq \|q - p_i\|, \forall j \neq i\} \tag{3.48}$$

Definition 136. *The **Voronoi partition** V_i^* of agent i, for the anisotropic case, is*

$$V_i^* = \left\{q \in \mathbf{Q}, \|q - p_i\|_{L_i} \leq \|q - p_i\|_{L_i}, \forall j \neq i\right\} \tag{3.49}$$

The anisotropic Voronoi partition is determined not only by the sensor's position but also by the sensor's orientation [158]

Definition 137. ***Centroidal Voronoi configuration:*** *Given the set of points $\mathbf{P} \in \mathbf{Q}$, $C_{V_i^*}$ is the center of mass (centroid) of an anisotropic Voronoi partition. A Voronoi tessellation is called an anisotropic centroidal Voronoi configuration if*

$$p_i = C_{V_i}; \forall i \tag{3.50}$$

That is, the points P serve as generators and also centroids for the anisotropic Voronoi tessellations.

One problem with a simple centroidal Voronoi transformation algorithm is that it requires a convex and obstacle-free environment to ensure that the motion is always possible to the centroids; otherwise, it may not converge to a steady state.

Alternatively, distributed coverage control on discrete spaces can be studied in a game theoretic framework. Sensors with variable footprints can achieve power-aware optimal coverage on a discretized space, and a team of heterogeneous mobile agents can be driven on a graph to maximize the number of covered nodes [312]. A multi-UAV coverage solution for a rectilinear polygonal environment focused on assigning partitions of the environment proportionally based on each UAV's capabilities. The algorithm divided the free space into simple regions, and focused on selecting a per-region coverage pattern that minimized the number of turns.

Algorithm 34 Lloyd's Method

1. Given a polygon P and a set of k generating points at positions z_i.
2. Loop
   a. Construct the Voronoi diagram $\{V_i\}$ for the points $\{z_i\}$.
   b. Compute the centroid $\{c_i\}$ for each Voronoi cell $\{V_i\}$.
   c. Set each point $\{z_i\}$ to the position of its associated centroid $\{c_i\}$.
   d. If this new set of points meets some convergence criterion, terminate.
3. end Loop

3.4.3.4.1 Optimal Coverage

Optimal coverage algorithms for general time-varying density functions can be used to influence a team of UAVs, making possible human operators interaction with large teams of UAVs. The density function could represent the probability of a lost person being at a certain point in an area, for a search and rescue mission. In order to talk about optimal coverage, a cost ϕ must be associated with a UAV configuration that describes how well a given area is being covered.

Problem 138. *The coverage problem involves placing n UAVs in $D \subset \mathbb{R}^2$ where $p_i \in D, i = 1, \ldots, n$ is the position of the ith UAV. Moreover, the domain itself is divided into regions of dominance $P_1, \ldots, P_n$, forming a proper partition of D, where the ith UAV has to cover P_i. The associated density function $\phi(q,t)$ assumed to be bounded and continuously differentiable in both arguments captures the relative importance of a point $q \in D$ at time t.*

For non-slowly varying density function, timing information must be included in the motion of the UAVs, for example, as in [221]:

$$\frac{d}{dt}\left(\sum_{i=1}^{n}\int_{v_i}\|q-c_i\|^2\,\phi(q,t)dq\right)=0 \tag{3.51}$$

The mass m_i and center of mass c_i of the ith Voronoi cell V_i is

$$m_{i,t}(p,t)=\int_{V_i}\dot{\phi}(q,t)dq$$

$$c_{i,t}=\frac{1}{m_{i,t}}\left(\int_{V_i}q\dot{\phi}(q,t)dq\right) \tag{3.52}$$

The solution can be given by:

$$\dot{p}_i=c_{i,t}-\left(k+\frac{m_{i,t}}{m_{i,0}}\right)(p_i-c_{i,0}) \tag{3.53}$$

3.4.3.4.2 Zermelo-Voronoi Diagram

In many applications of UAV, significant insight can be gleaned from data structures associated with **Voronoi-like partitioning** [33]. A typical application can be the following: given a number of landing sites, divide the area into distinct non-overlapping cells (one for each landing site) such that the corresponding site in the cell is the closest one (in terms of time) to land for any UAV flying over this cell in the presence of winds. A similar application that fits in the same framework is the task of subdividing the plane into **guard/safety** zones such that a **guard/rescue aircraft** residing within each particular zone can reach all points in its assigned zone faster than any other guard/rescuer outside its zone. This is the **generalized minimum distance problems** where the relevant metric is the minimum intercept or arrival time. Area surveillance missions can also be addressed using a frequency-based approach where the objective implies to optimize the elapsed time between two consecutive visits to any position known as the refresh time [5].

The construction of **generalized Voronoi diagrams** with time as the distance metric is in general a difficult task for two reasons:

1. The distance metric is not symmetric, and it may not be expressible in closed form.
2. Such problems fall under the general case of partition problems for which the UAV's dynamics must be taken into account.

It is assumed that the UAV's motion is affected by the presence of temporally varying winds. Since the generalized distance of this Voronoi-like partition problem is the minimum time to go of the Zermelo problem, this partition of the configuration space is known as **Zermelo–Voronoi diagram** (ZVD). This problem deals with a special partition of the Euclidean plane with respect to a generalized distance function. The characterization of this Voronoi-like partition takes into account the proximity relations between a UAV that travels in the presence of winds and the set of Voronoi generators. The question of determining the generator from a given set which is the closest in terms of arrival time, to the agent at a particular instant of time, reduces the problem of determining the set of the Zermelo–Voronoi partition that the UAV resides in at the given instant of time: this is the **point location problem**. The **dynamic Voronoi diagram problem** associates the standard Voronoi diagram with a time-varying transformation as in the case of time-varying winds. The **dual ZVD** problem leads to a partition problem similar to the ZVD with the difference that the generalized distance of the dual ZVD is the minimum time of the Zermelo problem from a Voronoi generator to a point in the plane. The minimum time of the Zermelo navigation problem is not a symmetric function with respect to the initial and final configurations.

3.4.3.4.3   Environment with Polygonal Obstacles

Solutions of the coverage control problem based on partitioning the mission space overlook the fact that the overall sensing performance may be improved by sharing the observations made by multiple sensors. In addition, many approaches assume uniform sensing quality and an unlimited sensing range. A number of solution techniques are also based on a centralized controller, which is inconsistent with the distributed communication and computation structure of sensor networks. Moreover, the combinatorial complexity of the problem constrains the application of such schemes to limited-size networks. Finally, another issue that appears to be neglected is the movement of sensors, which not only impacts sensing performance but also influences wireless communication: because of the limited on-board power and computational capacity, a sensor network is required not only to sense but also to collect and transmit data as well. For this reason, both sensing quality and communication performance need to be jointly considered when controlling the deployment of sensors.

Problem 139. *Coverage with polygonal obstacles: The mission space $\Omega \subset \mathbb{R}^2$ is a non-self-intersecting polygon. The mission space may contain obstacles which can interfere with the movement of the sensor nodes and the propagation of event signals. The boundaries of these obstacles are modeled as m non-self-intersecting polygons properly contained in Ω.*

$$\max_{s} \int_{\Omega} R(x)P(x,s)dx$$

subject to $s_i \in F, i = 1, \ldots, N$ where

$$P(x,s) = 1 - \prod_{i=1}^{N} \hat{p}_i(x,s_i)$$

and

$$\hat{p}_i(x,s_i) = \left\{ \begin{array}{ll} p_i(x,s_i) & \text{if } x \in V(s_i) \\ \tilde{p}_i(x,s_i) & \text{if } x \in \bar{V}(s_i) \end{array} \right\} \tag{3.54}$$

with

$$p_i(x,s_i) = p_0 \exp^{-\lambda_i \|x-s_i\|}$$

and $\tilde{p}_i(x,s_i) \leq p_i(x,s_i)$

A gradient-based motion control scheme is developed in [314] to maximize the joint detection probability of random events in such mission spaces, taking into account the discontinuities that are introduced by obstacles in the sensing probability models. The optimization scheme requires only local information at each node. A modified objective function is also proposed, which allows a more balanced coverage when necessary.

3.4.3.4.4   Area Coverage in Non-Convex Environment

Non-convex domains pose non-convex optimization problems with non-convex constraints. An approach based on the Lloyd algorithm and the tangent bug algorithm is presented in [66]. The tangent bug algorithm is a local path planner with obstacle avoidance behavior. The control strategy is composed of two layers of abstraction:

1. The Lloyd algorithm provides goal updates based on successive computation of Voronoi regions and their centroids on the upper level. The geodesic distance helps in keeping the UAV inside the environment region on its way to the target. The geodesic distance measure calculates the paths along the boundaries and avoids the obstacles.

2. The tangent bug algorithm plans the UAV path to the next centroid target position on the lower level. It is a simple but efficient sensor-based planner capable of handling unknown environments by using a range sensor. This algorithm shows both characteristic behaviors:
   a. Motion-toward-target, a form of gradient descent.
   b. Boundary-following, a form of exploration of obstacle boundary.

The description of the implemented navigation algorithm is detailed in Algorithm 35, while the tangent bug subroutine is given in Algorithm 36. The proposed control strategy computes the Lloyd algorithm using only the virtual generators.

Algorithm 35 Coverage for Non-Convex Environment

1. Set of UAVs $i = 1, \ldots, n$ with initial positions p_i^s in environment Q and each UAV is provided with:
   a. Localization and knowledge of Q
   b. Voronoi region computation
   c. subroutine tangent bug algorithm
2. initialize at time $T_i^s = 0, g_i^{\text{real}} \leftarrow p_i^s, g_i^{\text{virt}} \leftarrow p_i^s$
3. Loop

   a. Acquire positions p_i and $\left\{ g_j^{\text{virt}} \right\}_{j=1}^{k}, j \neq i$ of k neighbors
   b. Construct the local Voronoi region V_i associated with g_i^{virt}
   c. Compute the mass centroid C_{V_i} of the Voronoi region, and update virtual target position $t_i^{\text{virt}} \leftarrow C_{V_i}$
   d. Run tangent bug algorithm
4. end Loop
5. compute the final Voronoi region associated with g_i^{real}

Algorithm 36 Tangent Bug

1. Set of n UAVs in environment Q and each UAV $i = 1, \ldots, n$ is provided with:
   a. Obstacle avoidance: sensing and computation
   b. Virtual target t_i^{virt} and $var \leftarrow t_i^{\text{virt}}$
2. Loop 2
   a. If V_i is an obstacle boundary Voronoi region, then
   b. project t_i^{virt} to point p_i^* onto ∂Q and set $var \leftarrow p_i^*$
   c. end if
   d. Update real target position $t_i^{\text{real}} \leftarrow var$
   e. Execute next motion step toward real target t_i^{real}, apply obstacle avoidance to drive to next position p_i
   f. Update real generator position $g_i^{\text{real}} \leftarrow p_i$
   g. simulate next motion step toward virtual target t_i^{virt} and update virtual generator position g_i^{virt}
3. end Loop
4. return virtual generator g_i^{virt}

In distributed mobile sensing applications, networks of agents that are heterogeneous respecting actuation as well as body and sensory footprint are often modeled by recourse to power diagrams: generalized Voronoi diagrams with additive weights. In [25], power diagrams are used for identifying collision-free multi-robot configurations, and a constrained optimization framework is proposed combining coverage control and collision avoidance for fully actuated disk-shaped robots.

Most weighted Voronoi cells assume the correct weightings are known a priori. In [246], an algorithm is proposed to adapt trust weightings online using only comparisons between UAV's sensor measurements and those of its neighbors. A measure of sensor discrepancy is integrated into a cost function for the team, and used to derive an adaptation law for each UAV to change its trust weightings on line, while simultaneously performing a Voronoi-based coverage control algorithm. The weightings serve as an adaptive way to assess trust between UAVs and improve the overall sensing quality of the team.

Given a convex- or non-convex-shaped area $A \subset \mathbb{R}^2$ decomposed approximately by a finite set of regular cells $C = \{c_1, \ldots, c_n\}$ such that $A \approx \bigcup_{c \in C} c$; a coverage trajectory P with a finite set of continuous way-points p, which can be written as $P = \bigcup_{p \in P} p$, where the way-points correspond to the centroid of a corresponding cell, and consequently, a cell corresponds to an image sample; thus, $\dim(P) = \dim(C)$, a team of UAVs with attitude and position control and capable of way-point navigation, each UAV is characterized by a position in $[X, Y, Z]$ and orientation $[\phi, \theta, \psi]$. The variable of interest to minimize is the number of turns performed in P, which corresponds to the number of rotations made by a UAV around the z-axis (yaw movements). The area coverage problem oriented to mosaicking missions can be abstractly described as follows [301]:

Problem 140. *For each quad-rotor of the team, an optimal trajectory can be computed by:*

$$min_\psi \left(J(\psi) = K_1 \times \sum_{i=1}^m \psi_k^i + K_2 \qquad k \in \{0, \pi/4, \pi/2, 3\pi/4\} \right) \qquad (3.55)$$

where $\psi_{\pm 3\pi/4} > \psi_{\pm \pi/2} > \psi_{\pm \pi/4} > \psi_0$ and $K_i \in \mathbb{R}$ are weights such that $K_2 > K_1$.

The geometric nature of the coverage problem necessitates the proper partitioning of the space among the nodes, using a proper metric so that each node's control action should be dependent on the part of the space instead of requiring global knowledge of the state of the network. Assuming homogeneous sensor networks, the region A can be partitioned and assigned among the n nodes into polygonal cells, $V_i, i = 1, \ldots, n$. The tessellation is based on the standard Euclidean metric defined as:

$$V_i = \{q \in A; \|q - p_i\| \leq \|q - p_j\|, j = 1, \ldots, n\} \qquad (3.56)$$

indicating assignment of points in A to the nearest node in the Euclidean sense.

Definition 141. *The **geodesic Voronoi partitioning** is defined as:*

$$V_i^g = \{q \in A; d(q - p_i) \leq d(q - p_j), j = 1, \ldots, n\} \qquad (3.57)$$

It assigns parts of the space under surveillance among the nodes based on the geodesic rather than the Euclidean distance. In general, the boundaries of the geodesic Voronoi cells are comprised of line segments and hyperbolas. The geodesic Voronoi diagram consists of a tessellation of the environment A, while each geodesic Voronoi cell V_i^g is always a compact set, contrary to the cells produced by the Euclidean ones, when dealing with non-convex environments [293].

3.4.3.4.5 Expanding Grid Coverage

In this paragraph, an approach is considered to coordinate a team of UAVs without a central supervision, by using only local interactions between the UAVs [17]. When this decentralized approach is used, much of the communication overhead is saved, the hardware of the UAVs can be simple and better modularity is achieved. A properly designed system should be readily scalable, achieving **reliability** and **robustness** through **redundancy**. A team must cover an unknown region in the grid that possibly expands over time. This problem is strongly related to the problem of distributed search after mobile or evading targets. In general, most of the techniques used for the task of a distributed coverage use some sort of cellular decomposition [208].

In the **cooperative cleaners** case study, a team of robots cooperate in order to reach the common goal of cleaning a dirty floor. A cleaning robot has a bounded amount of memory; it can only observe the state of the floor in its immediate surroundings and decide on its movement based on these observations. Consequently, the overall topology of the floor contamination is unknown to the robots. The robots use an indirect means of communication based on signals and sensing, and the desired goal of cleaning the whole floor is thus an emerging property of multi-robot cooperation.

1. In the **static cooperative cleaners problem**, the dirty shape of the floor does not grow due to spreading of contamination. The shape of the floor is a region in $\mathbb{R}^2$ represented as an undirected graph G. Let V be the set of vertices in G. Each element in V is a tile of the floor and is represented as a pair $v = (x, y)$. Let E be the set of edges in G, and each edge is a pair of vertices (v, w) such that v and w are connected through a 4-neighbor relation. The dirty floor F_t is a subgraph of G, where t represents the time. In the initial state, G is assumed to be a single connected component without holes or obstacles, and $F_0 = G$. All the cleaning robots are identical, and there is no explicit communication between the robots (only broadcast and sensing actions in the local environment are allowed). All the cleaning robots start and finish their task in the same vertex. In addition, the whole system should support fault tolerance: even if almost all the robots cease to work before completion of the mission, the remaining ones will eventually complete the mission [291].

2. In the **dynamic cooperative cleaners problem**, the time is discrete. Let the undirected graph $G(V, E)$ denote a 2D integer grid $\mathbb{Z}^2$ whose vertices have a binary property called contamination. Let $\text{Cont}_t(v)$ be the contamination state of the vertex v at time t, taking either the value 1 or 0. Let F_t be the contamination state of the vertex at time t,

$$F_t = v \in G | \text{Cont}_t(v) = 1 \tag{3.58}$$

F_0 is assumed to be a single connected component. This algorithm will preserve this property along its evolution. Let a team of k UAVs that can move on the grid G be placed at time t_0 on F_0 at point $P_0 \in F_t$. Each UAV is equipped with a sensor capable of telling the contamination status of all vertices in the digital sphere of diameter 7. In all the vertices, their **Manhattan distance** from the UAV is equal to or smaller than 3. A UAV is also aware of other UAVs that are located in these vertices, and all of them agree on a common direction. Each vertex may contain any number of UAVs simultaneously. When a UAV moves to a vertex v, it has the possibility of cleaning this tile (i.e., causing $\text{Cont}(v)$ to become 0). The UAVs do not have any prior knowledge of the shape or size of the subgraph F_0 except that it is a single and simply connected component. The contaminated region F_t is assumed to be surrounded at its boundary by a rubber-like elastic barrier, dynamically reshaping itself to fit the evolution of the contaminated region over time. This barrier is intended to guarantee the preservation of the simple connectivity of F_t crucial for the operation of the UAVs, due to their limited memory. When a UAV cleans a contaminated vertex, the barrier retreats in order to fit the void previously occupied by the cleaned vertex. In every step, the contamination spreads; that is, if $t = nd$ for some positive integer n, then

$$\forall v \in F_t, \forall u \in \{4 - \text{Neighbors}(v)\}, \text{Cont}_{t+1}(u) = 1 \tag{3.59}$$

Here, the term $4 - \text{Neighbors}(v)$ simply means that the four vertices are adjacent to vertex v. While the contamination spreads, the elastic barrier stretches while preserving the simple connectivity of the region. For the UAVs traveling along the vertices of F, the barrier signals the boundary of the contaminated region. The UAV's goal is to clean G by eliminating the contamination entirely. No central control is allowed and the system is fully decentralized: all the UAVs are identical, and no explicit communication between the UAVs is allowed. An important advantage of this approach in addition to the simplicity of the UAVs is **fault tolerance**; even if some UAVs are no longer in good flying condition before completion, the remaining ones will eventually complete the mission, if possible. A cleaning algorithm is proposed in [17], for exploring and cleaning an unknown contaminated sub-grid F, expanding every d time steps. This algorithm is based on a constant traversal of the contaminated region, preserving the connectivity until the region is cleaned entirely. Until the conditions of completion of mission are met, each UAV goes through the following sequence of commands. The pseudocode is presented in Algorithm 37.

Algorithm 37 Cleaning Algorithm

1. First, each UAV calculates its desired destination at the current time.
2. Then, each UAV calculates whether it should give a priority to another UAV located at the same vertex and wishes to move to the same destination.
3. When two or more UAVs are located at the same vertex and wish to move toward the same direction, the UAV that had entered the vertex first gets to leave the vertex, while the other UAVs wait.
4. Before actually moving, each UAV that had obtained a permission to move must now locally synchronize its movement with its neighbors, in order to avoid simultaneous movements that may damage the connectivity of the region.
5. When a UAV is not delayed by any other agent, it executes its desired movement.

Remark 142. *Any waiting UAVs may become active again if the conditions change. More details on this implementation can be found in [17].*

3.4.3.4.6 Coverage Control

Three related problems of UAV movement in arbitrary dimensions are coverage, search and navigation. A UAV is asked to accomplish a motion-related task in an unknown environment whose geometry is learned by the UAV during navigation [68]. In the standard coverage control problem, the goal of the UAV team is to reach asymptotically a configuration where the agent positions $\lim_{t\to\inf} p_i(t); i \in [n]$ minimize the following performance measure capturing the quality of coverage of certain events:

$$E_n(p) = E\left[\min_{i\in[n]} f(\|p_i - Z\|)\right] \tag{3.60}$$

where $f : \mathbb{R}_{\geq 0} \to \mathbb{R}_{\geq 0}$ is an increasing continuously differentiable function. The random variable Z represents the location of an event of interest occurring in the workspace. To interpret relation (3.60), the cost of servicing an event at location z with a UAV at location π is measured by $f(\|p_i - Z\|)$, and an event must be serviced by the UAV closest to the location of this event. For example, in monitoring applications, $f(\|p_i - Z\|)$ can measure the degradation of the sensing performance with the distance to the event. In VRPs, this cost might be the time it takes a UAV to travel to the event location, i.e., $f(\|p_i - Z\|) = \|p_i - Z\|/v_i$, assuming enough time between successive events [222]. Typically, partitioning is done so as to optimize a cost function, which measures the quality of service provided over all of the regions. Coverage control additionally optimizes the positioning of UAVs inside a region. Many algorithms for partitioning and coverage control in robotic networks build on Lloyd's algorithm on optimal quantizer selection through **centering and partitioning**. The basic pseudocode is presented in Algorithm 34. There are also multiagent partitioning algorithms built on **market principles** or auctions. Distributed Lloyd methods are built around separate partitioning and centering steps [113].

3.4.3.4.7   Way-Point Coverage of Time-Varying Graphs

Dynamic graphs represent the UAV's highly dynamic networks. The **dynamic map visitation problem** (DMVP) is considered, in which a team of UAVs must visit a collection of critical locations as quickly as possible, in an environment that may change rapidly and unpredictably during the UAVs' navigation. The DMVP applies formulations of highly dynamic graphs or **time-varying graphs** (TVGs), to a graph navigation problem. When incorporating dynamics into DMVP, there are many options for how to constrain or model the dynamics of the graph. Dynamics can be deterministic or stochastic. The deterministic approach is also relevant for situations in which some prediction of changes is feasible, and the graph must be connected at all times. Indeed, for complete map visitation to be possible, every critical location must be eventually reachable. However, in UAVs' application environments, at any given time, the way-point graph may be disconnected. There are three classes of TVGs, each of which places constraints on edge dynamics: edges must reappear eventually, edges must appear within some time bound, and edge appearances are periodic. These classes have proven to be critical to the TVG's taxonomy.

Definition 143. *A **dynamic graph** is a five-tuple $G = (V, E, \tau, \rho, \Xi)$, where $\tau \subset \mathbb{T}$ is the lifetime of the system, the presence function $\rho(e,t) = 1$ means that edge $e \in E$ is available at time $t \in \tau$, and the latency function $\Xi(e,t)$ gives the time it takes to cross e if starting at time t. The graph $G = (V,E)$ is called the underlying graph of G, with $|V| = n$.*

The discrete case in which $\mathbb{T} = \mathbb{N}$, edges are undirected is considered, and all edges have uniform travel cost $\Xi(e,t) = 1$ at all times. If agent a is at u, and edge (u,v) is available at time τ, then agent a can take (u,v) during this time step, visiting v at time $\tau + 1$. As a traverses G, a both visits and covers the vertices in its traversal. A temporal subgraph of a TVG G results from restricting the lifetime τ of G to some $\tau' \subset \tau$. A static snapshot is a temporal subgraph throughout which the availability of each edge does not change; i.e., edges are static.

Definition 144. $J = \{(e_1, t_1), \ldots, (e_k, t_k)\}$ *is a **journey** $\Leftrightarrow$ $\{e_1, \ldots, e_k\}$ is a walk in G (called the **underlying walk** of J), $\rho(e_i, t_i) = 1$ and $t_{i+1} \geq t_i + \Xi(e_i, t_i)$ for all $i < k$. The **topological length** of J is k, the number of edges traversed. The temporal length is the duration of the journey: (arrival date) - (departure date).*

Problem 145. *Given a dynamic graph G and a set of starting locations S for k agents in G, the TVG foremost coverage or **DMVP** is the task of finding journeys starting at time 0 for each of these k agents such that every node in V is in some journey, and the maximum temporal length among all k journeys is minimized. The decision variant asks whether these journeys can be found such that no journey ends after time t. Let $T = \sum_{i=1}^{m} t_i$.*

For the DMVP minimization problem (G, S) and the corresponding decision problem $(G; S; t)$, input is viewed as a sequence of graphs G_i each represented as an adjacency matrix, with an associated integer duration t_i, i.e., $G = (G_1, t_1), (G_2, t_2), \ldots, (G_m, t_m)$, where G_1 appears initially at time zero [1].

3.4.3.4.8   Multi-UAV Persistent Coverage

Persistent coverage differs from static and dynamic coverage in that the coverage of the environment persistently decays and the UAVs have to continually move to maintain the desired level; i.e., it requires repetition and redundant actions. Therefore, in this case, the task can generally never be completed. Dynamic coverage is defined by the use of a mobile sensor network and results from the consistent mobility of sensors. While sensors move around, locations that were uncovered at the beginning will be covered at a later time; therefore, a wider area is covered over time, and intruders that might never be discovered in a fixed sensor network can now be detected by mobile sensors.

Remark 146. *The main difference between multiagent persistent monitoring and dynamic coverage lies in that dynamic coverage task is completed when all points attain satisfactory coverage level, while the persistent monitoring would last forever.*

The **coverage level** can be seen as the quality of a measurement, and in this sense, persistent coverage is often addressed as persistent surveillance or environmental monitoring, especially with UAVs. This approach is more dynamic, flexible and suitable for multiple source localization, but requires resolving many challenging technical problems such as endurance, planning, coordination, communication, cooperation and navigation of all the vehicles [38]. The solutions to the persistent coverage problem intend to derive results, which are applicable for infinite time. In [243], this problem is formulated in discrete time due to the fact that a distributed system requires discrete communications. The UAVs form a network defined by a communication graph $G^{\text{com}}(k) = (V(k), E(k))$. The vertices $V(k)$ of the graph are the positions $p_i(k) \in Q$ of UAV $i, i = 1, \ldots, N$ at time $k \geq 1$. An edge $(i,j) \in E(k)$ if $\|p_i(k) - p_j(k)\| \leq r^{\text{com}}$, where r^{com} is the communication radius, i.e., the maximum distance between two UAVs at which they can communicate. The neighbors of the UAV i at instant k are $N_i(k) = \{j = 1, \ldots, N; (i,j) \in E(k)\}$. The coverage of the environment, or coverage function or global map, is modeled with a time-varying field $Z(q,k)$.

Problem 147. *Persistent coverage: Let $Q \subset \mathbb{R}^2$ be a bounded environment to be persistently covered by a team of N UAVs, assumed to be holonomic:*

$$p_i(k) = p_i(k-1) + u_i(k-1) \quad \text{where } \|u_i(k)\| \leq u^{\text{max}}$$

The aim of the UAV team is to maintain a desired coverage level $Z^*(q) > 0, \forall q \in Q$. The coverage function at each time instant k is given by Equation (3.61):

$$Z(q,k) = d(q)Z(q,k-1) + \alpha(k) \tag{3.61}$$

where $0 < d(q) < 1$ is the decay gain. The UAV increases the value of the coverage by $\alpha(k)$.

In the first step, at each communication time k, each UAV generates its map-to-communicate $Z_i^{\text{com}}(k)$ as:

$$Z_i^{\text{com}}(k) = dZ_i(k) + \alpha_i(k) \tag{3.62}$$

Each UAV sends its map-to-communicate to its neighbors and receives their maps. With this information, the first step of the update is performed by dividing the map into two parts: the coverage area $\Omega_i(k)$ and the rest of the map. The UAV updates each region according to the following equation:

$$
\begin{aligned}
Z_i^-(k) &= Z_i^{\text{com}}(k) + \sum_{j \in N_i(k)} \left(\max(Z_j^{\text{com}}(k) - dZ_i(k-1), 0) \right) \quad \forall q \in \Omega_i(k) \\
Z_i^-(k) &= \max \left(Z_j^{\text{com}}(k), dZ_i(k-1) \right) \quad \forall q \notin \Omega_i(k)
\end{aligned}
\tag{3.63}
$$

To counteract the error of estimation, a second updating step is executed. At first, each UAV extracts the region of its coverage area that is overlapped with another UAV and then sends its coverage function in this region to its neighbors:

$$\beta_i(k) = \beta_i(q, p_i(k)) = \alpha_i(q, p_i(k)) \quad \forall q \in \Omega_i^0(k) \tag{3.64}$$

where $\Omega_i^0(k) = \{q \in \Omega_i(k) \bigcup \Omega_j(k) | j \in N_i(k)\}$ is the overlapped area of UAV i with its neighbors. The UAVs exchange the overlapped productions with their neighbors, and with the received ones, they perform the final update:

$$
Z_i(k) = \left\{
\begin{array}{ll}
Z_i^-(k) & \forall q \in \Omega_i(k) \\
Z_i^-(k) - \max_{j \in N_i(k)} b_j(k) + \sum_{j \in N_i(k)} \beta_j(k) & \forall q \notin \Omega_i(k)
\end{array}
\right\}
\tag{3.65}
$$

The final step adds the contributions that are not first considered and ends the estimation, as shown in Algorithm 38.

Algorithm 38 Local Map Update

1. Calculate map-to-communicate $Z_i^{com}(k)$ with relation (3.62)
2. Communicate map to neighbors
3. Update local map $Z_i^-(k)$ with relation (3.63)
4. Extract overlapped production $\beta(q, p_i(k))$ with relation (3.64)
5. Communicate region to neighbors
6. Update local map $Z_i(k)$ with relation (3.65)

The problem of dispatching UAVs with cameras to monitor road traffic in a large city can be considered. Often, UAVs have limited range and can stay in air only for a limited amount of time. Traffic events such as congestion tend to have strong local correlations; i.e., if the vehicle density at an intersection is high, the same is likely true at intersections that are close-by. Therefore, sequentially visiting intersections following the road network's topological structure may offer little incremental information. As UAVs are not restricted to travel along roads, routes with carefully selected, not necessarily adjacent, intersections can potentially offer much better overall traffic information per unit of traveled distance. Under such settings, the following question then naturally arises: how to plan the best tours for the UAVs so that they can collect the maximum amount of traffic information per flight? Due to spatial and temporal variations, such fields can be highly complex and dynamic. However, in applications involving large spatial domains, the underlying spatial domain often does not change. The observation allows to work with the premise that nearby nodes have mostly time-invariant spatial correlations, even though the overall field may change significantly over time. Exploiting these correlations, at any given time, it becomes possible to infer the field's value at a certain node from the values of adjacent nodes [313].

In [275], a multi-vehicle sampling algorithm generates trajectories for non-uniform coverage of a nonstationary spatio-temporal field characterized by varying spatial and temporal de-correlation scales. The sampling algorithm uses a nonlinear coordinate transformation that renders the field locally stationary so that existing multi-vehicle control algorithm can be used to provide uniform coverage. When transformed back to original coordinates, the sampling trajectories are concentrated in regions of short spatial and temporal de-correlation scales. Employing a 2D planner, a 3D structure is approximated using multiple 2D layers, treated individually.

3.4.3.4.8.1 Continuous target coverage Design of coordination algorithms for non-convex environments requires tackling several issues such as signal attenuation or visibility loss because of the obstacles. Different kinds of sensing devices have been classified in Euclidean/Geodesic footprint (non-visibility) and visibility sensors. A transformation of non-convex domains into convex ones was one of the first techniques. Then, the real trajectories of the UAVs are obtained through the inverse transformation. For a frontier-based exploration of a non-convex environment, the latter is transformed into a star-shaped domain. Another solution based on Lloyd's algorithm and a path planning method for deploying a group of nodes in a concave environment can also be presented. However, this method is not effective for all types of environments, since it maximizes the coverage of the convex hull of the allowable environment rather than the coverage of the environment itself. In another solution, each UAV moves at a direction determined by the repulsive forces received by other UAVs and/or obstacles. However, this control scheme may lead to sub-optimal topologies. This fact motivated the incorporation of an attractive force to the centroid of the respective Voronoi cell. The geodesic Voronoi diagram allows deployment in non-convex environments. The algorithm assumes that the sensing performance degrades according to the square of the geodesic, rather than the Euclidean distance. In another control strategy, each node is assumed to move to the projection of the geometric centroid of its geodesic Voronoi cell onto the boundary of the environment, the domain of interest considered is unknown, and therefore, an **entropy metric** is used as a density function that allows the nodes to explore and cover the area at the same time. However, in the field of area coverage in non-convex environments, the visibility-based Voronoi diagram was introduced and an algorithm based on Lloyd's algorithm is implemented for a team of UAVs with unlimited range, omnidirectional visibility-based sensors. The 3D case of that problem is also addressed for a team of UAVs equipped with omnidirectional visibility sensors of infinite sensing range. Non-smooth optimization techniques for the coordination of a homogeneous UAV swarm by minimizing the sensing uncertainty, given that the performance of the visibility-based sensors is reduced according to the square of the distance [202].

3.4.3.4.8.2 Partition of environment coverage The distributed territory partitioning problem for UAV networks consists of designing individual control and communication laws such that the team will divide a space into territories. Typically, partitioning optimizes a cost function, which measures the quality of service provided by the team. Coverage control additionally optimizes the positioning of UAVs inside a territory. A distributed coverage control algorithm can be described for a team of UAVs to optimize the response time of the team to service requests in an environment represented by a graph. Optimality is defined with reference to a cost function, which depends on the locations of the UAVs and geodesic distances in the graph. As with all multi-UAV coordination applications, the challenge comes from reducing the communication requirements: the proposed algorithm requires only gossip communication, i.e., asynchronous and unreliable pairwise communication [112].

Problem 148. *Given a team of N UAVs with limited sensing and communication capabilities, and a discretized environment, partition the environment into smaller regions and assign one region to each agent. The goal is to optimize the quality of coverage, as measured by a cost function, which depends on the current partition and the positions of the agents.*

Coverage algorithms for UAV deployment and environment partitioning can be described as dynamical systems on a space of partitions. **Centering and partitioning** Lloyd's algorithm is an approach to facility location and environment partitioning problems. The Lloyd's algorithm computes centroidal Voronoi partitions as optimal configurations of an important class of objective functions, called multicenter functions.

Definition 149. *An **N-partition** of Q, denoted by $v = (v_i)_{i=1}^N$, is an ordered collection of N subsets of Q with the following properties:*

1. $\cup_{i\in\{1,...,N\}} v_i = Q$
2. *$int(v_i) \cap int(v_j)$ is empty for all $i, j \in \{1,...,N\}$ with $i \neq j$ and*
3. *each set $v_i, i \in \{i = 1,...,N\}$, is closed and has non-empty interior*

The set of $N-$partitions of Q is denoted by V_N.

Definition 150. *Let Q be a totally ordered set, and $p_i \in \mathbb{C}$, the **generalized centroid** of p_i, is defined as follows:*

$$Cd(p_i) = min\{argmin_{i=1..N}H_1(h, p_i)\} \tag{3.66}$$

where $H_1(h, p_i)$, the one-center function, is defined as follows:

$$H_1(h, p_i) = \sum_{k \in p_i} d_{p_i}(h, k)\phi_k \tag{3.67}$$

where ϕ_k is a bounded positive weight function.

Let $p = (p_1, ..., p_N) \in Q^N$ denote the position of N UAVs in the environment Q. Given a team of N UAVs and an N-partition, each UAV is naturally in one-to-one correspondence with a component of the partition; v_i is the dominance region of UAV $i \in \{1, ..., N\}$. On Q, a density function is defined to be a bounded measurable positive function: $Q \to \mathbb{R}_{\geq 0}$ and a performance function to be a locally Lipschitz, monotone increasing and convex function $f : \mathbb{R}_{\geq 0} \to \mathbb{R}_{\geq 0}$. With these notions, the multicenter function is defined as follows:

$$H_{\text{multicenter}}(v, p) = \sum_{i=1}^{N} \int_{v_i} f(||p_i - q||)\phi(q)dq \qquad (3.68)$$

This function is well defined because closed sets are measurable. The objective function $H_{\text{multicenter}}$ must be minimized with respect to both the partition v and the locations p.

Remark 151. *The distributed coverage law, based upon Lloyd's algorithm, has some important limitations: it is applicable only to UAV networks with synchronized and reliable communication along all edges of the Delaunay graph.*

In order to generate the Delaunay graph, some representative points are chosen on the boundaries of the computational domain. For this given set of boundary points, there exists a unique triangulation for 2D cases or tetrahedralization for 3D cases. The grid generated in such a way is defined as a Delaunay graph for the given moving grid problem. The graph covers the whole computational domain for the given configuration, including the interior elements. Such triangulation or tetrahedralization is unique, maximizing the minimum angle of a triangle or tetrahedral. The Delaunay graph provides a unique mapping from the given boundary points to a coarse unstructured grid. There exists a pre-determined common communication schedule for all UAVs and, at each communication round, each UAV must simultaneously and reliably communicate its position. Some questions arise, which are as follows:

1. Is it possible to optimize UAVs' positions and environment partition with asynchronous, unreliable and delayed communication?
2. What if the communication model is that of gossiping agents, i.e., a model in which only a pair of UAVs can communicate at any time?

A **partition-based gossip** approach is proposed, in which the robots' positions essentially play no role and where instead dominance regions are iteratively updated as in Algorithm 39. When two agents with distinct centroids communicate, their dominance regions evolve as follows: the union of the two dominance regions is divided into two new dominance regions by the hyperplane bisecting the segment between the two centroids [71].

Algorithm 39 Gossip Coverage Algorithm

For all $t \in \mathbb{Z}_{\geq 0}$, each agent $i \in 1, ..., N$ maintains in memory a dominance region $v_i(t)$. The collection $(v_1(0), ..., v_N(0))$ is an arbitrary polygonal N-partition of Q. At each $t \in \mathbb{Z}_{\geq 0}$, a pair of communicating regions, $v_i(t)$ and $v_j(t)$, is selected by a deterministic or stochastic process to be determined. Every agent $k \notin i, j$ sets $v_k(t+1) := v_k(t)$. Agents i and j perform the following tasks:

1. Agent i transmits to agent j its dominance region $v_i(t)$, and vice versa
2. Both agents compute the centroids $C_d(v_i(t))$ and $C_d(v_j(t))$
3. If $C_d(v_i(t)) = C_d(v_j(t))$, then
4. $v_i(t+1) := v_i(t)$ and $v_j(t+1) := v_j(t)$
5. Else
6. $v_i(t+1) := (v_i(t) \cup v_j(t)) \cup H_{\text{bisector}}(C_d(v_i(t)); C_d(v_j(t)))$
7. $v_j(t+1) := (v_i(t) \cup v_j(t)) \cup H_{\text{bisector}}(C_d(v_j(t)); C_d(v_i(t)))$

3.4.3.4.9   *Detection of Coverage Holes*

Delaunay triangulation can be constructed to discover the topological properties of a network. Delaunay triangulation is an important data structure in computational geometry, which satisfies the empty circle property: for each side in Delaunay triangulation, a circle passing through the end points of this side can be determined without enclosing other points [219]. The proposed method of detecting and localizing coverage holes consists of four phases, which are as follows:

1. **Detection of coverage holes:** Every node detects whether coverage holes exist around it based on a hole detection algorithm.
2. **Merging of coverage holes:** A merging method of holes can be provided to present the global view of a coverage hole by indicating the location and shape of an isolated coverage hole.
3. **Size estimation of local coverage holes:** The inscribed empty circles are used for the estimation of the size of every local coverage hole.
4. **Tree description:** For each isolated coverage hole, line segments are used to connect the centers of each pair of inscribed empty circles. If a separated tree can be recognized, the corresponding coverage hole that contains the tree can be exclusively determined.

3.4.3.4.10   *Probabilistic Approach*

Let S be the set of all the nodes representing the potential locations the UAV needs to visit. In each problem instance, only the nodes that belong to a subset s need to be visited. The probability that the set s requires a visit is given by $p(s)$. If an a priori tour τ is available, $L_\tau(s)$ represents the length of the tour in which all the nodes in s are visited according to the order of the a priori tour, skipping nodes not requiring a visit; $E[L_\tau]$ represents the expected length of the a priori tour τ

$$E[L_\tau(s)] = \sum_{s \in S} p(s) L_\tau(s) \tag{3.69}$$

and $E[L_{\text{PTSP}}] = \min_\tau [E[L_\tau]]$ represents the expected length of optimal a priori tour. $L_{\text{TSP}}(s)$ represents the length of the optimal TSP over nodes in s, and $E[\Sigma]$ represents the expected length of the tour produced using a re-optimization technique in which the optimal TSP tour is produced after the problem instance is known:

$$E[\Sigma] = \sum_s p(s) L_{\text{TSP}}(s) \tag{3.70}$$

The optimal probabilistic TSP solution over n nodes is supposed to be known; this is an a priori tour. A parameter $\beta > 1$ is selected. The nodes are clustered according to their order in the a priori tour and their coverage probabilities. The pseudocode of the clustering is given in Algorithm 40, where $G_i, i = 1,..,m$ represents the ith obtained team. After that, the groups must be routed: the median locations $Y_1, Y_2, ..., Y_m$ are the representatives of the groups $G_1, G_2, ..., G_m$, and the **Christofides heuristics** are used to construct a tour over all the representatives $Y_1, Y_2, ..., Y_m$. The Christofides

---

Algorithm 40 Cluster Algorithm

---

1. Let $m = \max\left\{ \left\lfloor \frac{\Sigma_i p_i}{\beta} \right\rfloor, 1 \right\}$ where the floor function $\lfloor x \rfloor$ represents the largest integer not exceeding x
2. If $\Sigma_i p_i \leq \beta$, all the nodes are clustered in one group; otherwise, follow step 3
3. Select k nodes until $\Sigma_i p_i < \frac{1}{m} \Sigma p_i$ and $\Sigma_{i=1}^{j} p_i \geq \frac{1}{m} \Sigma_{i=1}^{j} p_i$. If $\Sigma_i p_i > \frac{1}{m} \Sigma p_i$, X_j is split into two nodes X'_j, X''_j. The coverage probability for X'_k is $\frac{1}{m} \Sigma_i p_i - \Sigma_{i=1}^{j-1} p_i$, while the coverage probability for X''_k is $\frac{1}{m} \Sigma_{i=1}^{j} p_i - \Sigma_i p_i$. Repeat this procedure until m groups of nodes are obtained.

---

heuristic relies on the development of a minimum spanning tree and then solves a matching problem. Then, each node is connected to its representative and forms a loop or uses sweep Algorithms 32 and 40 to construct an a priori tour within each group.

3.5 CONCLUSION

The first part of this chapter presents operational research basics such as the TSP, postperson problem (Chinese and rural) and knapsack problem. These approaches are at the root of orienteering and coverage missions.

The second part of this chapter presents the UAV orienteering problem as a generalization of the TSP and the knapsack problem. This is the problem of routing a UAV over POI.

The third part of this chapter presents the fundamental ideas in the coverage mission. It is how to cover all points in a given barrier/perimeter/area using one or many UAVs with different sets of sensors.

4 Deployment, Patrolling and Foraging

4.1 INTRODUCTION

In this chapter, operations concern generic robotic problems such as **deployment, patrolling** and **foraging**. That is a core area of robotics, artificial intelligence and operations research.

First, about deployment, finding a distribution of a group of UAVs in a known or unknown environment is one of the challenges in multi-robot systems. One challenge is how to balance the workload among UAVs or partition the environment into regions and assign a UAV to each of the regions. The main issue of target coverage is to complete monitoring and information collection of the target coverage through the scheduling and deployment of distributed sensor nodes within the coverage area. Target coverage research includes low power consumption, real time, objective coverage ability, versatility and connectivity of algorithm.

Second, about patrolling, large populations of UAVs are ideal for coverage of a large geographic area. They can maintain coverage of the environment after the dispersion is complete. The size of the environment that can be covered is proportional to the population size. After deployment of the mapping UAVs, controlling the patrolling UAVs amounts to a coverage control problem:

1. How should the UAVs move in order to ensure small worst-case latency in patrolling all areas of the surveyed environment? This may require only simple local strategies that do not involve complicated protocols or computations for coordinating the motion of the mobile components, while still achieving complete coverage, with small latency to surveyed locations. Complete coverage of an unknown region can be achieved by performing a structured exploration by a multi-UAV system with bearing only low-resolution sensors.
2. What local policies should be used for patrolling the cell or region? A natural choice for this task is **least recently visited** approach (LRV), in which each cell keeps track of the time elapsed since its last visit from a patrolling UAV. The patrolling UAV policy directs it to move to the adjacent cell with the smallest latency. This amounts to tracking the visit times of regions [227].

Third, foraging represents the broad class of problems integrating robotic exploration, navigation and object identification, manipulation and transport. In multi-robot systems, foraging is a canonical problem for the study of robot–robot cooperation. The anchoring problem is an important aspect of the connection between symbolic and sensory-based processes in autonomous robotic systems. Anchoring is in fact the problem of how to create, and maintain in time, the connection between the symbol and the signal-level representations of the same physical object.

4.2 AERIAL DEPLOYMENT

The deployment problem considers the number of needed UAVs for a specific situation (e.g., objective, scenario, constraints) and their initial locations. The **deployment problem** is deciding the number of UAVs and where they will be initially located before performing the mission using their control strategy. Among different topics, covering an area is an important objective in a deployment problem. In this problem, the environment is partitioned into regions, and each UAV of the team should be responsible for covering the events happening inside its assigned region. Placing the UAVs in pre-determined regions on the basis of simple geometric structures, such as triangular, square, rhombus and hexagon lattice, is simple when the environment is known [255]. The environment may change over time, and thus, deployment must be repeated periodically in order to re-distribute the UAVs. The **coverage** and **connectivity** are one of the fundamental considerations in deployment strategies. A survey of the coverage and connectivity by considering deployment strategies, sleep scheduling mechanism and adjustable coverage radius is presented in [288]. The problem of multi-agent deployment over a source can also be explored. Based on the heat partial differential equation and extremum seeking, a source seeking control algorithm can be designed to deploy the group of agents around the source.

4.2.1 DEPLOYMENT PROBLEM

To measure the performance of any specific solution, a deployment function represents the quality of the UAV's distribution over the field. Such a function might be defined based on the distance of UAVs to the points in the environment, which must be minimized. Therefore, the deployment problem can be translated to a minimization problem [15]:

1. One category of multi-UAV deployment control schemes is based on **artificial potential field** or **virtual force fields**. Moreover, graph-based nonlinear feedback control laws in a group of mobile agents can be used to make the graph stay connected at all times.
2. A second category is based on common coverage control approach through the definition of feedback control laws defined with respect to the centroids of Voronoi regions. A distributed approach for optimally deploying a team of UAVs in a domain can be based on the Lloyd algorithm. Each UAV follows a control law, which is a gradient descent algorithm that minimizes the functional encoding the quality of the deployment. In a non-convex environment, geodesic distance metric to address non-convex region instead of Euclidean distance can be used.

4.2.1.1 Deployment Methodology

According to the environment, the state-of-the-art in a deployment problem is divided into continuous and discrete setups. One of the useful pre-processing techniques is modeling the workspace as a form of graph, and then, the problem becomes one of graph partitioning. In the graph representation, the deployment problem can be interpreted as a **p-median problem**, in which a limited number of facilities will be assigned to customers. The **p-median problem** is a basic discrete location problem. A problem is classified to belong to the location field if some decision regarding the position of new facilities has to be made. In general, the objective or goal of the location problem is related to the distance between new facilities and other elements of the space where they have to be positioned.

Location models may be divided into three groups: continuous, discrete and network models [233]. Large distances between nodes weaken the communication links, lower the throughput and increase energy consumption. In optimal node placement for sensor deployment, several heuristics have been proposed to find sub-optimal solutions for a fixed topology, assessing the quality of candidate positions on a structural quality metric such as distance and network connectivity. On the other hand, some schemes have advocated dynamic adjustment of nodes location since the optimality of the initial positions may become void during the operation of the network depending on the network state and various external factors [285]. For a centralized coverage algorithm in a discrete environment, an applied **spanning tree covering** (STC) can be applied on the corresponding grid cell in order to partition the graph and direct the UAVs to cover the environment. The multi-objective problem can be formulated as a mixed linear integer programming with two objectives: finding the best position in the environment to deploy the UAVs on them and minimizing the length of UAVs paths from the initial to the end position. Target overlapping areas and a greedy algorithm are used in [273] to design an optimal deployment algorithm based on goal weights, and as a result the greedy algorithm realizes optimal coverage monitoring on discrete targets and ensures the connectivity of node-monitoring network. Unlike round coverage model of a 2D plane, coverage model of 3D space chooses node position as center of the sphere and the perceived distance as sphere radius.

Voronoi-based coverage control uniquely combines both deployment and allocation in an intrinsically distributed manner via gradient descent (the **move-to-centroid law**) down a utility function minimizing the expected event sensing cost to adaptively achieve a centroidal Voronoi configuration. The application to homogeneous point UAVs and heterogeneous groups of UAVs with various sensori-motor capabilities differ by recourse to power diagrams: **generalized Voronoi diagrams with additive weights**. Voronoi-based coverage control involves collision avoidance for point UAVs since UAVs move in their pairwise disjoint Voronoi cells, but an additional collision avoidance strategy is mandatory for safe navigation of finite size UAVs. Existing work on combining coverage control and collision avoidance generally uses either heuristic approaches based on **repulsive fields** and **reciprocal velocity obstacles** causing UAVs to converge to configurations far from optimal sensing configurations; or the projection of a vector field whenever a UAV reaches the boundary of its partition cell introducing a source of discontinuity [24].

4.2.1.1.1   Taxonomy for Task Allocation

Cooperation is the **task assignment problem** (TAP) which assigns a finite number of agents to complete a finite number of tasks as efficiently as possible. This problem can be solved with a centralized or decentralized solution. Task allocation for an individual agent is relatively simple; the difficulty occurs when a decentralized algorithm is used for consensus between all UAVs. The auction-based solution is the **consensus-based auction algorithm** (CBAA), which solves the TAP for single agent tasks that are defined as tasks that require a single agent to complete. The CBAA lets agents make bids for tasks and provides a system for decentralized consensus on assignments, giving a conflict-free solution. The consensus-based bundle algorithm solves an extension of the TAP where agents queue up tasks they will complete: individual agents take available tasks and compute every permutation given their current queue of tasks, where the highest rewarded permutation becomes their bid for that task. In this way, agents continually remove and revise new tasks, as other agents find they can create a more valuable sequence with that task. This algorithm can be extended to allow multiagent tasks requiring agents to cooperate in completing individual tasks [174].

The following useful definitions are taken from [211]:

Definition 152. *A task t is **decomposable** if it can be represented as a set of sub-tasks σ_t for which satisfying some specified combination (ρ_t) of sub-tasks in σ_t satisfies t. The combination of sub-tasks that satisfy t can be represented by a set of relationships ρ that may include constraints between sub-tasks or rules about which or how many sub-tasks are required. The pair (σ_t,ρ_t) is also called a **decomposition** of t.*

The term **decomposition** can also be used to refer to the process of decomposing a task.

Definition 153. *A task t is **multiply decomposable** if there is more than one possible decomposition of t.*

Definition 154. *An **elemental (or atomic) task** is a task that is not decomposable.*

Definition 155. *A **decomposable simple task** is a task that can be decomposed into elemental or decomposable simple sub-tasks, provided that there exists no decomposition of the task that is multi(agent)-allocatable.*

Definition 156. *A **simple task** is either an elemental task or a decomposable simple task.*

Definition 157. *A **compound task** t is a task that can be decomposed into a set of simple or compound sub-tasks with the requirement that there is exactly one fixed full decomposition for t (i.e., a compound task may not have any multiply decomposable tasks at any decomposition step).*

Definition 158. *A **complex task** is a multiply decomposable task for which there exists at least one decomposition that is a set of multi[agent]-allocatable sub-tasks. Each sub-task in a complex task's decomposition may be simple, compound or complex.*

The degree of interdependence is represented with a single categorical variable with four values:

1. **No Dependencies:** These are task allocation problems with simple or compound tasks that have independent agent-task utilities. The effective utility of an agent for a task does not depend on any other tasks or agents in the system.
2. **In-schedule dependencies:** These are task allocation problems with simple or compound tasks for which the agent-task utilities have intraschedule dependencies. The effective utility of an agent for a task depends on what other tasks that agent is performing. Constraints may exist between tasks on a single agent's schedule or might affect the overall schedule of the agent.
3. **Cross-schedule dependencies:** These are task allocation problems with simple or compound tasks for which the agent-task utilities have interschedule dependencies. The effective utility of an agent for a task depends not only on its own schedule, but also on the schedules of other agents in the system. For this class, allowable dependencies are simple dependencies in that the task decomposition can be optimally pre-determined prior to task allocation. Constraints may exist between the schedules of different agents.
4. **Complex dependencies:** The agent-task utilities have interschedule dependencies for complex tasks, in addition to any in-schedule and cross-schedule dependencies for simple or compound tasks. The effective utility of an agent for a task depends on the schedules of other agents in the system in a manner that is determined by the particular chosen task decomposition. Thus, the optimal task decomposition cannot be decided prior to task allocation, but must be determined concurrently with task allocation. Furthermore, constraints may exist between the schedules of different agents.

4.2.1.1.2 Communication Constraints

Cooperative UAVs have to move to complete their tasks while maintaining communication among themselves without the aid of a communication infrastructure. Initially deploying and ensuring a mobile ad-hoc network (MANET) in real and complex environments is difficult since the strength of the connection between two UAVs can change rapidly in time or even disappear. A wrong decision about the number of UAVs and their initial location may greatly jeopardize the mission. A coverage algorithm evaluation using both centralized and random initial deployments concludes that the algorithm convergence was slower using a random initial deployment but tended to lead to better overall coverage for sparse topologies. In real situations, it is necessary to ensure several constraints of the system. If the network supports multi-hop connectivity, this kind of constraints may significantly increase the complexity of the random distribution since it would depend not only on the communication constraints, but also on the number of UAVs and their own position. Moreover, **random deployment** may cause unbalanced deployment, thus increasing the number of needed UAVs and energy depletion. A solution that can cover the deployment area within the maximum coverage time allowed is iteratively determined by varying the number and size of groups based on heuristics. An initial deployment in which UAVs start from a compact configuration works well when the main purpose is to spread the UAVs within area coverage scenarios. **Fault tolerance** in multi-UAVs system can be summarized by **bi-connectivity**, meaning that each pair of nodes in the network has at least two disjoint routes between them. Therefore, the failure at any single node does not partition the network. Despite the positive results provided by bi-connected networks, complementary strategies can be introduced such as **attraction forces**, **redundancy** or **transmission power adaptation** to maintain MANET connectivity. Furthermore, in **wireless sensor networks** (WSN), the bi-connectivity feature can be generalized to multi-connectivity, or **k-connectivity**, $k \in \mathbf{N}$ [94].

4.2.1.1.3   UAVs Deployment as a Decision Problem

The team comprises a set of N UAVs operating in either single or multiple layers such that each UAV has a radio range R. Each UAV is capable of handling a set of K users making continuous request from a particular demand area. The number of requests S_r comes with an arrival rate λ and mean packet size of each service request is $1/\mu$. The deployment model of the UAVs in the heterogeneous networks may consider two aspects:

1. Single layer with multiple UAVs: The number of UAVs is decided on the basis of provisioning of the connectivity with the macro-base station and the number of user requests from a particular demand area. For the single-layer model, the number of UAVs for connection between the base station and the demand area is calculated as $|N| = Z/R$ where Z is the distance of excessive demand area from the base station and R is the radius range. For the full capacity link between the UAVs and the base station, $|N| = S_r/S_u$ where S_u is the number of service requests a single UAV can handle.
2. Multiple layers model with multiple UAVs in each layer: The altitude is taken into account to distinguish between the UAVs of different layers. Each base station has a limited number of UAVs. In this model, the upper layer UAVs act as the main pivot between the lower layer UAVs and the base station. These pivot UAVs can support a number of UAVs by acting as an aerial base station for them. This model is very useful in case of UAV failures or base station failures.

In this decision-based approach, the base station divides the complete area into the priority zones on the basis of the user requests from them. The upper range of the user requests is accounted by the upper limit of the requests a UAV can handle as defined in the network model. For an area A, let $A_1, A_2, \ldots, A_{|B|}$ be the demand areas generating requests for extra users. Now, the complete request areas are assigned a priority value based on the number of requests pending in the particular zone provided that the number of requests pending is always less than or equal to the number of requests supported by a single UAV. The decision of the network topology is totally based on the priority value. The part of the area which is not included as the demand zone will either be handled using the multi-layer model or be assigned a number of UAVs for direct connectivity with the base station [264].

The problem of locating and routing UAV–UGV teams with a specific objective function can be formulated as an mixed-integer linear problem with the aim of maximization of the total score collected from visited points of interest by flight routes of UAVs. These routes are originated from assigned base stations determined simultaneously along with the visit order of points of interest [282]. Ideally, in order to obtain the minimum cost mission plan, these decisions would be made via a single optimization problem. To tackle the complexity of the optimization problem, given mission parameters, a multi-stage optimization algorithm can compute a mission plan with locations for each of UAV to visit:

1. In the first stage, an optimal deployment location problem for ground vehicles is formulated as a mixed-integer linear programming problem, with the objective to minimize the summation of the distances from each ground vehicle location to the associated aid requests.
2. In the second stage, optimal routes for ground vehicles from their initial locations to the destinations are determined such that the total travel time is minimized. Optimal trajectories for the UAVs along the locations of aid requests are determined. Once the mission plan has been determined, the way-points are sent to the respective autonomous vehicles. The way-points serve as input to the vehicle guidance and so enable automated deployment given the ability to operate autonomously for each of the vehicles.

4.2.1.2   Deployment Strategies

4.2.1.2.1   Initial Deployment

The initial deployment of UAVs from the ground and the re-deployment of UAVs once an area is searched are investigated to reduce energy costs and search time. Three strategies are compared that are **scalable** and decentralized, and require low computational and communication resources. The strategies exploit environment information to reduce unnecessary motion, and reduce diminishing returns and interference between UAVs:

1. **Linear-temporal incremental deployment** (LTID): This strategy deploys UAVs one at a time with a fixed time interval between consecutive launches. Longer inter-launch intervals (λ) slow deployment, but decrease the number of concurrent UAVs. This reduces spatial interference and unnecessary flight by exploiting environmental information acquired from the expanding network. Once a sub-area of the environment has been searched, UAVs re-deploy as explorers to new unexplored areas. Before this re-deployment commences, there may be multiple explorers flying into this sub-area where they are not required, which is reduced with longer inter-launch intervals. Thus, LTID reduces energy consumption by reducing interference and unnecessary movement.

2. **Single incremental deployment** (SID): This strategy is similar to LTID and deploys one UAV at a time, but waits for the previous UAV to become a beacon before launching the next. Single incremental deployment reduces unnecessary flight time because the next UAV will only deploy once the beacon network has sensed the environment and perceived if and where a new beacon is required. Thereby, explorers always fly directly to the desired deployment location. To implement SID, the network communicates if an explorer is flying. This can be achieved by propagating local messages across the beacon network. Beacons signal to the whole team if they perceive a flying explorer and UAVs only deploy if no signal is received. To ensure only a single UAV deploys at a time, random timeouts are used. When no flying explorer signal is present, UAVs wait a short random time period. If after this period there is no flying explorer signal, the UAV can deploy.

3. **Adaptive group size** (AGS): This strategy adapts the density of UAVs, initially rapidly deploying UAVs, every 2–3 s. Explorers measure the density of neighboring UAVs using their relative-positioning sensor and probabilistically land if the density is higher than a predefined threshold. This decreases the ratio of UAVs, diminishing returns and interference. UAVs which have landed launch again when there are no UAVs flying in the vicinity.

4.2.1.2.2   Optimal Deployment at Constant Altitude

The flying zone can be represented as a discretized parallelepiped of height $h_{\min} \leq h \leq h_{\max}$, width $y_{\max}$ and length $x_{\max}$, where U is the set of available UAVs, with the coordinates (x_u, y_u, h_u) of each UAV u and T is the set of targets to be monitored, $t_i = (X_{t_i}, Y_{t_i})$ with the distance $D_{t_i}^{x_u, y_u} = \sqrt{(X_{t_i} - x_u)^2 + (Y_{t_i} - y_u)^2}$. The visibility θ of each UAV u is represented by a disk with radius r^{h_u}. Two main decisions must be taken:

1. The decision variables are given by

$$\delta_{xyh}^u = \left\{ \begin{array}{cc} 1 & \text{if the UAV}u \text{ is located at } (x,y,h) \\ 0 & \text{otherwise} \end{array} \right\} \tag{4.1}$$

2. The targets $t_i \in T$ to be monitored by a UAV u:

$$\delta_{t_i}^u = \left\{ \begin{array}{cc} 1 & \text{if the target } t_i \text{ is observed by the UAV}u \\ 0 & \text{otherwise} \end{array} \right\} \tag{4.2}$$

The objective is to monitor all the targets with at least one UAV, minimizing either the number of UAVs or the total energy consumption. Each UAV consumes the energy:

$$E = (\beta + \alpha k)t + P_{\max}(k/s) \tag{4.3}$$

where α is a motor speed multiplier and β is the minimum power needed to hover over the ground, t being the operating time.

Problem 159. *The deployment problem can be stated with the objective:*

$$\min f(\delta) = \sum_{(x,y,h)} \sum_{u \in U} \delta^u_{xyh} + E \tag{4.4}$$

subject to

$$\sum_{x,y,h} \delta^u_{x,y,h} \leq 1 \quad \forall u \in U \tag{4.5}$$

This constraint ensures that the UAV u is located at most one position.

$$\gamma^u_{t_i} \leq \sum_{x,y,h} \delta^u_{x,y,h} \left(\frac{r^{h_u}}{D^{uxy}_{t_i}} \right) \quad \forall u \in U, t_i \in T \tag{4.6}$$

This condition is used to set the value of the variable $\gamma^u_{t_i}$ which can have the value 0 or 1 depending on the radius

$$\sum_{u \in U} \gamma^u_{t_i}, \forall t_i \in T \tag{4.7}$$

Each target is observed by at least one UAV.

Heuristics can be presented to solve this nonlinear mixed-integer optimization problem [289].

4.2.1.2.3 Generalized Discrete Lloyd Descent

A multiagent system is composed of interconnected subsystems or agents. In control of UAVs, the aim is to obtain a coordinated behavior of the overall system through local interactions among the agents. Communication among the agents often occurs over a wireless medium with finite capacity [7]. Some definitions useful for the sequel are introduced as follows:

Definition 160. *A **landmark** is an abstraction of a point or a small area of interest that must be kept under observation and that may be of a larger surface. A landmark is formally defined as the tuple $\ell = (q, \hat{m})$ where $q \in \mathbb{R}^3$ is the position of the landmark, and $\hat{m} \in S^2$ (unit sphere) is the orientation of the landmark. More specifically, $\hat{m}$ is a direction that characterizes the orientation of the landmark.*

Definition 161. *A **mobile sensor** is a tuple $s = (p, \hat{n}, f)$ where $q \in \mathbb{R}^3$ is the position of the sensor, $\hat{n} \in S^2$ is the orientation and $f : (\mathbb{R}^3 \times S^2)^2 \to \mathbb{R}^+$ is the footprint of the sensor. The orientation of a sensor is the unit vector corresponding to the direction where the sensor is pointing. The footprint of the mobile sensor is a function that describes the sensor's perception of the surrounding environment.*

The coverage of a finite set of landmarks attained by a set of mobile sensors is defined with respect to a partition of the landmarks among the sensors, and it is given by the sum of the coverages attained by each sensor for its subset of landmarks.

Definition 162. *Coverage of a finite set of landmarks attained by a team of mobile sensors: Consider a team of mobile sensors $S = (s_1, \ldots, s_N)$, a finite set of landmarks $L = \{\ell_1, \ldots, \ell_N\}$ and a* *partition* $P = \{P_1, \ldots, P_N\}$ *so that each subset* L_i *is assigned to the sensor* s_i, *the coverage of the set L attained by the team S with respect to the partition P as the sum of the coverage of* L_i *attained by* s_i *for* $i \in \{1, \ldots, N\}$:

$$cov(S, P) = \sum_{i=1}^{N} cov(s_i, L_i) = \sum_{i=1}^{N} \sum_{\ell \in L} per(s_i, \ell_i) \tag{4.8}$$

where

$$per(s_i, \ell_i) = f(p_i, \hat{n}_i, q_i, \hat{m}_i) \tag{4.9}$$

Problem 163. *The objective is to find a partition* $P = \{P_1, \ldots, P_N\}$ *and the positions and orientations of the sensors, such that the coverage of the set L attained by a team with respect to the partition is minimized.*

Algorithm 41 aims at progressively adjusting iteratively the positions and orientations of the sensors as well as the partition P. The coverage is improved by adjusting iteratively the positions and orientations of the sensors, as well as the partition P. The partition P is improved by considering a pair of sensors s_i and s_j at each partition and rearranging the landmarks in $L_i \cup L_j$ so that each landmark is assigned to the sensor between s_i and s_j. A lower value of coverage corresponds to a better coverage.

Algorithm 41 Generalized Discrete Lloyd Descent

1. Assign the mobile sensors $S = \{s_1, \ldots, s_N\}$ with $s_i = (p_i, \hat{n}_i, f_i)$
2. Assign the landmarks $L = \{\ell_1, \ldots, \ell_N\}$
3. Assign a partition $P = \{P_1, \ldots, P_N\}$
4. Assign $\varepsilon > 0$ and set $Z_i \{s_i\}$ for $i \in \{1, \ldots, N\}$
5. while $Z_i \neq \emptyset$ for some $i \in \{1, \ldots, N\}$ do
6. pick s_i such that Z_i is not empty
7. pick $s_i \in Z_i$
8. for $\ell \in L_i$ do
9. if $per(s_j, \ell) < per(s_j, \ell) - \varepsilon$ then
10. transfer ℓ from L_i to L_j
11. end if
12. end for
13. if one or more landmarks have been transferred then
14. $Z_i \leftarrow S \backslash \{s_i\}$
15. $Z_j \leftarrow S \{s_j\}$
16. $(p_i, \hat{n}_i) \leftarrow optcov(s_i, L_i, \Omega_i)$
17. $(p_j, \hat{n}_j) \leftarrow optcov(s_j, L_j, \Omega_j)$
18. else
19. remove
20. end if
21. end while

4.2.1.2.4 *Location Problem on the Plane*

This paragraph considers the problem of locating M facilities on the unit square so as to minimize the maximal demand faced by any facility subject to closest assignments and coverage constraints. By minimizing the demand faced by the facilities, the difference in demand rates between the busiest and the least busy facilities is the **equitable location problem** (ELP).

Definition 164. *A location vector x represents an* **equitable facility configuration** *(EFC) if the demand rates to all facilities are the same, i.e., if* $\lambda_{\max} = \Lambda/M$, *where* Λ *is the total demand rate.*

Let x_j be a vector denoting the location of facility $j, x_j \in P$, let $I_x^j = 1$ if the jth facility is the closest one to x and $I_x^j = 0$ otherwise, and let $R(x_j) = \max_{x \in P} \|xx_j\| I_x^j$ be the maximum travel distance of customers' assigned to facility j. Assume that at each $x \in P$ customers' demand rate is $\lambda(x)$ such that $\int_{x \in P} \lambda(x) dx = \Lambda < \infty$, thus $\lambda_{x_j} = \int_{x \in P} I_x^j \lambda(x) dx$ is the arrival rate to the jth facility. Let r be an exogenous given distance, which is the maximum distance allowed from a customer to a facility, and let ε be the smallest distance allowed between distinct facilities. The Voronoi region associated with the ith facility is denoted by V_i. Then, the demand rate to this facility is

$$\lambda_{x_i} \int_{x \in V_i} dx \cdot dy \qquad \forall i = 1, \dots, M \tag{4.10}$$

Assuming the Voronoi diagram is given, the ELP can be formulated as follows.

Problem 165. **Equitable location problem** *Let P be a space* $P \subset \mathbb{R}^2$ *equipped with some norm* $\|.\|$ *and* $M > 0$ *denote the number of facilities. Given M facilities, the closest assignment constraints with a given distance norm, divides the plane to M areas using a Voronoi diagram of this distance norm, find the equitable location configuration.*

$$min \ \lambda^{\max} \tag{4.11}$$

subject to

$$\begin{aligned} \lambda_{x_i} &\leq \lambda^{\max} \quad \forall i = 1, \dots, M \\ \|x_i, x_j\| &\geq \varepsilon \quad \forall i, j = 1, \dots, M, i \neq j \\ \|x_i, x\| &\leq r \quad \forall i = 1, \dots, M, x \in V_i \end{aligned} \tag{4.12}$$

When the values of M and r are small, $ELP(M)$ may be infeasible and therefore no feasible EFC that is feasible exists. In [36], sufficient conditions for the existence of EFC are deduced, given the closest assignment constraints.

Remark 166. *The deterministic feasible equitable facility location is a major component in the* **stochastic capacity and facility location problem** *(SCFLP).*

The SCFLP focuses on three sources of uncertainty: the timing, location and actual amount of the demand generated by the customers. This problem optimizes three types of decision variables:

1. The number of facilities to be located
2. The location of the facilities
3. The service capacity of each facility.

The approach to solve the SCFLP is based on the following. For a given M, the ELP is solved to provide optimal location for the facilities and maximal demand rate to a facility, $\lambda_{\max}$. The solution of ELP on the unit square is investigated, with a uniform demand using Voronoi diagrams [240].

4.2.1.2.5 *Team-Based Optimization Scheme*

The locational optimization function can be translated to maximizing the sensing performance:

$$H(P, Q) = \sum_{i=1}^{N} \int_{W_i} f(\|q - p_i\|) \Phi(q) dq \tag{4.13}$$

where for n teams, $N = \sum_{t=1}^{n} n_t$, and P is the set of all UAVs. The ith UAV is assigned to the region W_i, and the cost function H is minimized by finding the optimum locations of the UAVs and their assigned regions W_i whose union is Q. A team-based partition of the agents considering agents as a collection of multiple teams pursuing their assigned task or objective is presented in [2]. The optimization problem is broken into two interconnected functions such that the solution to each problem represents the optimum configuration of the teams and their associated agents. Let $L = (\ell_1, \ell_2, \ldots, \ell_n)$ define the set of teams where each $\ell_t, t = 1, \ldots, n$ represents the nucleus of team t function of the agent's position in the associated team $\ell_t = g(p_{t_1}, \ldots, p_{t_{n_t}})$. The polytope Q is partitioned into a set of Voronoi cells: $V(L) = \{V_1, \ldots, V_n\}$ considered as the optimal partitioning for a set of agents with fixed locations at a given space as

$$V_t = \{q \in Q, \|q - \ell_t\| \leq \|q - \ell_s\|\} \tag{4.14}$$

The obtained Voronoi cells associated with the nuclei of the teams are then considered as the convex polytopes set to deploy their associated agents. The Voronoi partitions $V_t(P_t) = \{V_{t_1}, \ldots, V_{t_{n_t}}\}$ generated by the agents $p_{t_1}, \ldots, p_{t_{n_t}}$ belonging to the ith team are defined as

$$V_{t_m} = \{q \in V_t, \|q - p_{t_m}\| \leq \|q - p_{t_r}\|\} \tag{4.15}$$

where p_{t_m} denotes the location of the mth in the tth team such that $m \in \{1, \ldots, n_t\}$. The basic characteristics of the **Voronoi partitions** are

1. **Associated mass:**

$$M_{V_{t_m}} = \int_{V_{t_m}} \Phi(q) dq \tag{4.16}$$

2. **Centroid:**

$$C_{V_{t_m}} = \frac{1}{M_{V_{t_m}}} \int_{V_{t_m}} q\Phi(q) dq \tag{4.17}$$

3. **Polar moment of inertia:**

$$J_{V_{t_m}, p_{t_m}} = \int_{V_{t_m}} \|q - p_{t_m}\|^2 \Phi(q) dq \tag{4.18}$$

The characteristics of the team's Voronoi cells can be deduced as

$$M_{V_t} = \sum_{m=1}^{n_t} M_{V_{t_m}} \tag{4.19}$$

$$C_{V_t} = \frac{1}{M_{V_t}} \int_{V_t} q\Phi(q) dq \tag{4.20}$$

The **nucleus of the team** is a function of the agents' position:

$$\ell_t = \frac{\sum_{m=1}^{n_t} M_{V_{t_m}} p_{t_m}}{\sum_{m=1} n_t M_{V_{t_m}}} \tag{4.21}$$

It is a representative of the agents' position in the team and can be considered as the collective position of the agents for drawing the Voronoi diagram of the teams V_t. The deployment task can be addressed by solving a two-level optimization problem.

1. The first function to be minimized represents the cost associated with partitioning the main space into partitions related to the teams of agents.

$$G_t(P_t, Q_t) = \sum_{i=1}^{n_t} \int_{Q_{t_m}} f(\|q - p_{t_m}\|) \Phi(q) dq \tag{4.22}$$

2. The solution of the second optimization problem results in deploying the agents in an optimum way inside the teams:

$$G(L, Q) = \sum_{i=1}^{n} \int_{Q_t} f(\|q - \ell_t\|) \Phi(q) dq \tag{4.23}$$

where the sensing performance is $f(\|q - p_i\|)$.

The extension of Lloyd algorithm can be used to solve this problem.

4.2.1.2.6 Cooperative Task Allocation

UAV coordination has to be reached in the presence of multiple uncertainties in the environment and in the communication channels. The specific application considered in this paragraph involves the decentralized assignment of tasks in which UAVs can only receive delayed measurements from other vehicles, and the environment disturbances are able to disrupt the planned sequence of actions [142]. UAVs are required to perform different tasks on stationary targets with known locations. The optimal and conflict-free decentralized computation of assignment of tasks in the presence of communication delays, measurement noise and wind disturbance is considered. **Conflict-free** means correct assignment of tasks to UAV in which a given task needs to be assigned to one and only one UAV and the tasks on the same target need to be performed in a certain order. UAVs will communicate only with a subset of UAVs or neighbors. Each UAV will estimate the position of every other UAV and obtain a list of assignments for the group. Local estimated positions as well as the estimates of other UAV positions are affected by zero mean white sensor noise. A **coordinated assignment plan** is required in order to guarantee that every task will be performed only once. This is to prevent that the same task on the same target is carried over by two different vehicles or that conflict in the plan occurs such that a given task is never performed. The cost function to be minimized is the cumulative distance the UAVs travel in order to perform all required tasks:

$$J = \sum_{i=1}^{N_u} D_i > 0 \tag{4.24}$$

A group of N_u UAVs, a set of targets $\{1, 2, \ldots, N_t\}$ and N_m the number of tasks to be performed at each target are considered. Each task is associated with an integer value, $N_s = N_t N_m$ is the number of single assignments and $S = \{1, 2, \ldots, N_s\}$ represents the set of stages. A decision variable $g_{i,j,k} \in \{0, 1\}$ is defined such that it is 1 if UAV $i \in U$ performs a task on target $j \in T$ at stage $k \in S$ and it is zero otherwise. The set of assignments, up to stage k, is represented by the list $G_k = \{\bar{g}_1, \bar{g}_2, \ldots, \bar{g}_k\}$ where $\bar{g}_k = [i, j]$ such that $g_{i,j,k} = 1$. The formulation of the cooperative multiple task assignment can be expressed as follows:

$$\min\left(J = \sum_{i=1}^{N_u} \sum_{j=1}^{N_t} \sum_{k=1}^{N_s} d_{i,j,k}^{G_k} g_{i,j,k}\right) \tag{4.25}$$

subject to

$$\sum_{i=1}^{N_u} \sum_{j=1}^{N_t} g_{i,j,k} = 1 \quad k \in S \tag{4.26}$$

$$\sum_{i=1}^{N_u} \sum_{lk=1}^{N_s} g_{i,j,k} = N_m \quad j \in T \tag{4.27}$$

The constraints guarantee that exactly one task is assigned at any given stage and that on each target, exactly N_m tasks are performed. A specific order on the tasks performed on the same target significantly increases the complexity of this optimization problem. To run a decentralized task allocation algorithm based on distance-to-task costs, each UAV needs an estimate of current positions of teammates. It is also desired to avoid continuous or frequent inter-UAV communication. At a reasonable increased computation cost, each UAV will implement models of all UAV dynamics, including itself. The assignment at each stage is found based on a cost matrix that evaluates the expected cost of each UAV to perform each one of the current tasks. Because of communication delays and other uncertainties, UAVs may arrive at different assignment plans. An algorithm for estimation and resolution of possible conflicts is based on generating new events when entries of the cost matrix are close to the minimum at any given stage of the optimization problem. To resolve an estimated conflict, UAVs bid on their best task at that particular stage. Their bids represent their cumulative cost on the task they are bidding on. Because these are real numbers representing the expected distance to travel to perform previous tasks and the conflicted task, the probability of UAVs bidding exactly the same cost is very low. This approach results in a trade-off between reducing inter-UAV communication and achieving a conflict-free assignment plan.

4.2.2 MOBILE SENSOR NETWORK

Various schemes have been proposed for the deployment of **mobile sensor nodes** (MSNs), with optimal utilization of resources. The following classification of sensor deployment techniques has been proposed in [263]:

1. Based on placement strategy
   a. Random
   b. Deterministic
2. Based on usage
   a. Barrier
   b. Blanket
   c. Area
   d. Target oriented
3. Based on deployment domain
   a. Open area (wide regions)
   b. Indoor.

4.2.2.1 Aerial Networks

In 2D space scenarios, the maximal coverage problem can be mapped to a circle packing formulation. The problem turns into the sphere packing problem in 3D, and the strategies designed for 2D become NP-hard in 3D. The problem of coverage in 3D space is often a critical part of the scenario for the observation of an environment. The number of nodes and their locations are restricted by the investigated environment and the reception range of node. Moreover, the dynamic UAV network topology and flight must be handled efficiently considering the communication constraints of the UAVs. In [10], a node positioning strategy for UAV networks is proposed with a wireless sensor and actor network structure according to different capabilities of the nodes in the network. The positioning algorithm utilizes the **valence shell electron pair repulsion** (VSEPR) theory of chemistry, based on the correlation between molecular geometry and the number of atoms in a molecule. By using the rules of VSEPR theory, the actor nodes in the proposed approach use a lightweight and distributed algorithm to form a self-organizing network around a central UAV, which has the role of the sink.

4.2.2.1.1 Minimization of Sensor Number

The objective of the **sweep coverage problem** is to minimize the number of sensors required in order to guarantee sweep coverage for a given set of points of interest on a plane. Use of both static and mobile sensors can be more effective, in terms of energy consumption.

Definition 167. *T-sweep coverage A point is said to be **T-sweep** covered if and only if at least one mobile sensor visits the point within every T time period, where T is called sweep period of the point.*

The inputs of the algorithm proposed in [153] are the graph G, speed v, sweep period T and energy consumption per unit time for static and mobile sensors λ and μ, respectively. The output of the algorithm is the number of mobile and static sensors with the deployment locations for the static sensors and the movement schedule for the mobile sensors.

4.2.2.1.2 Evasion Paths

In **minimal sensor network** problems, one is given only local data measured by many weak sensors but tries to answer a global question. The mobile sensor network problem is considered where sensors are ball-shaped. A sensor cannot measure its location but knows when it overlaps a nearby sensor. An evasion path exists if a **moving intruder** can avoid being detected by the sensors. The evasion problem can also be described as a **pursuit-evasion problem** in which the domain is continuous and bounded, there are multiple sensors searching for intruders and an intruder moves continuously and with arbitrary speed. The motions of the sensors are not controlled; the sensors wander continuously but arbitrarily. The locations of the sensors cannot be measured but instead know only their time-varying connectivity data. Using this information, it is important to determine whether it is possible for an intruder to avoid the sensors. Both the region covered by the sensors and the uncovered region change with time. Zigzag persistent homology provides a condition that can be computed in a streaming fashion [78]. The technical basis for zigzag persistence comes from the theory of graph representations. However, no method with time-varying connectivity data as input can give necessary and sufficient conditions for the existence of an evasion path. The existence of an evasion path depends not only on the type of the region covered by sensors but also on its embedding in space-time. Both the region covered by the sensors and the uncovered region change with time. For planar sensors that also measure weak rotation and distance information, necessary and sufficient conditions for the existence of an evasion path are provided in [8].

4.2.2.1.3 Blanket Coverage

In the category of blanket coverage problems, the main objective is to maximize the total detection area. This coverage problem is defined as how to position or deploy the sensors in a particular **region of interest** (ROI) so that coverage percentage is maximized and coverage holes are minimized. The deployment of nodes can be done either randomly or deterministically. The deterministic deployment of nodes can be considered because sensor network coverage can be improved by carefully planning position of sensors in the ROI prior to their deployment. Grid-based sensor networks divide the ROI into square cells, and sensors can be placed at the center of the square cell in order to maximize the coverage, and also the number of sensors required for placing inside the square cell is less than the number of sensors required for placing at the intersection of the grids. The sensors can be placed at the center of the square cell. In case of grid-based deployment, problem of coverage of sensor field reduces to the problem of coverage of one cell and its neighbor because of symmetry of cells [266].

Problem 168. *Given N mobile nodes with isotropic radial of range R_s and isotropic radio communication of range R_c, how should they deploy themselves so that the resulting configuration maximizes the net sensor coverage of the network with the constraint that each node has at least K neighbors?*

Definition 169. *Two nodes are considered **neighbors** if the Euclidean distance between them is less than or equal to the communication range R_c.*

Three metrics are introduced to evaluate the performance of the deployment algorithm [248]:

1. The **normalized per-node coverage** defined as:

$$\text{cov} = \frac{\text{Net area covered by the network}}{N\pi R_s^2} \tag{4.28}$$

2. The **percentage of nodes** in the network that have at least K neighbors.
3. The **average degree** of the network.

In this deployment algorithm, **virtual forces** are constructed between nodes so that each node can attract or repel its neighbors. The forces are of two kinds. The first causes the nodes to repel each other to increase their coverage, and the second constrains the degree of nodes by making them attract each other when they are on the verge of being disconnected. By using a combination of these forces, each node maximizes its coverage while maintaining a degree of at least K.

4.2.2.1.4 Optimal UAV Deployment for Stationary Nodes

In order to find the optimal locations of UAVs functioning as communication relays at a fixed position between stationary nodes, the performance index for network connectivity is used. As the number of nodes increases, network complexity increases. Thus, the concept of **minimum spanning tree** can be used to obtain the highest probability of a successful transmission using minimum possible links. A **spanning tree** is a subgraph that is itself a tree connecting all the vertices of the graph together. For a successful transmission, the weight of each graph node is

$$W_{ij} = -log P_r^{ij} \tag{4.29}$$

The smaller the weight, the higher the probability of a successful transmission. If the positions of the nodes are given, the minimum spanning tree can be constructed with the weight W_{ij}. The performance index for the global message connectivity can be set as

$$J = \sum_{i=1}^{n} \sum_{j=1}^{n} \mathbf{A}_{ij} W_{ij} \tag{4.30}$$

where $\mathbf{A} \in \mathbb{R}^{n \times n}$ represents the adjacency matrix of the minimum spanning tree for a given configuration. The implementation of this deployment optimization can be centralized for a stationary environment [206].

4.2.2.1.5 Task Execution

In a multi-UAVs system, a mission can be divided into different tasks and a number of specialized UAVs can be introduced to solve each task concurrently. These tasks may be known by the UAVs before task execution stage or may dynamically appear during task execution. In an exploration mission, the aim is to locate and visit a number of pre-determined targets in a partially unknown terrain. The challenge is then how to assign these targets to the UAVs in order to optimize an overall system objective required by the mission. Multi-UAV **task allocation** problems are often solved in a distributed manner using **market-based algorithms**, while **auction algorithms** are efficient in terms of both computation and communication. The information of the UAVs and tasks can be compressed into numerical bids and computed in parallel by each UAV. For **single-assignment problems** where each UAV can handle at most one task, the single-item auctions can be used where UAVs bid on tasks that are auctioned off individually. The UAV with the highest bid wins the task and then has to finish it. However, for **multi-assignment problems** where each UAV is able to handle several tasks, they belong to the class of combinatorial optimization problem. Strong synergies exist between the tasks for bidders. It is considered that a set of tasks have a positive synergy for a UAV if their combined cost for executing them together is less than the sum of their individual cost incurred by doing them separately, and conversely for a negative synergy. A near-optimal allocation of a set of tasks to UAVs uses single-round combinatorial auctions, calculating the bid for every UAV, based on bundles of tasks rather than individual tasks. The bid for each UAV to hold a bundle is computed through the smallest path cost that needs to visit all tasks in the bundle from the UAV's current location.

Parallel single-item auctions treat each task independent of other tasks and every UAV bids for each task in parallel. Such mechanism has its computation and communication efficiencies while it leads to highly sub-optimal solutions since it does not account for any synergies between tasks. On balance, the sequential (multi-round) single-item auctions provide the advantages of solution quality from single-round combinatorial auctions, and computation and communication efficiencies from parallel single-item auctions. It works in a multi-round manner and in each round every UAV places bid on the unallocated tasks. The bid is computed as the smallest cost increase resulted from winning the task and the UAV with the overall smallest bid is allocated the corresponding task. The process is repeated until all the tasks have been allocated [287].

4.2.2.2 Visual Coverage

The visual coverage problem differs in several respects from the standard coverage control.

1. While the standard coverage assumes **isotropic sensors**, the camera sensor has **anisotropic property**.
2. The image acquisition process of a camera sensor involves a nonlinear projection from the 3D world to 2D image plane, which is significant especially in the monitoring problem in the 3D world.
3. A camera sensor does not provide any physical quantity, while temperature or radiation sensors trivially sample a scalar field describing importance of each point in the environment.

Thus, computer vision techniques must be integrated with control scheme to extract the meaning of the sensed data [128]. A visual coverage problem is considered under the situation where vision sensors with controllable orientations are distributed over the 3D space to monitor 2D environment. In this case, the control variables, i.e., the rotation matrices must be constrained on the Lie group $SO(3)$. The problem is directly formulated as an optimization on $SO(3)$, and the gradient descent algorithm is applied on matrix manifolds. A vision sensor has an image plane containing the sensing array, whose pixels provide the numbers reflecting the amount of light incident. The objective function to be minimized by sensor i is defined by a sensing performance function and a density function at a point $q \in E$. The **sensing performance function** describes the quality of the acquired data about $q \in E$, and the **sensing density function** indicates the relative importance of $q \in E$. The function is accumulated only at the center of the pixels projected onto the environment E in order to reflect the discretized nature of the vision sensors. The gradient is derived assuming that the image density describing relative importance over the image is given in the form of the mixed Gaussian function. The gradient descent approach is a standard approach to coverage control; the rotation is updated in the direction of this gradient [162].

A **visual sensor network** (VSN) consists of a number of self-configurable visual sensors with adjustable spherical sectors of limited angle: **field of view** that is meant to cover a number of targets randomly positioned over a deployment area. One of the fundamental problems of VSNs is to cover a maximum number of targets using the minimum number of sensors. VSNs can be classified in two different categories:

1. **Over-provisioned systems** when the number of sensors is sufficient to cover all the targets. In this coverage task, the number of cameras must be minimized besides maximizing coverage.
2. **Under-provisioned systems** when this number is insufficient, the target coverage should be maximized regardless of the number of cameras being used.

Two approaches can be considered [286]:

1. **Sensor-oriented approach:** One can look into the cameras and determine the exact coverage count in different field of views of each camera. While counting the coverage, the overlapping regions of the neighboring cameras must also be considered to exclude the possibility of redundant coverage.
2. **Target-oriented approach:** One can look into the targets first. Some targets might be located in a difficult corner of the deployment area and could be covered only by a single camera. In order to maximize the target coverage, those targets need to be covered first. The targets must be prioritized based on their coverage vulnerability, and then, one must select a minimal set of cameras that can cover targets in their order of priorities.

The distance at which a UAV camera sensor should be positioned from the target is

$$h = H\frac{f}{d} \qquad (4.31)$$

where f is the focal length, h is the camera sensor height, d is the distance from the sensor to the far plane and H is the height of the far plane of the frustum. A **view frustum** is a 3D volume that defines how models are projected from camera space to projection space. Objects must be positioned within the 3D volume to be visible. Points beyond this distance are considered invisible to the camera.

Definition 170. *The **functional coverage** is defined as the ratio of the area covered by the set of cameras to the area to be reconstructed.*

In order to determine the area covered by a given set of cameras, a viewing space for each camera is defined according to geographical position and the distance from the camera to the far plane. The terrain points that are visible to each individual camera are determined by an **occlusion test**, involving reflecting the point cloud in the frustum onto a spherical surface away from the camera. Any points that are not reflected are not included in the frustum; however, the entire set of visible points is included. The distance from the sensor to the object of interest determines the resolution of the final model. Several variables and sets of constraints are applied simultaneously in the process of adjusting each camera to find the optimal coordinates, position, altitude, and optimal flight path [254].

Problem 171. *Schedule of the active periods of sensors: Given*

1. *A set of targets $T = \{t_1, t_2, \ldots, t_m\}$ and their corresponding weights $W = \{w_1, w_2, \ldots, w_m\}$,*
2. *A set of homogeneous cameras $S = \{s_1, s_2, \ldots, s_n\}$ randomly deployed in a 2D plane,*
3. *A subset $F = \{S_{i,j}, 1 \le i \le n, 1 \le j \le q\}$ subsets $S_{i,j}s \subseteq T$ computed by an identifiability test,*

4. *the required coverage level C_L, specific to a given mission where*

$$\max_{k \in \{1,...,m\}} w_k \leq C_L \leq \sum_{k \in \{1,...,m\}} w_k \tag{4.32}$$

The problem is to schedule the active periods of each camera such that the sum of weights of all targets which are effectively covered is at least C_L at any time and the network lifetime is maximized.

This problem can be divided into two sub-problems: determining the direction (active sensor) of each node and assigning the sleep–wake-up schedule to it. A heuristic has been proposed in [166] to solve this problem.

4.2.2.2.1 Voronoi Approach for the Visibility Landmark

The problem in this paragraph is to consider a team of UAVs $a_i \in \mathbb{A}$, where $i = 1, \ldots, m$ and $m \in \mathbb{N}$, with a set of poses $\mathbf{A} = (A_1, A_2, \ldots, A_m)$. A set of m disjoint partitions of the set of landmarks $\mathbf{S}$ is $\mathbf{P} = (\mathbf{P}_1, \mathbf{P}_2, \ldots, \mathbf{P}_m)$. In this scenario, each UAV a_i is responsible of a subset $\mathbf{P}_j$ for $j = 1, \ldots, m$ with $\mathbf{P}_i \cap \mathbf{P}_j = \emptyset$. The visibility of a landmark is calculated with respect to the pose $\mathbf{A}_i$ of the UAV a_i that is responsible for this landmark [270]. The coverage score of a_i is given by

$$C(\mathbf{A}_i, \mathbf{P}_i) = \sum_{s_k \in \mathbf{P}_i} \text{vis}(\mathbf{A}_i, s_k) \tag{4.33}$$

where the visibility of a point $s \in \mathbb{R}^2$ with respect to an agent a with generic pose $\mathbf{A}$ can be derived as

$$\text{vis}(A, s) = \text{vis}_{\mathbf{I}_3} \left(\left(\mathbf{A}^{-1} \tilde{s} \right)_{xy} \right) \tag{4.34}$$

where the homogeneous coordinates of s are $\tilde{s} = [s_x, s_y, 1]^T$ and

$$\text{vis}_{\mathbf{I}_3}(s) = \left\{ \begin{array}{ll} 0 & s \in \mathbf{R}_1 \\ s_x & s \in \mathbf{R}_2 \\ s_x / \|s\|^2 & s \in \mathbf{R}_3 \end{array} \right\}$$

while the partition of $\mathbb{R}^2$ is

$$\mathbf{R}_1 = \left\{ s \in \mathbb{R}^2 : s_x \leq 0 \right\},$$
$$\mathbf{R}_2 = \left\{ s \in \mathbb{R}^2 : s_x > 0 \text{ and } \|s\| \leq 1 \right\},$$
$$\mathbf{R}_3 = \left\{ s \in \mathbb{R}^2 : s_x > 0 \text{ and } \|s\| > 1 \right\}.$$

The coverage score of the whole team of UAVs is calculated as the coverage score of each UAV as follows:

$$C(\mathbf{A}, \mathbf{P}) = \sum_{i=1}^{m} C(\mathbf{A}_i, \mathbf{P}_i) \tag{4.35}$$

In order for the algorithm to take the landmarks partitions to a Voronoi configuration, while improving the visibility of each landmark and thus the overall coverage score, it has to be designed for handling non-trivial interaction between the agents, as well as simultaneous communications between different pairs of agents. After the initialization procedure has finished, all the agents first calculate the visibility of each landmark in their own set $\mathbf{P}_i$ from their current positions, then optimize the pose in order to maximize the coverage score on that set. The optimization process takes place every time the set of landmarks $\mathbf{P}$ of an agent changes, i.e., every time two agents trade some landmark successfully as in Algorithm 42.

Algorithm 42 Pose Optimization Procedure a_i

1. $old_{score} \leftarrow$ Old coverage score
2. $new_{score} \leftarrow$ New coverage score
3. $p \leftarrow$ current position of the agent
4. $\psi \leftarrow$ current orientation of the agent
5. $p_n \leftarrow$ optimized position of the agent
6. $\psi_n \leftarrow$ optimized orientation of the agent
7. $(p_n, \psi_n) \leftarrow$ optimization routine on (p, ψ)
8. for $s_k \in \mathbf{P}$ do
9. calculate visibility of s_k from (p_n, ψ_n)
10. calculate the coverage score of the agent from (p_n, ψ_n)
11. if $new_{score} > old_{score}$ then
12. new way-point $\leftarrow (p_n, \psi_n)$
13. else
14. new way-point $\leftarrow (p, \psi)$
15. end if
16. end for

The trading procedure is the main part of the coverage task, presented in Algorithm 43. This part involves actual communication between agents, with exchange of information about current pose of each agent and on the partition of landmarks currently owned by each UAV. A UAV can start its trading routine only if it has reached the last way-point generated by the optimization algorithm. The way-point is considered reached if the distance between the way-point and the current position of the UAV is below a certain threshold. In order to maintain consistence in the information about the ongoing coverage mission, a trading procedure can involve only two UAVs at a time.

Algorithm 43 Trading Algorithm for Client Agent a_i

1. $o_w \leftarrow$ last way-point generated by the optimization routine 42
2. $q_c \leftarrow$ client agent
3. $q_s \leftarrow$ server agent
4. $A_c \leftarrow$ the pose of q_c
5. $\mathbf{P}_c \leftarrow$ the landmark partition of q_c
6. $\mathbf{Q}_{c_{in}} \leftarrow$ initial $\mathbf{Q}_c$
7. $state_c \leftarrow (A_c, \mathbf{P}_c)$
8. if o_w is reached then
9. pick an item $q_i \in \mathbf{Q}_c$
10. $q_i \leftarrow q_s$
11. send to q_s state
12. if q_s is available and $n \neq 0$ landmarks can be traded then
13. call the optimization routine 42
14. $\mathbf{Q}_c = \mathbf{Q}_c \, q_s$
15. else if $\mathbf{Q}_i = \emptyset$ then
16. all possible trading completed
17. end if
18. end if

4.2.2.2.2   Visibility Coverage in Polygonal Environments with Hole

A visual coverage problem can be considered under the situation where vision sensors with controllable orientations are distributed over the 3D space to monitor a 2D environment. In the case, the control variables, i.e., the rotation matrices must be constrained on the **Lie group** SO(3). The problem is directly formulated as an optimization on SO(3) and the gradient descent algorithm is applied on matrix manifolds. A vision sensor has an image plane containing the sensing array, whose pixels provide the numbers reflecting the amount of light incident. The objective function to be minimized by sensor i is defined by a **sensing performance function** and a **density function** at a point $q \in E$. The sensing performance function describes the quality of the acquired data about $q \in E$ and the density function indicates the relative importance of $q \in E$. The function is accumulated only at the center of the pixels projected onto the environment E in order to reflect the discretized nature of the vision sensors. The gradient is derived assuming that the image density describing relative importance over the image is given in the form of the mixed Gaussian function [162].

In another scenario, UAVs begin deployment from a common point, possess no prior knowledge of the environment and operate only under **line-of-sight sensing and communication**. The objective of the deployment is for the agents to achieve **full visibility coverage** of the environment while maintaining line-of-sight connectivity with each other. This is achieved by incrementally partitioning the environment into distinct regions, each completely visible from some UAV. Approaches to visibility coverage problems can be divided into two categories:

1. Those where the environment is known a priori, in the **art gallery problem**, one seeks the smallest set of guards such that every point in a polygon is visible to some guard.
2. Those where the environment must be discovered, **simultaneous localization and mapping** (SLAM) techniques explore and build a map of the entire environment and then use a centralized procedure to decide where to send agents. For example, deployment locations can be chosen by a human user after an initial map has been built. Waiting for a complete map of the entire environment to be built before placing agents may not be desirable.

Problem 172. *The **Distributed Visibility-Based Deployment Problem with Connectivity** is stated as follows: design a distributed algorithm for a network of UAVs to deploy into an unmapped environment such that from their final positions every point in the environment is visible from some UAV. The UAVs begin deployment from a common point, their visibility graph $G_{vis,E}(P)$ is to remain connected, and they are to operate using only information from local sensing and line-of-sight communication. Each UAV is able to sense its visibility gaps and relative positions of objects within line-of-sight. Additionally, the following main assumptions are made:*

1. *The environment* **E** *is static and consists of a simple polygonal outer boundary together with disjoint simple polygonal holes. Simple means that each polygon has a single boundary component, its boundary does not intersect itself and the number of edges is finite.*
2. *UAVs are identical except for their unique identifiers $(0,\ldots,N-1)$.*
3. *UAVs do not obstruct visibility or movement of other UAVs.*
4. *UAVs are able to locally establish a common reference frame.*
5. *There are no communication errors nor packet losses.*

A centralized algorithm to incrementally partition the environment E into a finite set of openly disjoint star-convex polygonal cells is used. The algorithm operates by choosing at each step a new vantage point on the frontier of the uncovered region of the environment, then computing a cell to be covered by that **vantage point** (each vantage point is in the kernel of its corresponding cell). The frontier is pushed as more and more vantage point–cell pairs are added until eventually the entire environment is covered. The vantage point–cell pairs form a directed rooted tree structure called the partition tree [238].

4.2.2.3 Wireless Sensor Network

A **wireless sensor network** (WSN) is a distributed system of sensor nodes interconnected over wireless links. Sensors gather data about the physical world and transmit these data to a central through single-hop or multi-hop communications. Wireless sensor nodes have integrated batteries with limited energy resources. Sensor nodes can be either thrown in as a mass or placed one by one during deployment. **Deployment planning** requires consideration of several objectives such as energy consumption, sensing coverage, network lifetime and network connectivity. Often these objectives conflict with one another, and operational trade-offs must be established during network design. In pre-determined deployment, the locations of the nodes are specified. This type of deployment is used when sensors are expensive or their operation is meaningfully affected by their position. Self-deployment is proposed as a technique assuming the sensors' own mobility. **Potential fields** or approaches based on **virtual force** are used to spread sensors out from a compact or random initial configuration to cover an unknown area [3].

4.2.2.3.1 Wireless Sensor Network Deployment Strategy

A major design step is to selectively decide the locations of the sensors in order to maximize the covered area of the targeted region. There are different sensor network deployment strategies: sensor capabilities, base station options, initial energy assignment, sensor locations and traffic generation pattern are the key parameters used to describe WSN deployment strategies. Most strategies with fixed sinks try to optimize routing/transmission power. They try to maximize coverage while keeping the number of sensors at a minimum. Another strategy is the use of redundant sensors at some points for redundancy. Deployment strategies can be introduced as

1. Coverage maximizing
2. Connectivity maximizing
3. Energy efficiency and lifetime optimization
4. Multi-objective deployment.

A common method is as follows [300]:

1. During deployment, UAVs can measure the average of the **received signal strength indicator** (RSSI) values, and then they can stop and place wireless sensor nodes where the average of measured RSSI values falls below a pre-determined threshold.
2. Determining a fixed distance between each sensor node.

The important contribution for implementing autonomous deployment on the multi-UAV system is saving time. In [22], the meaning of the time is duration of the complication deployment task.

4.2.2.3.2 Service Optimization in a Convex Field

Let Q be a convex field in $\mathbb{R}^2$ with a group of n agents randomly distributed in Q where the position of the ith agent is denoted by p_i. Let $\Phi : Q \to \mathbb{R}^+$ be a priority function representing the likelihood of an event taking place at any arbitrary point $q \in Q$. This function can reflect a measure of the relative importance of different points in the field. Let the strictly increasing convex function be $f_i : \mathbb{R} \to \mathbb{R}^+, f_i(q) = \alpha_i \|p_i - q\|^2, i = 1, \ldots, n$ denoting the cost of serving an event taking place at point q by the ith agent, α_i are pre-specified strictly positive coefficients. The cost function can encode the travel time or the energy consumption required to serve any point in the field. Let S denote a set of n distinct weighted nodes $\{(S_1, w_1), \ldots, (S_n, w_n)\}$ in a 2D field where $w_i > 0$ is the weighting factor associated with the node $(S_i, w_i), i = 1, \ldots, n$.

Definition 173. *The **weighted distance** of a point q from the node $(S_i, w_i), i = 1, \ldots, n$ is defined as*

$$d_w(q, S_i) = \frac{d(q, S_i)}{w_i} \qquad (4.36)$$

where $d(q, S_i)$ denotes the Euclidean distance between q and S_i.

Problem 174. *Service optimization problem: The objective is to develop an appropriate agent-deployment algorithm such that the following cost function is minimized:*

$$H(P, W) = \sum_{i=1}^{n} \int_{W_i} f_i(q) \Phi(q) dq \qquad (4.37)$$

where the set $W = \{W_1, W_2, \ldots, W_n\}$ represents a partition of the field Q into n regions, where the ith agent is in charge of serving all points in region W_i.

Minimizing the above cost function implies maximizing the overall efficiency of the network. When f_i are agent-dependent, the conventional Voronoi partitioning is no longer optimal. The **multiplicatively weighted (MW) Voronoi diagram partition** is such that each region contains only one node, which is the nearest node, in the sense of weighted distance to any point inside the region. It is described as follows:

$$\Pi_i = \left\{ q \in \mathbb{R}^2; d_w(q, S_i) \leq d_w(q, S_j), j = 1, \ldots, i-1, i+1, \ldots, n \right\} \qquad (4.38)$$

Definition 175. *Given two points A and B in a 2D plane and a positive constant k, the **Apollonian circle** $\Omega_{S_i, S_j, w_i, w_j}$ is the locus of any point E such that $AE/BE = k$.*

The smallest region generated by the Apollonian circles and containing the ith node is the ith MW-Voronoi diagram.

Definition 176. *The mass and center of mass of a region W_i with respect to the priority function Φ are, respectively, defined as follows:*

$$M_{W_i} = \int_{W_i} \Phi(q) dq \qquad (4.39)$$

$$C_{W_i} = \frac{1}{M_{W_i}} \int_{W_i} q \Phi(q) dq \qquad (4.40)$$

A distributed coverage control law for double-integrator dynamics is proposed in [262,305].

4.3   PATROLLING

Patrolling is a standard way of addressing security threats. Securing citizens, infrastructures, forests, as well as wild animals is a critical concern around the world.

Definition 177. *Patrolling is the activity of going around or through an area at regular intervals for security purposes. The goal of patrolling is to repeatedly visit a set of positions and often to minimize the downtime between visits.*

However, security resources are often limited or expensive preventing full security coverage at all the times. Instead, these limited security resources can be allocated and scheduled efficiently while simultaneously taking into account adversary responses to the security coverage and potential uncertainty over such preferences and capabilities. It involves one or multiple decision-makers reasoning about the course of actions to achieve so as to cover an environment as quickly as possible [69].

The environment to patrol is commonly abstracted through a navigation graph, and numerous works explore spanning trees or graph partitioning to compute minimal-cost cycles that assign efficient routes for each UAV in the patrolling mission. **Auctions and market-based coordination, task allocation, artificial forces, Gaussian processes** theory, **evolutionary algorithms** and **linear programming** modeling are also popular. Agents may bid to exchange vertices of the patrol graph to increase overall patrol performance. Based on a topological representation of the environment and using global/centralized information, optimal patrolling could eventually be obtained for the single UAV case by solving the traveling salesperson problem. For the multi-UAVs situation, optimal performance depends on the environment topology and the number of UAVs. Theoretically superior performance can be obtained by either optimal **k-way** graph partitioning, especially for high number of agents or graphs with unbalanced edges; or having all UAVs following the same traveling salesperson problem cycle, equally distributed in time and space, especially for low number of agents or balanced graphs. Both the traveling salesperson problem and the graph partitioning problem are NP-hard. Therefore, solving these problems is non-trivial, particularly in sparse topologies, which is the case of most real-world environments [249].

For applications of multi-UAVs **persistent monitoring**, the goal can be to patrol the whole mission domain while driving the uncertainty of all targets in the mission domain to zero [268]. The uncertainty at each target point can be assumed to evolve nonlinearly in time. Given a closed path, multi-UAVs persistent monitoring with the minimum patrol period can be achieved by optimizing the UAV's moving speed and initial locations on the path [280].

Recent work in patrolling can be classified as [272]:

1. **Offline versus online:** Offline algorithm computes patrols before sensors are deployed, while online algorithm controls the sensor's motion during operation and is able to revise patrols after the environment has changed.
2. **Finite versus infinite:** Finite planning horizon algorithm computes patrols that maximize reward over finite horizon, while infinite horizon maximize an expected sum of rewards over an infinite horizon.
3. **Controlling patrolling versus single traversal:** This is dynamic environment monitoring versus a one snapshot of an environment.
4. **Strategic versus non-strategic patrolling**.
5. **Spatial or spatio-temporal dynamics**.

4.3.1 PERIMETER PATROL

The application of UAVs in perimeter surveillance missions can be assumed as a patrolling mission along defined paths: the perimeter [256]. The main challenges involved with target localization include maintaining tracks of all potential targets within sensor coverage region and developing platform trajectories, such that the target localization error is minimized. Maintaining the sensor's field of view over the target is extremely challenging due to certain factors such as limited UAV maneuverability, payload range constraints, and environmental disturbances like wind or turbulence.. The function of tracking multiple targets in an aerial surveillance mission is further affected due to much more complex sources of uncertainties such as false alarms, ambiguity in data association, lower detection probability, the target sudden maneuvers, closed spaced target trajectories with multiple miss detections, etc.

The patrolling problem has a graphical structure: V is the vertex set of that graph and E the edges set. Let L be an $|V| \times |V|$ matrix in which L_{ij} is a real number that represents the time required to go travel from i to j if $[i, j] \in E$ and is infinite otherwise. Each vertex i has a non-negative importance weight w_i. **Idleness** can be used as a performance measure. The idleness of vertex i, noted τ_i represents the time since the last visit of a UAV to that vertex. The idleness is 0 if and only if a UAV is currently at vertex i and $\tau_{i+1} = \tau_i + \Delta t$ if there are no visits to i in the time interval $(t, t + \Delta t)$. As idleness is an unbounded quantity, exponential idleness is used $k_i^t = b\tau_i^t$ with $0 < b < 1$. It can be seen as the expected value of a **Bernoulli random variable**, and k_i^t is the probability that this random variable is 1 at time t. b is the rate at which k_i decays over time. The probability evolves as $k_i^{t+\Delta t} = k_i^t b\Delta t$ if there are no visits to i during time interval $(t, t + \Delta t)$. If a UAV with noisy observations visits i at time t, idleness becomes 0 with probability $b < (1 - a) \leq 1$, where a is the probability that the idleness does not become 0 when a UAV visits a vertex. If n UAVs visit vertex i at time $t + \Delta t$ and there were no visits since time t, then

$$k_i^{t+\Delta t} = k_i^t b\Delta t + 1 - a^n \tag{4.41}$$

An instance of the patrolling problem is a tuple $\langle L, w, a, b \rangle$ consisting respectively of the matrix L of edge lengths, the vector w of importance weights and parameters a, b.

In patrol missions, the status of some sites must be monitored for events [229]. If a UAV must be close to a location to monitor it correctly and the number of UAVs does not allow covering each site simultaneously, a path planning problem arises.

Problem 178. *How should the UAVs visit the locations in order to make sure that the information about all locations is as accurate as possible?*

One patrolling algorithm is based on a graph patrolling formulation on which agents use reinforcement learning on a particular MDP. The MDP is defined over a countably infinite state space, assuming that as UAVs communicate by leaving messages on the nodes of the graph, leading to unrealistic communication model. Reactive algorithms, such as the **ant colony** approach, have been shown to perform well; however, this approach relies on the simplistic communication models relying on pheromones. When all locations are equally important, the shortest **Hamiltonian circuit** is an optimal solution for a single UAV. Multiagent strategies using a unique cycle are the best whatever the graph is. However, as some locations may be more important than others, not visiting the less important ones from time to time may be advantageous.

In [200,201], the following base perimeter **patrol problem** is addressed:

Problem 179. *A UAV (or more) and a remotely located operator cooperatively perform the task of perimeter patrol. **Alert stations** consisting of **unattended ground sensors** (UGS) are located at key locations along the perimeter. Upon detection of an incursion in its sector, an alert is flagged by the UGSs. The statistics of the alert arrival process are assumed known. A camera-equipped UAV is on continuous patrol along the perimeter and is tasked with inspecting unattended ground sensors with alerts. Once the UAV reaches a triggered unattended ground sensor, it captures imagery of the vicinity until the controller dictates it to move on.*

There are m alerts stations/sites on the perimeter where a nearby breaching of the perimeter by an intruder is flagged by an UGS. Upon detection of an incursion in its sector, an alert is flagged by the UGS. Camera-equipped UAVs are on a continuous patrol along the perimeter and their task is to inspect UGS with alerts. To determine whether an incursion flagged by a UGS is a false alarm or a real threat, a patrolling UAV flies to the alert site to investigate the alert. The longer a UAV loiters at an alert site, the more information it gathers; however, this also increases the delay in responding to other alerts. The decision problem for the UAV is to determine the **dwell time** so as to maximize the expected payoff. This perimeter patrol problem falls in the domain of discrete time controlled queuing systems. A **stochastic dynamic programming** approach may be employed to obtain optimal policies for the patrolling UAV. The customers are the flagged UGS/alerts waiting to be serviced, and the UAVs are the servers. Only unit/single buffer queuing is considered, for the UGS either flags an alert or it does not. Once it flags an alert, its state does not change even if additional triggering events were to occur, until the flag is reset by a loitering UAV. Thus, this perimeter patrol problem constitutes a **multi-queue multi-server**, unit buffer queuing system with deterministic interstate travel and service times. Because the UAVs are constantly on patrol or are servicing a triggered unattended ground sensor, the framework considered here is analogous to a cyclic polling system. The basic model of a **cyclic polling system** consists of separate queues with independent Poisson arrivals served by a single server in cyclic order. A related problem is the **dynamic traveling repairmen problem**, where the stations are not restricted to being on a line segment or a closed path. One is interested from a patrolling perspective in the optimal service time in addition to the dynamic scheduling of the server's movement. The basic question then would be to decide how long the server/UAV should dwell at a triggered alert station/UGS as well as in which direction is a bidirectional server.

One objective may be to maximize the information gained and at the same time reduce the expected response time to alerts elsewhere [230]. The problem is simplified by considering discrete time evolution equations for a finite fixed number m of UGS locations. The UAV has access to real-time information about the status of alerts at each alert station. Because the UAV is constantly on patrol and is servicing a triggered UGS, the problem is a **cyclic polling system** in the domain of discrete time controlled queuing system. The patrolled perimeter is a simple closed curve with $N \geq m$ nodes that are spatially uniformly separated of which m are the alert stations (UGS locations). The objective is to find a suitable policy that simultaneously minimizes the service delay and maximizes the information gained upon loitering. A stochastic optimal control problem is thus considered. A **Markov decision process** is solved in order to determine the optimal control policy [67]. However, its large size renders exact dynamic programming methods intractable. Therefore, a state aggregation-based approximate linear programming method is used instead, to construct provably good sub-optimal patrol policies. The state space is partitioned and the optimal cost-to-go or value function is restricted to be a constant over each partition. The resulting restricted system of linear inequalities embeds a family of Markov chains of lower dimension, one of which can be used to construct a lower bound on the optimal value function. The perimeter patrol problem exhibits a special structure that enables tractable linear programming formulation for the lower bound [27].

Remark 180. *Scalable swarm robotics [150] and game-theoretic [276] approaches allow taking into account adversary responses to the security strategy. The issue of designing security patrolling strategies using UAVs while providing strong performance guarantees requires guarantees on the level of security.* **Partially observable stochastic patrolling game** *(POSG) considers a general framework for modeling security patrolling problems with UAVs [161]. The goal of solving POSG is to find a patrol strategy that achieves the best security performance. In this game-theoretic framework, two teams are considered: defenders and attackers. The security game proceeds by alternating between defender and attacker decisions [247,283,284]. However, this security game repeats until the attackers get arrested. Other salient features include uncertain action effects and partial observability about both the state of the world and what each team does or plans to do.*

In multi-UAV Markov decision process, the problem is assumed to be fully observable; that is, each UAV has the same complete information to make its decision. In the patrolling problem, however, the actions of each UAV have a concurrent effect on the environment. These actions have also different durations. Concurrency in decision processes is modeled with a **generalized MDP**. Such decision processes generalize **multi-UAV MDP** to continuous time with asynchronous events. The state variables for this problem describe the position of each UAV and the idleness of each vertex. If the total number of UAVs is N, the state space is

$$S = V^N \times [0,1]^{|V|} \qquad (4.42)$$

Given some states $s = (v,k) \in S$, v_i is the position of the ith UAV and k_i the idleness of the ith vertex. At various time points, called **decision epochs**, the UAV must choose an action. The actions from which a UAV can choose depend on the structure of the graph and on its position: if a UAV is at vertex v, it can choose its action from $A_v = \{u : [v,u] \in E\}$. If a UAV occurs at time $t^{i+1} = t^i + L_{vu}$

and $v^t = v$ while $t \in [t^i, t^{i+1}]$ and $v^t = u$ as soon as $t = t^{i+1}$, the problem is concurrent because the decision epochs of all UAVs can be interleaved arbitrarily, each component k_i of k and the number of UAVs n. Let $\{y^j\}_j$ be the non-decreasing sequence of decision epochs, and n_i^j is the number of UAVs arriving at vertex i at time t^j, $\Delta t^j = \{t^{j+1} - t^j\}$:

$$k_i^{t^{j+1}} = k_i^{t^j} a^{n_i^{j+1}} b^{\Delta t^j} + 1 - a^{n_i^{j+1}} \tag{4.43}$$

The reward R is defined in terms of k, the rate at which reward is gained is given by

$$dR = w^T k^t dt \tag{4.44}$$

The discounted value function for a generalized Markov decision process is defined as

$$V^\pi(s) = E\left[\int_0^\infty \gamma^t dR\right] = E\left[\gamma^{t^j} \int_0^{\Delta t^j} \gamma^t w^T k^t dt\right] = E\left[\gamma^{t^j} w^T k^{t^j} \frac{(b\gamma)^{\Delta t^j} - 1}{\ln(b\gamma)}\right] \tag{4.45}$$

where $\gamma \in (0, 1]$ is the discount factor.

Online planning has the advantage that it solves equation (4.45) only for the current state, in contrast to offline algorithms that do so for all states. The patrolling problem is simpler to solve online than offline. **Anytime error minimization search** (AEMS) performs a heuristic search in the state space. The search procedure proceeds using a typical **branch-and-bound** scheme. Since the exact long-term value of any state is not exactly known, it is approximated using lower and upper bounds, guiding the expansion of the search tree by greedily reducing the error on the estimated value of the root node. In the patrolling problem, actions have the same interpretation as in a partially observable setting, whereas observations are the travel durations. In anytime error minimization search, the error is defined using the upper bound and the lower bound on the value of some state. Let $s \in S$ be a state, $L(s) \le V(s) \le U(s)$, where $L(s), U(s)$ are, respectively, the lower and upper bounds and $V(s)$ the actual value of s. Given some search tree T, whose set of leaf nodes is noted $F(T)$, the bounds for the root node are estimated recursively. When a UAV must choose an action, the action of maximum lower bound is chosen. A lower bound for the value of any state is the value of following any policy from that state. An upper bound is usually obtained by relaxing problem constraints, assuming that UAVs are ubiquitous (they can be in more than one location at the same time). Whenever a UAV reaches a vertex, it instantaneously multiplies itself and starts heading to adjacent unvisited locations. This bound estimates the shortest time that a team of UAVs would take to cover the entire graph and estimates through the discount factor and upper bound on the maximum reward obtainable. This bound implicitly assumes that the optimal policy does not require having more than one UAV at any vertex.

Extending anytime error minimization search to asynchronous multi-UAV is simple: whenever a node is expanded, there is a branch for every joint action and observation. **Asynchronicity** is handled with state augmentation. The state is now (s, η) where η_i is the time remaining before the next decision epoch of UAV i. At anytime t, the next decision epoch happens at time $t + \min_i\{\eta_i\}$. The expand operation adds actions and observations for any UAV for which $\eta = 0$. Whenever UAV i performs an action of duration Δt, η_i is assigned Δt. Otherwise, η_i is updated according to its depth in the search tree. Anytime error minimization search can be used to perform online planning for any subset of UAVs. The complexity being exponential in the number of UAVs, these UAVs are thus coordinated locally and a partial order is defined among UAVs. A UAV is said to be greater than (respectively less than) another UAV if it must choose its policy before (respectively after). The UAVs compute their policy according to that order. Once a UAV knows the policies of all greater UAVs, it proceeds to compute its policy and then communicates it to the lesser UAVs. Whenever a UAV selects its policy, it chooses the best policy given the policy of greater UAVs. A useful property of this coordination algorithm is that if the UAVs use an online anytime planner, then it is also anytime and online. A **fallback strategy** is to ignore the presence of the greater UAVs until their policy has been received.

4.3.2 AREA COOPERATIVE PATROLLING

This section considers a problem where the area of interest has to be patrolled by a team of multiple UAVs from a frequency-based approach. These approaches assume that the probability of event appearing is equal along the whole area and are useful when there is not a priori information about the problem. The objective is to cover any position in the area with the maximum frequency, which is equivalent to minimize the elapsed time between each two consecutive visits to any position or refresh time. A solution where each position in the path is visited with the same period is the optimal solution to cover a path with a set of UAVs. Different cooperative patrolling strategies may be analyzed and compared from the elapsed time (frequency-based) and latency time criteria. The area partitioning strategy allows the system to obtain theoretically the optimal performance from frequency-based approach (**elapsed time criterion**) if the sub-areas are sized correctly. Surveillance scenarios usually involve large areas where the communications among the UAVs and with the control stations cannot be guaranteed. In order to improve the robustness and scalability of the whole system, a distributed coordination control is the most efficient option to approach the surveillance missions with multiple UAVs. The distributed coordination method should allow the multi-UAV system to converge to a common cooperative patrolling strategy from local decisions and asynchronous communications among pair of them. Some authors propose algorithms based on peer-to-peer (or one-to-one) coordination. The algorithms based on the one-to-one coordination assume a different problem to be solved by each pair of contacting UAVs. Each pair of contacting UAVs addresses a reduced version of the whole problem considering just their own information. These algorithms require low information storage capabilities for the UAVs because they just have to store their own local information. Moreover, this technique has proved to converge to the desired solution, but its convergence time complexity is increased quadratically with the total number of UAVs. On the other hand, the algorithms based on the coordination variables assume that the problem can be totally described by a limited set of variables (the coordination variables) and that, using these variables, each UAV can solve independently the whole problem [6].

4.3.2.1 Multiple Depot Multi-TSP

The multi-UAV patrolling can be cast as a multiple depot **multiple traveling salesmen problem** (MTSP), where a set of m UAVs, located at different initial positions, must visit a set of n target locations and come back to their depots. The main objective is to find an efficient assignment of the target locations to the team of UAVs such that all the targets are covered by exactly one UAV and the cost is minimal. A distributed solution based on a market-based approach is presented in [212], involving the cooperation of the UAVs to incrementally allocate targets and remove possible overlap. In each step, a UAV moves and attempts to improve its solution while communicating with its neighbors. The approach consists of four main phases: an initial target allocation followed by a tour construction, then negotiation of conflicting targets and finally solution improvement.

The MTSP arises when several Hamiltonian tours are considered. It can in general be defined as follows:

Problem 181. *Multiple traveling salesmen problem: Given a set of nodes, let there be m salesmen located at a single-depot node. The remaining nodes that are to be visited are called intermediate nodes. Then, the MTSP consists of finding tours for all m salesmen, who all start and end at the depot, such that each intermediate node is visited exactly once and the total cost of visiting all nodes is minimized.*

The cost metric can be defined in terms of distance and time [39]:

1. **Single versus multiple depots:** In the single-depot case, all UAVs start from and end their tours at a single point. On the other hand, if there exist multiple depots with a number of

UAVs located at each, the UAVs can either return to their original depot after completing their tour or return to any depot with the restriction that the initial number of UAVs at each depot remains the same after all the travel. The former is referred to as the fixed destination case, whereas the latter is named as the non-fixed destination case.

2. **Number of UAVs:** The number of UAVs in the problem may be a bounded variable or fixed a priori.
3. **Fixed charges:** When the number of UAVs in the problem is not fixed, then each UAV usually has an associated fixed cost incurring whenever this UAV is used in the solution.
4. **Time windows:** In this variation, certain nodes need to be visited in specific time periods, named as time windows. This is the **multiple traveling salesperson problem with time windows** (MTSPTW).
5. **Other special restrictions:** These restrictions may consist of bounds on the number of nodes each UAV visits, the maximum or minimum distance a UAV travels or other special constraints.

The MTSP can be defined by a complete graph $G(V, A)$ where $V \{0\}$ is the set of nodes to be visited, node 0 is the depot and A is the set of arcs connecting the n nodes to be visited. A c_{ijk} value is associated with each arc (i, j) which is traversed by the salesperson k. The matrix C is symmetric when an undirected graph is considered and asymmetric otherwise. The traveled distance of a tour is calculated as the sum of the lengths of the arcs belonging to the tour. In addition, the travel time of a salesperson for each arc (i, j) of A is given by t_{ijk}. The travel time of a tour is calculated as the sum of the lengths of the arcs belonging to the tour. Given an integer m, the problem consists of finding m tours passing through the depot node. Each node must be visited by a tour only once so that the total traveled distance is minimized and the working times of the traveling salesmen are similar. The multi-objective version of the MTSP can be formulated as a multi-objective mixed-integer linear model with two objective functions:

1. Minimization of the distance traveled by all the salesmen
2. Similar working times of the salesmen.

A multi-objective optimization problem can be formulated through a mathematical model defined by a set of h objective functions which must be minimized or maximized subject to a set of m inequality constraints, a set of ℓ equality constraints and lower and upper limits in the k decision variables. With each arc (i, j) of G?? is associated a binary variable which takes the value of 1 if arc (i, j) was traversed by the salesperson k?? into the final solution and 0 otherwise [62].

Problem 182. *The optimization problem can be formulated as*

$$\min Z_1 = \sum_{k=1}^{m} \sum_{i=1}^{n} \sum_{j=1}^{n} c_{ijk} x_{ijk} \tag{4.46}$$

$$\min Z_2 = \sum_{k=1}^{m} \left| t_{avg} - t_k \right| \tag{4.47}$$

where t_{avg} is the average travel times of the tours defined as

$$t_{avg} = \frac{1}{m} \sum_{k=1}^{m} \sum_{i=1}^{n} \sum_{j=1}^{n} t_{ijk} x_{ijk} \tag{4.48}$$

and t_k is the time spent on each tour $k, k = 1, \ldots, m$:

$$t_k = \sum_{i=1}^{n} \sum_{j=1}^{n} t_{ijk} x_{ijk} \tag{4.49}$$

The following set of constraints must be also considered, with the binary decision variables $x_{ijk} \in \{0,1\}$:

$$\sum_{k=1}^{m} \sum_{i=1}^{n} x_{ijk} = 1, \forall j = 1, \ldots, n \tag{4.50}$$

$$\sum_{k=1}^{m} \sum_{j=1}^{n} x_{ijk} = 1, \forall i = 1, \ldots, n \tag{4.51}$$

These two constraints are related to the degree restrictions and ensure that each node is visited only once by a single UAV, except the depot node.

$$\sum_{k=1}^{m} \sum_{i=1}^{n} x_{i0k} = m \tag{4.52}$$

$$\sum_{k=1}^{m} \sum_{j=1}^{n} x_{0jk} = m \tag{4.53}$$

These two constraints ensure that each UAV must leave and return to the depot node.

$$\sum_{i \in S} \sum_{j \notin S} x_{ijk} \geq 1, \forall k = 1, \ldots, m, \forall S \subseteq V; 0 \in S \tag{4.54}$$

This constraint is the connectivity constraint, avoiding sub-tours generation on the final solution.

Many approaches can be proposed to solve this problem such as the computational intelligence with **genetic algorithm** [207], **artificial bee colony** algorithm [281] and **a set covering approach** [35].

4.3.2.2 Exploration

There are different strategies proposed for robotic exploration [277]:

1. **Frontier-based exploration:** A group of UAVs are directed to regions on the boundary between unexplored space and the space known to be open. In this strategy, each UAV builds and maintains its global map. During the mission, each UAV shares perceptual information with the others.
2. **Role-based exploration:** UAVs are categorized in two groups:
   a. **Relays:** They maintain the connection between explorers and a base station responsible for controlling the mission. If relays discover information about the zone while relaying, this information is added to the exploration team knowledge.
   b. **Explorers:** They explore an unknown zone and communicate their findings to a relay in previously agreed rendezvous points. They use frontier-based exploration strategy.
3. **Market-driven exploration:** The exploration task is divided into sub-tasks. In this strategy, UAVs place bids on the sub-tasks, and when communication strength is factored into the bids, UAVs avoid going beyond communication range.

The design of a multi-target exploration/visiting strategy for a team of UAVs in a cluttered environment is able to

1. Allow visiting multiple targets at once (for increasing the efficiency of the exploration),
2. Always guaranteeing connectivity maintenance of the group despite some typical sensing/communication constraints representative of real-world situations,
3. Without requiring presence of central nodes or processing units
4. Without requiring that all the targets are known at the beginning of the task.

Designing a decentralized strategy that combines multi-target exploration and **continuous connectivity** maintenance imposes often antithetical constraints. A fixed topology and centralized method which, using a virtual chain of mobile antennas, may be able to maintain the communication link between a ground station and a single UAV visiting a given sequence of target points. A similar problem is addressed by resorting to a partially centralized method where a linear programming problem is solved at every step of motion in order to mix the derivative of the second smallest eigenvalue of a weighted **Laplacian** or algebraic connectivity, or **Fiedler eigenvalue** and the **k-connectivity** of the system. A line-of-sight communication model is considered where a centralized approach, based on polygonal decomposition of the known environment, is used to address the problem of deploying a group of UAVs while achieving periodical connectivity. The case of periodical connectivity optimally solves the problem of patrolling a set of points to be visited as often as possible [236]. Triangulations in [217] provide complete coverage with only basic local geometry. The underlying topological structure of a triangulation allows to exploit its dual graph for mapping and routing, with performance guarantees for these purposes. The maximum edge length of a triangle is bounded by the communications range of the UAVs. If the number of available UAVs is not bounded a priori, the problem of minimizing their number for covering all of the region is known as the **minimum relay triangulation problem** (MRTP); if their number is fixed, the objective is to maximize the covered area, which is known as the **maximum area triangulation problem** (MATP). Both problems have been studied both for the offline scenario, in which the region is fully known, and the online scenario, where the region is not known in advance.

4.3.2.2.1 *Covering in Minimal Time*

For the exploration and rescue application, an important goal is covering the whole environment to find possible victims in a minimal time. Compared with single-UAV systems, multi-UAV systems not only speed up the exploration process but also have advantages of robustness and redundancy. In addition, multi-UAV systems can finish tasks that the single UAV systems cannot do. In order to utilize the advantages of multi-UAV systems, it is essential to have a coordinated algorithm which realizes **moving target selection** for each UAV while considering the environment status and the moving targets of other UAVs. A primary method for multi-UAV coordination is **market economy-based approach**. All UAVs in the team trade tasks and resources with each other to maximize individual profits. In the **threshold-based approach**, each UAV has to handle the closest event, and without duplicating another UAV's work, the stimulus $\sigma(r, e)$ produced by an event e for UAV r can be proposed as

$$\sigma(r, e) = \frac{1}{d(r, e)} \quad \theta_e = \frac{1}{D_r} \quad p_e = \frac{\sigma(r, e)^n}{\sigma(r, e)^n + \theta_e^n} \tag{4.55}$$

In [199], the best performance was found when the threshold θ_e for every event e is equal to the inverse of expected distance D_r between the UAVs. This threshold value is consistent with the best reserve price for the market.

4.3.2.3 Coordination in a Unknown Environment

A **distributed bidding algorithm** (DBA) can be proposed to coordinate the multi-UAV systems exploring unknown environments where a nearness measure was introduced in bidding to maintain all UAVs close to each other to overcome the shortage of communication range. On the other hand, a decision theoretic approach dispersed UAVs over an unknown environment through a utility calculation scheme that a UAV utility reduction was inversely related to distances to targets of other UAVs. For a known environment, task allocation is realized by repeating the auction algorithm when a UAV finished its task. More recently, an **energy consumption-based bidding scheme** was proposed. In the proposed approach, a hierarchical coordinated architecture for multi-UAV exploring an unknown environment is proposed where the lower level is developed by maximizing two kinds of new utility, respectively. The other kind of improvement is to deal with heterogeneity in multi-UAV systems. The method developed is that each UAV offers three bids from the aspects of the shortest-path length, moving speed and battery life, respectively. A back-propagation neural network-based approach was proposed to decide which UAV was the winner. The method used vectors to express capabilities of UAV and task, respectively. The auctioneer broadcasts a task and its capability vector, and each UAV bids for the task based on its capability. The task allocation was realized within the framework of **contract net protocol**. Finally, an index for describing exploration performance of each UAV is proposed for a heterogeneous multi-UAV system. However, the approaches mentioned above depend on a perfect communication among all teammate UAVs. Different to the existing results, the proposed algorithm in [98] improves the market economy-based approach for the purpose of dealing with both target selection and heterogeneity. **Colored Petri nets** were used to realize the complex and concurrent conversations between agents. According to whether there is an installed task detection device, all UAVs are categorized into initiators or helpers, respectively. The latter assisted the former to finish task when it matched the task requirements. The UAVs can change their teammate selection strategies to adapt to a dynamic environment.

4.3.2.3.1 Communication

Early exploration strategies were based on the principle of keeping the UAVs within communication range. Some authors propose market-driven exploration strategies, in which an exploration task is divided into sub-tasks and UAVs place bids on these sub-tasks. In these strategies, goal point selection is an important decision and aims to select unexplored regions for UAVs to visit. Here, bids are based on specific values such as traveling cost to a particular location and expected information gain. Although a market-driven exploration approach does not rely on perfect communication and is still functional with zero communication, when communication strength is factored into these bids, UAVs avoid going beyond the communication range. Goal point selection strategies are

1. **Random goal point** selection is the simplest strategy, in which goal points are chosen randomly. If the area that surrounds the goal point has already been visited, it is discarded. This strategy has been effectively used in practice.
2. In the **greedy exploration strategy**, a goal point in the closest unexplored region is selected as a candidate, and this strategy is effective for single UAV exploration.
3. **Frontier-based exploration** is an exploration strategy with a key idea of gaining the most new information by moving to the boundary between open space and uncharted territory. In this strategy, UAVs share perceptual information while maintaining separate global maps. Even if UAVs are dispersed to the maximum extent which their communication ranges allow, unexplored spaces may remain.

A solution to deal with the limited communication during multiple UAV exploration is rendezvous points in which shared knowledge is communicated. **Clustering approach** was proposed by some authors, in which groups of UAVs stay close together while exploring the zone. Another strategy using the same principle is **role-based exploration**. In this strategy, UAVs are categorized into two groups: explorers and relays. While explorers explore an unknown zone by using frontier-based exploration strategy and communicate their findings to a relay in rendezvous points, relays maintain the connection between the base station responsible for the mission and the explorers. Role-based exploration strategy offers a solution to connectivity-related issues in large environments at the expense of additional UAVs responsible for messaging. Though most exploration strategies are successful in maintaining connectivity during an exploration mission, their performances are limited due to the constraint of having to keep the UAVs within communication range [277].

4.3.2.3.2 *Ergodic Trajectories*

In the context of exploration, **ergodic trajectory optimization** computes control laws that drive a dynamic system along trajectories such that the amount of time spent in regions of the state space is proportional to the expected information gain in those regions. Using **ergodicity** as a metric encodes both exploration and exploitation, both needed for non-myopic search when variance is high and convexity is lost, as well as myopic search when variance is low and the problem is convex. By encoding these needs into a metric, generalization to nonlinear dynamics is possible using tools from optimal control. Different dynamical systems can achieve nearly identical estimation performance using **ergodic exploration of distributed information**. The ability to actively explore and respond to uncertain scenarios is critical in enabling UAVs to function autonomously. Active sensing or sensor path planning refers to control of sensor parameters, such as position, to acquire information or reduce uncertainty. Planning for search/exploration is challenging as the planning step necessarily depends not only on the sensor being used but also on the parameter being estimated, such as target location versus target size. Methods for representing and updating the estimate and associated uncertainty and the belief state to determine expected information are, therefore, required. The best choice for representing and updating the belief state for a given application depends on UAV dynamics, sensor physics, and the estimation task (modeling a field versus target localization). Ergodic theory relates the time-averaged behavior of a system to the space of all possible states of the system. Ergodicity can be used to compare the statistics of a search trajectory to a map of **expected information density** (EID). An efficient exploration strategy should spend more time exploring regions of space with higher expected information, where useful measurements are most likely to be found. The UAV should not, however, only visit the highest information region but distribute the amount of time spent searching proportional to the overall expected information density. This is the key distinction between using **ergodicity** as an objective and previous work in **active sensing** (e.g., information maximization). The ergodic metric encodes the idea that measurements should be distributed among regions of highly expected information. Information maximizing strategies otherwise require heuristics in order to force subsequent measurements away from previously sampled regions so as not to only sample the information maxima [232].

4.3.2.3.3 *Human–UAV Collaboration*

To realize effective **human–UAV collaboration** (HAC) under uncertainty, in the search task, a UAV is capable of searching the mission domain autonomously by utilizing its cameras, thus allowing the exploration of a potentially dangerous environment that would otherwise be unsafe for a human to enter. However, due to the limitations in sensing capabilities and image processing, it is difficult for a UAV to confirm a found object of interest (e.g., to distinguish between human, animal, and mechanical movements). This might lead to errors (missed detections or false alarms) and hence low overall task performance. It is therefore necessary to have timely and more effective manual processing of these camera images. When situation demands, the UAV will send an intervention request to the human operator through a human–machine interface. However, keeping human monitoring for a long duration will probably make the operator cognitively overloaded due to decrements in vigilance, which consequently degrades the overall performance. This is especially true in the case when the human operator is required to collaborate with multiple UAVs in a large-scale task domain. Hence, a reasonable solution is to switch between the manual and autonomous operations to balance human workload while guaranteeing a desired level of task efficiency and accuracy. This type of mixed autonomous operation and tele-operation is called tele-autonomous operation. Existing decision-making approaches such as Bayesian methods and data-driven techniques for process monitoring and control almost always seek optimal solutions based on measurements in the presence of uncertainty [278].

4.3.2.3.4   Uncertain Terrain

UAVs often fly in a complicated environment. Many threats such as hills, trees, and other aircrafts can be fatal in causing the UAV to crash. These threats, in general, can only be detected within a limited range from a single UAV. However, by sharing information with other UAVs, these threats can be detected over a longer distance. Furthermore, an effective path for navigation should be smooth, should provide an escape route and must be computationally efficient. In the previous work on path planning for a single UAV, Voronoi graph search and visibility graph search have been proven to be effective only in a simple environment. They are not real-time and also lead to fatal failure when map information is not entirely available, e.g., when obstacles are not detected. Path planning of multiple UAVs concentrates on the **collaborative framework, collaborative strategies** and **consistency**. The Voronoi graph search and the A* or Dijkstra algorithms plan a global path for multiple UAVs to simultaneously reach the target in an exhaustive procedure [19,20,228]. Path planning of multiple UAVs can also be addressed from the perspective of reinforcement learning. Q-learning is a way to solve the path planning problem. The basic idea of Q-learning is to obtain the optimal control strategy from the delayed rewards according to the observed state of the environment in a learning map and to make a control strategy to select the action to achieve the purpose.

Modeling of probabilistic risk exposure to obstacles can be presented as follows. It is necessary for a UAV to keep a certain distance from regions of high risk to ensure safe flying. So the measure of probabilistic risk exposure to obstacles can be seen as a continuous distribution function. For example, consider the case where an obstacle is at position (x_i, y_i, z_i). The measure of the risk is denoted by F_i, in which the parameters are related to the dimension of the planning space. In the 3D space,

$$F_i(x, y, z) = \frac{1}{\sqrt{2\pi}\sigma_i} \exp\left(-\frac{d_i^2}{2\sigma_i}\right) \tag{4.56}$$

where

$$d_i = \sqrt{(x - x_i)^2 + (y - y_i)^2 + (z - z_i)^2} \tag{4.57}$$

σ_i is an adjustable parameter. The probabilistic risk of the area where UAVs cannot fly over can be represented as a very big value. Furthermore, when more than one obstacle exists on the map, the probabilistic risk at position (x, y, z) can be calculated as

$$F(x, y, z) = 1 - \prod_{i=1}^{M} [1 - f_i(x, y, z)] \tag{4.58}$$

The key idea of cooperative and geometric learning algorithm is to calculate the cost matrix $\mathbf{G}$ which can be used to find the optimal path from a starting point to a target point in terms of distance and integral risk. Each element in the matrix $\mathbf{G}$ is defined to be the sum of costs from its position to a target point. The process underlying path planning in an unknown environment can be modeled as a controlled Markov chain or an MDP. In the case of an unknown environment, the transition probabilities of the underlying process are not known and have to be estimated while planning paths in the environment. This corresponds to **adaptive control of an MDP**. The state space is defined by the ordered pair (s, q) where s represents the system state and q denotes the environment state. The system state is assumed to be known perfectly, while the environment state may be noise corrupted. It is assumed that the terrain does not change throughout the duration of the path planning maneuvers. If the environment at any system state can be modeled as a stationary random process, the method developed in [80] can be applied. The system state is assumed to be finite (for example, in 2D, eight states: North, East, South, West, North-East, North-West, South-East, South-West; in 3D, 26 possible states), N is the number of system states, D the number of environment states, M the number of controls and U the set of control actions.

Problem 183. *Let* $\mathbb{F}^T = \{(s_0, q_0), u_0, \ldots, (s_{t-1}, q_{t-1}), u_{t-1}\}$ *represent the history of the process till t; then*

$$ProbProb\left((s_t, q_t)|\mathbb{F}^T\right) = Prob\left((s_t, q_t)|(s_{t-1}, q_{t-1}), u_{t-1}\right) \tag{4.59}$$

The dynamics of the system can be represented probabilistically through the probability density function $Prob\left((s_t, q_t)|(s_{t-1}, q_{t-1}), u_{t-1}\right)$ where the function represents the probability of the system making a transition to state (s_t, q_t) given that the current state is (s_{t-1}, q_{t-1}) and the current control is u_{t-1}.

The current system state s_t is assumed to be independent of the past environment state q_{t-1}:

$$\text{Prob}\left(s_t|(s_{t-1}, q_{t-1}), u_{t-1}\right) = \text{Prob}\left(s_t|s_{t-1}, u_{t-1}\right) \tag{4.60}$$

The above assumption is satisfied if the UAV is controllable. The current environment state q_t is dependent only on the current system state:

$$\text{Prob}\left((s_t, q_t)|(s_{t-1}, q_{t-1}), u_{t-1}\right) = \text{Prob}\left(s_t|s_{t-1}, u_{t-1}\right)\text{Prob}\left(q_t|s_t\right) \tag{4.61}$$

The probability distribution function quantifies the localization and the control uncertainties inherent in the system and is assumed to be known beforehand. The terrain/environment uncertainty $p(q_t, s_t)$ is unknown, and successive estimates $p_t(q_t, s_t)$ are made of the uncertainty as the planning proceeds to completion. The question is then how to use this increasing information of the terrain in order to plan better paths. In fact, the goal of path planning can be framed as an **infinite horizon discounted stochastic optimization problem**. The above assumption defines the goal of path planning in an uncertain terrain, i.e., the average value of the total discounted cost incurred by the system needs to be minimized. Adaptive control involves controlling an uncertain system while simultaneously obtaining better estimates of the system parameters. The problem of path planning can be envisaged as one of adaptive control of an uncertain MDP; i.e., the transition probabilities are not known. In such a scenario, the strategy of adaptive control is to use the policy that is optimal with respect to the current estimate of the system, since it corresponds to the current knowledge of the system that is being controlled and referred to as the certainty equivalence principle in adaptive control. This is the equivalent of the **persistent excitation condition** in adaptive control which seeks to tackle the exploration/exploitation trade-off inherent in every adaptive control problem. More details on this implementation can be found in [80].

4.3.2.3.5 *Rescue in Uncertain Adversarial Environment*

The rescue path planning problem is a variant of a resource-and-risk-constrained shortest-path planning. Nonlinearity characterizing problem complexity introduces a composite measure of performance coupling rescue UAV **survivability**, and path length subject to multiple side constraints makes problem solving very challenging [42]. Examples may be hurricane or earthquake rescue tasks [16]. The simplest rescue path planning problem in an uncertain adversarial environment involves a UAV planning a trajectory over a certain 2D area avoiding obstacles and threats in order to rescue stranded individuals while minimizing a combination of traveled distance and threat level exposure of an equivalent survivability measure subject to a variety of temporal, itinerary and/or survivability constraints such as deadline or survivability threshold. The UAV is assumed to move from a base station (source site s) to a rescue service point. The adversarial environment is pictured as a grid or lattice composed of N cells. Cells are colored using a spectrum of shades to depict various threat levels (or survivability) ranging from white for safe cells to red for physical obstruction or zero for survivability cells. Modeled as a directed graph $G(V,A)$, the grid specifies a set of vertices or nodes representing individual cells connected to one another (center to center) through a set A of arcs (i, j). Arcs define neighborhood relationship or possible moves between adjacent cells i, j. Typically, in a 2D environment, a non-boundary node i has eight outgoing arcs and eight incoming arcs.

The rescuing UAV is assumed to move one step at a time. Map representation uses a discretized probabilistic threat risk exposure of survivability. The grid is initially built from the probabilistic threat presence $p_T(\ell)$ estimated from a previous search task. Threat risk exposure, UAV destruction or detection risk contribution $p_{\text{dest}}(\ell, j)$ in cell j from threat presence in cell ℓ for a UAV traveling over a given cell is then derived from the map as

$$risk_j = 1 - \underbrace{1 - p_{\text{dest}}(\ell, j) p_T(\ell)}_{survivability_j} \tag{4.62}$$

or

$$risk_j = 1 - p_{s_j} \tag{4.63}$$

where p_{s_j} is the UAV probability to survive cell j visit.

The basic rescuing path planning problem consists in minimizing traveled distance and threat risk exposure measures. Given user preferences or threat risk exposure over a traveled distance, a UAV must move from a source node s to a destination site d over an N-cell grid environment characterized by a survivability distribution P_s, progressively constructing a path by making decisions x_{ij}, binary decision variables on visiting arcs (i, j). The ideal solution would be to find the shortest and safest paths separating source s and destination d sites subject to various constraints. The problem can be formulated as follows:

Problem 184.

$$\min_{x_{ij}} (F = (1 - \alpha)L + \alpha(1 - S)) \tag{4.64}$$

where

$$0 \leq L \leq 1 \ and \ 0 \leq \alpha \leq 1 \tag{4.65}$$

with

$$L = \frac{\sum_{(i,j)\in A} d_{ij} x_{ij} - L_{\min}}{L_{\max} - L_{\min}} \tag{4.66}$$

and

$$S = \prod_{(i,j)\in A} p_{sj} \tag{4.67}$$

subject to

$$\sum_{j\in V} x_{sj} = 1 \quad path \ starts \ at \ source \ node, s \tag{4.68}$$

$$\sum_{j\in V} x_{jd} = 1 \quad path \ ends \ at \ destination \ node, d \tag{4.69}$$

$$\sum_{i\in V} x_{ij} \leq 1 \quad at \ most \ one \ visit \ per \ node \tag{4.70}$$

$$t_i^s + d_{ij} - K(1 - x_{ij}) \leq t_j^s \quad (i, j) \in A, k \in \mathbb{N} \quad elimination \ of \ disjoint \ sub\text{-}tours \tag{4.71}$$

$$\sum_{j\in V} x_{ij} - \sum_{j\in V} x_{ji} = 0, \forall i \in V/\{s,d\}, (i, j) \in A \quad flow \ constraint/continuity \tag{4.72}$$

$$\sum_{(i,j)\in A} d_{ij} x_{ij} \leq L_{\max}, \forall i \in V/s, d, (i, j) \in A \quad itinerary \ constraint/deadline \tag{4.73}$$

$$- \sum_{(i,j)\in A} x_{ij} \log(p_{sj} \leq \log(S_m), S_m \in [0,1] \quad survivability \ threshold \tag{4.74}$$

$$x_{ij} = 0 \forall (i, j) \notin A \tag{4.75}$$

$$x_{ij} \in \{0, 1\}, t_i^s \geq 0 for \ i \in V \tag{4.76}$$

*where F is the objective function, α is a user-defined threat exposure, S_m is the minimum surviv-
ability probability, P_s survival probability distribution over the N-cell grid, x_{ij} is a binary decision
variable representing the UAV positive ($x_{ij} = 1$) or negative ($x_{ij} = 0$) decision to travel along arc
(i, j), d_{ij} is the distance separating centroids from cells i and j, $L_{\max}$ is user-defined maximum
distance or $= 2\sqrt{N}$ and $L_{\min} = \sqrt{2N}$. They are, respectively, the maximal and minimal possible
traveled distances from s to d, and S is the overall UAV survivability on path completion.*

The **survivability-biased risk-free path** $α = 1$ attempts to minimize any threat exposure at the
expense of extra travel cost, whereas the strictly distance-biased route $α = 0$ slightly moves around
the deadly core of the threat to satisfy minimal survivability requirements. This problem is then
simplified to a quadratic objective function approximation through logarithmic survivability function transformation. It consists in replacing S by an approximate function $\bar{S}$ having the property to
make the resulting approximate quadratic program formulation capture key elements of the original
problem [42].

One computationally low-cost and efficient online complete coverage path planning approach
for an unknown environment is presented in [205]. The coverage task is performed by following
an online **boustrophedon** motion along with an efficient backtracking technique called **two-way
proximity search** (TWPS) to reduce total coverage time. The proposed algorithm generates the
shortest possible path for backtracking. For the complete coverage path planning, it is assumed that
the environment must be closed, where all the regions are connected so that any accessible position
within the environment is reachable by UAV.

4.4 FORAGING

Foraging robots are mobile robots capable of searching for and, when found, transporting objects to one or more collection points. Foraging robots may be single robots operating individually or multiple robots operating collectively. Single foraging robots may be remotely tele-operated
or semi-autonomous; multiple foraging robots are more likely to be fully autonomous systems.
The study of foraging strategies has led to stochastic optimization methods such as **ant colony
optimization** (ACO), **particle swarm optimization** (PSO), **bacterial foraging optimization algorithm** (BFOA), **artificial bee colony** (ABC) and **information foraging theory** (IFT), among
others [82].

4.4.1 PROBLEM FORMULATION

4.4.1.1 Abstract Model

A **finite state machine** (FSM) can represent a foraging UAV. In the model, the UAV is always in
one of four states [279]:

1. **Searching:** In this state, the UAV is physically moving through the search space using
   its sensors to locate and recognize the target items. At this level of abstraction, it could,
   for instance, wander at random, or it could employ a systematic strategy such as moving
   alternately left and right in a search pattern. The fact that the UAV has to search follows
   from the real-world assumptions that either the UAV's sensors are of short range or the
   items are hidden (behind occluding obstacles for instance); in either event, the UAV cannot
   find items simply by staying in one place and scanning the whole environment with its
   sensors. Object identification or recognition could require one of a wide range of sensors
   and techniques. When the UAV finds an item, it changes state from searching to grabbing.
   If the UAV fails to find the target item, then it remains in the searching state forever;
   searching is therefore the default state.

2. **Grabbing:** In this state, the UAV physically captures and grabs the item ready to transport it back to the home region. Here a single UAV is capable of grabbing and conveying an item. As soon as the item has been grabbed, the UAV will change state to homing.

3. **Homing:** In this state, the UAV must move, with its collected object, to a home or nest region. Homing clearly requires a number of stages: firstly, determination of the position of the home region relative to where the UAV is now; secondly, orientation toward that position and, thirdly, navigation to the home region. Again there are a number of strategies for homing: one would be to retrace the UAV's path back to the home region; another would be to home in on a beacon with a long-range beacon sensor. When the UAV has successfully reached the home region, it will change state to depositing.

4. **Depositing:** In this state, the UAV deposits or delivers the item in the home region and then immediately changes state to searching and hence resumes its search.

Successful object collection and retrieval can be formulated as follows:

$$F(O_i, t) = \left\{ \begin{array}{ll} 1 & \text{Object } O_i \text{ is in a sink at time } t \\ 0 & \text{Otherwise} \end{array} \right\} \tag{4.77}$$

If the foraging task is performance time limited and the objective is to maximize the number of objects foraged within fixed time T, then a performance metric for the number of objects collected in time T can be defined:

$$P = \sum_{i=1}^{N_0} F(O_i, t_0 + T) \tag{4.78}$$

where N_o is the number of objects available for collection and t_0 is the start time. A metric for the number of objects foraged per second $P_t = P/T$ is independent of the number of UAVs. In order to measure the performance improvement of multi-UAV foraging, the **normalized performance** P_m of a multi-UAV system

$$P_m = \frac{P}{N_r} \tag{4.79}$$

where N_r is the total number of UAVs. The **efficiency** of multi-UAV foraging is then the ratio P_m/P_s.

4.4.1.2　Continuous Foraging

In continuous foraging, UAVs visit locations in the environment to forage resources and deliver them to a home location. The resources replenish over time, and three models of resource replenishment can be considered: the first two are the **Bernoulli** and **Poisson** models where resources replenish probabilistically, which are suitable for scenarios with independently occurring resources. The third model is the **stochastic logistic model** where the rate of resource replenishment depends on the number of existing resources. The logistic model is suitable when the resources are populations of living things. The problem can be formulated as follows [220].

Problem 185. *Maximize the rate of resources foraged to l_0, the UAVs' home location after T time steps; i.e., maximize $\frac{v_{0,T}}{T}$ where $A = \{a_1, \ldots, a_n\}$ be the foraging UAVs; each UAV a_i has an associated speed V_i, maximum capacity c_i and payload $y_i \leq c_i$, i.e., the number of resources a_i is currently carrying. Let R be the reconnaissance UAV that performs information gathering and $L = \{l_0, \ldots, l_m\}$ be the set of locations, where $v_{j,t}$ be the number of resources available at location l_j at time step t. The number of resources at each location changes over time, in a Markovian way; i.e., the number of resources $v_{j,t}$ at a location l_j at time step t depends only on $v_{j,t-1}$. Let $\hat{v}_{j,t}^{(i)}$ be a_i's estimate of $v_{j,t}$. When a foraging UAV a_i arrives at a location $l_j (j > 0)$, $max(v_{j,t}, c_i - y_i)$ resources at l_j (i.e., all resources at l_j subject to the remaining capacity of a_i) are transferred to a_i, and a_i makes an observation $\hat{v}_{j,t}^{(i)}$ of the number of resources remaining.*

When a_i arrives at l_0, all y_i resources carried by a_i are transferred to l_0. Let $D : L \times L \to \mathbb{R}^+$ be the distance function of the locations. Thus, a UAV a_i takes $t(a_i, l_j, l_k) = \frac{D(l_j, l_k)}{V_i}$ time steps to move from location l_j to location l_k.

At every time step, the reconnaissance UAV R observes the number of resources at $M \leq m$ locations. This corresponds to the UAV being launched from l_0, instantaneously visiting M locations, observing the number of resources and sharing the information with all $a_i \in A$.

The important features of the system for online decentralized information gathering cited in [101] are as follows:

1. Tasks are distributed randomly within the environment. The spatial distribution of tasks is not known a priori by the UAVs. The UAVs perform a distributed search in the environment to discover the tasks. A UAV needs to move to the vicinity of a task to be able to discover it by perceiving/observing it on its sensors.

2. A single UAV is only capable of discovering and partially performing tasks but lacks the computational resources required to complete a task on its own. A task can be completed only if multiple UAVs collaborate to share their computational resources toward executing the task. Loosely coupled tasks are considered, and different UAVs collaborating with each other to perform the same task can execute those actions asynchronously and independently of each other.

3. To enlist the cooperation of other UAVs required to complete a task, a UAV that discovers a task communicates the task's information to other UAVs.

4. On completing its portion of execution of a task, a UAV communicates the progress of the task after its execution to other UAVs within its communication range. Those UAVs then consider their own commitments and selectively choose to visit the task to perform it. After completing its portion of a task, a UAV either continues to perform any other tasks it is already committed to or reverts to individually searching the environment for tasks if it does not have any other committed tasks.

Different multi-UAV aggregation strategies can be used. Each strategy is implemented using a heuristic function that calculates the suitability or priority of a task in the UAV's allocated task list. The different heuristic functions that have been used for this purpose are described below:

1. **Closest task first:** Each UAV selects a task from its task list that is closest to it. Distances are normalized over the sum of distances to all tasks in a UAV's task list to enable comparison between task distances. The value of this heuristic is very simple to calculate. However, it can lead all UAVs to prefer tasks that are close to them and consequently give a lower preference to tasks further away from them. Consequently, tasks that are further away from most of the UAVs can remain incomplete for long times, and more than the required number of UAVs can get allocated to tasks that are closer to most of the UAVs. Overall, the closest task first heuristic is not very efficient in spatially distributing the task load across the UAVs.

2. **Most starved task first:** To address the drawbacks of the closest task first heuristic and to balance the task load across the environment, a higher priority is given to tasks that have the least number of UAVs in their vicinity and are likely to be requiring more UAVs to complete them. The most starved task first heuristic does this by enabling each UAV to select a task from its task list that has the least number of UAVs in its vicinity. A potential drawback of the most starved task first heuristic is that UAVs are attracted toward a recently discovered task which is likely to have few UAVs near it. This can result in almost complete tasks being left incomplete because UAVs prefer more starved and possibly less complete tasks over a task that is almost nearing completion but has more UAVs (which possibly already visited the task) near it.

3. **Most starved, most complete task first:** The first heuristic extends the most starved task first heuristic by considering the number of UAVs still required to complete a task. While selecting a task, a UAV using this heuristic considers a product of the number of UAVs in the vicinity of the task as well as the progress of the task from the pheromone value associated with the task. Tasks that are nearing completion are given higher preference.

4. **Most proximal task first:** In the previous two heuristics, the task allocation algorithm only considers the effect of other UAVs on tasks in its allocated task list, but it does not consider the UAV's own relative position to the other UAVs. This can result in UAVs that are further away from the task allocating the task and moving toward it, only to be informed en route that other UAVs have completed the task before it. To alleviate this problem, the most proximal task first heuristic first determines how many other UAVs are closer to the task being considered than the UAV itself. It then selects the task that has the least number of UAVs closer to the task than itself and also is the nearest toward being completed.

4.4.1.3   Foraging Algorithms

4.4.1.3.1   *Levy Flight*

An observation made about foraging animals is that foragers when in no or limited prior knowledge of food show searching patterns that have special characteristics. The patterns are different from what can be seen in Brownian motion, the random walk by particles diffusing in a liquid. Foragers sometimes take long paths in just one direction. This strategy is found to be the key to the foragers' success in finding food rapidly in an unknown environment. In a target searching context, flight lengths, ℓ, are said to have the power-law distribution when all these lengths are drawn from a probability distribution of the form:

$$p(\ell) = C\ell^{-\mu} \tag{4.80}$$

where $p(\ell)$ is a probability density, μ is the scaling parameter and C is the normalization constant. A more general Levy distribution can be classified as a power-law distribution where the scaling parameter, μ, lies in the range $1 < \mu \leq 3$. Thus, a **Levy flight** consists of flight lengths that obey the power-law distribution with $1 < \mu \leq 3$. The following form for the continuous case can be given:

$$p(\ell) = \frac{\mu - 1}{\ell_{\min}} \left(\frac{\ell}{\ell_{\min}} \right)^{-\mu} \tag{4.81}$$

Assuming a 2D random search, the forager follows these rules:

1. The UAV moves from one point to another until it locates a target; i.e., this UAV keeps drawing lengths to follow from a power-law distribution with a fixed μ.
2. If the UAV locates a target in its vicinity, the sensor range of radius r_v, the UAV sends out the information to an operator outside.
3. Without stopping, the UAV moves on to its search; i.e., the next flight length is drawn from the power-law distribution with a predetermined value of scale factor μ. The angle UAV takes is drawn from a uniformly distributed set of range $[0, 2\pi]$.
4. The mapping of the region has not been specified; the UAV can revisit any region.

One flight length can be defined as the distance traveled by the forager between one point to another without stopping and changing the angle of its path. The forager may take several of such flight lengths to find one target [265].

When the target sites are sparsely and randomly distributed, the optimum strategy is a Levy flight. By performing Levy flight, a forager optimizes the number of targets encountered versus the traveled distance. Combining Levy flight algorithm with artificial potential field method allows us to perform an optimum bio-inspired foraging strategy. The **Levy flight** algorithm generates the length of the movement, while the artificial potential method improves the dispersion of the deployed UAVs by generating repulsion forces among them [274].

1. The **Levy probability distribution** has the following form:

$$p_\alpha(\ell) = \frac{1}{\ell} \int_0^{10} \exp\left(-\gamma q^\alpha\right) \cos(q\ell) dq \qquad (4.82)$$

where γ is the scaling factor and α determines the shape of the distribution such that different shapes of probability distribution in the tail region can be obtained. The distribution is symmetrical with respect to $\ell = 0$. The length of the Levy flight of the UAV during the foraging phase can be calculated as the limit of the following sequence (or n around 100):

$$z_n = \frac{1}{n^{1/\alpha}} \sum_{k=1}^n v_k \qquad (4.83)$$

requiring two independent random variables a, b having a normal Gaussian distribution from this nonlinear transformation $v = \frac{a}{|b|^{1/\alpha}}$

2. The basic idea of the potential field method is that the UAV is attracted toward searching target while being repulsed by the known obstacles. To improve the dispersion process during deployment, the repulsion forces among UAVs are used. A repulsive field can be proposed as

$$U_{\text{rep}}(q) = \left\{ \begin{array}{cc} \frac{1}{2} k_{\text{rep}} \left(\frac{1}{\rho(q)} - \frac{1}{\rho_0} \right) & \text{if } \rho(q) \geq \rho_0 \\ 0 & \text{elsewhere} \end{array} \right\} \qquad (4.84)$$

where k_{rep} is a scaling factor, $\rho(q)$ is the minimal distance from q to the adjacent UAV and ρ_0 the threshold value of the distance. This leads to the repulsive force

$$F_{\text{rep}}(q) = \left\{ \begin{array}{cc} k_{\text{rep}} \frac{1}{\rho^2(q)} \left(\frac{1}{\rho(q)} - \frac{1}{\rho_0} \right) & \text{if } \rho(q) \geq \rho_0 \\ 0 & \text{elsewhere} \end{array} \right\} \qquad (4.85)$$

This approach has the advantage that it does not require a centralized control or localization system and therefore scales the possibility to apply a very large number of UAVs. An alternative macroscopic model to the Levy flight model is the intermittent search model that views foraging in two alternating phases. In the first phase, the animal performs a local Brownian search, and in the second phase, the animal performs a ballistic relocation. In both the Levy flight and intermittent search models, the key macroscopic observation is that the animal performs a local exploration for some time and then moves to a far-off location. In animal foraging, the energy aggregated from a patch can be thought of as the reward from the patch, and the animal's objective is to maximize intake energy rate while minimizing expenditure in time and energy. In robotic foraging, the UAV searches an area, and the reward is the aggregated evidence. Analogous to the animal, the UAV's objective is typically to maximize evidence collected while minimizing expenditure of time and energy. The **multi-armed bandit problem** models the foraging objective in optimal foraging theory well, and the associated block allocation strategy captures the key features of popular macroscopic search models [271].

4.4.1.3.2   Fish School

The UAVs, just like fish, forage when in search of a target. The team of UAVs can behave/respond like a school of fish in search of food; thus the **fish prey algorithm** can be used. It uses the concept of adaptive networks and adds mobility as another attribute. The adaptive network is strongly connected and examined from the viewpoint of a group of nodes capable of learning and interacting with one another locally to accomplish distributed inference and processing challenges in real time.

The nodes relate locally with their neighbors within neighborhoods which are constantly changing due to the mobility of nodes. This leads to a network topology adaptive in nature. Each UAV can be represented as a fish in this school [116].

4.4.1.3.3 Greedy Rate

The Greedy rate algorithm actively replans the UAV a_i's destination based on the estimates $\hat{v}_{j,t}^{(i)}$. The Greedy rate algorithm is inspired by continuous area sweeping approaches, and its pseudocode is given by Algorithm 44.

Earmarking can be used to minimize communication among the foraging UAVs. Foraging UAVs tend to have limited computational power and communication bandwidth to minimize the computational and communication requirements. If communication among foraging UAVs was impossible, it would still be feasible for a foraging UAV to infer another UAV's destination by observing its direction of travel.

Algorithm 44 Next Destination of UAV a_i at Location l_α

1. if $c_i = y_i$ then
2. return l_0
3. end if
4. compute the rate if a_i heads home
5. $(r_{\text{best}}, l_{\text{best}}) \leftarrow \left(\frac{y_i}{t(a_i, l_\alpha, l_0)}, l_0 \right)$
6. compute the rate if a_i visits l_j then heads home
7. for all $l_j \in L$ such that $j > 0$ do
8. $e_j \leftarrow \sum_{a_k \in A \text{ heading to } l_j} (c_k - l_k)$
9. $y_i' \leftarrow \max \left(c_i, y_i + \max(0, \hat{v}_{j,t+t(a_i,l_\alpha,l_j)} - e_j) \right)$
10. $r' \leftarrow \frac{y_i'}{t(a_i, l_\alpha, l_j) + t(a_i, l_j, l_0)}$
11. if $r' > r_{\text{best}}$ then
12. $(r_{\text{best}}, l_{\text{best}}) \leftarrow (r', l_j)$
13. end if
14. end for
15. return l_{best}

4.4.1.3.4 Adaptive Sleep

The adaptive sleep algorithm is adapted from sustainable foraging. In Algorithm 45, each UAV a_i chooses a location l_α, and the UAV forages from the location when it has $K_\alpha/2$ resources, where K_α is the maximum number of resources at l_α.

The UAV sleeps until the estimated number of resources is at least $K_\alpha/2$ and the foraging UAVs coordinate on the locations, subject to communication range limits.

Algorithm 45 Compute if UAV a_i that is Assigned to Location l_α Should Sleep Further

1. if a_i is not at l_0 then
2. return false
3. end if
4. if $\hat{v}_{\alpha,t+t(a_i,l_l,l_\alpha)} < K_\alpha/2$
5. return true
6. else
7. return false
8. end if

4.4.1.4 Anchoring

Autonomous systems situated in the real world often need to recognize, track, and reason about various types of physical objects. In order to allow reasoning at a symbolic level, one must create and continuously maintain a correlation between symbols denoting physical objects and sensor data being collected about them, a process called anchoring. Anchoring must necessarily take place in any robotic system that comprises a symbolic reasoning component.

Definition 186. *Anchoring is the process of creating and maintaining the correspondence between symbols and sensor data that refer to the same physical objects.*

The anchoring problem is the problem of how to perform anchoring in an artificial system. Anchoring can be performed top-down, bottom-up or in both directions simultaneously [91]. The anchoring process must take the temporal dimension into account with the flow of continuously changing sensor input. An anchor can be seen as an internal model of a physical object that links together the symbol-level and sensor-level representations of that object. Uncertainty and ambiguity arise when anchoring is performed using real sensors which have intrinsic limitations and in an environment which cannot be optimized in order to reduce these limitations. Moreover, at the symbolic level, there are several ways to refer to objects. An anchor is a unique internal representation of an object o in the environment. At every moment t, $\alpha(t)$ contains a symbol meant to denote o, a percept generated by observing o and a signature meant to provide the current (best) estimate of the values of the observable properties of o [90]. The anchoring process is performed in an intelligent embedded system that comprises a **symbol system** σ and a **perceptual system** π. The symbol system manipulates individual symbols meant to denote physical objects. It also associates each individual symbol with a set of symbolic predicates that assert properties of the corresponding object. The perceptual system generates percepts from the observation of the physical objects. A **percept** is a structured collection of measurements assumed to originate from the same physical object.

For a UAV performing surveillance tasks of a road network, the symbol system consists of the planner; individual symbols denote cars and elements of the road network. The perceptual system is a reconfigurable active vision system able to extract information about car-like objects in aerial images, and they have attributes like position, width and color. The predicate grounding relation is a table that associates each predicate symbol with a fuzzy set of admissible values for the corresponding attribute. An anchor stores an individual symbol, the index of a region and an association list recording the current estimates of the object's attributes. Symbolic formalisms, such as chronicle recognition, require a consistent assignment of symbols, or identities, to the physical objects being reasoned about and the sensor data received about those objects. Image analysis provides a partial solution, with vision percepts having symbolic identities that persist over short intervals of time. However, changing visual conditions or objects temporarily being out of view leads to problems that image analysis often cannot handle. This is the task of the anchoring system, which also assists in object classification and in the extraction of higher-level attributes of an object. A geographic information system is used to determine whether an object is currently on a road. Concrete events corresponding to changes in such attributes and predicates provide sufficient information for the chronicle recognition system to determine when higher-level events such as reckless overtakes occur.

In the case study, anchoring links vision percepts from an object tracker to world objects, which are then linked to on-road objects. Link conditions are intended to demonstrate key concepts and could be elaborated to take more complex conditions into account. The temporal unit in the formulas is milliseconds, which is the temporal granularity used in the real system. When a traffic monitoring scenario is considered, traffic violations and other events to be detected should be represented formally and declaratively, which is a specific example of a general classification task which is common in robotic applications. This can be done using chronicle recognition, where each chronicle defines a parameterized class of complex events as a simple temporal network whose nodes correspond to occurrences of high-level qualitative events, and edges correspond to metric temporal constraints between event occurrences. Creating these high-level representations from low-level sensor data, such as video streams from the color and thermal cameras on-board the UAVs, involves extensive information processing within each sensor platform. Anchoring is a central process making symbolic reasoning about the external world possible by creating symbols referring to objects in the world based on processing of sensor data. The anchoring process is actually a set of concurrent link processes [164].

4.4.2   AERIAL MANIPULATION

Recently, UAVs have been employed in tasks such as grasping and manipulation as well as in cooperative transportation. An aerial manipulator, a multi-rotor UAV equipped with n degrees of freedom robotic arm, merges the versatility of multi-rotor UAV with the precision of a robotic arm. Multi-rotor UAVs have simple mechanical structures; they have the desirable capabilities of hovering and vertical take-off and landing. In particular, several aggressive maneuvers have been demonstrated by utilizing their high thrust-to-weight ratio. These properties are also particularly useful for autonomous load transportation. Small-size single or multiple autonomous vehicles can be considered for load transportation and deployment. Load transportation with multiple multi-rotors is useful when the load is heavy compared with the maximum thrust of a single multi-rotor or when additional redundancy is required for safety. But, this is challenging since dynamically coupled quadrotors should cooperate safely to transport load [177]. These are challenging issues in foraging since the system is characterized by unstable dynamics and the presence of the object causes non-trivial coupling effects where the dynamic model of the whole system, UAV plus manipulator, is devised. A Cartesian impedance control can be developed in such a way to cope with contact forces and external disturbances. The problem of motion control of the end effector of a manipulator mounted on a UAV can be tackled through a hierarchical control architecture. Namely, in the top layer, an inverse kinematics algorithm computes the motion references for the actuated variables, i.e., position and yaw angle of the UAV and joint variables for the manipulator, while in the bottom layer, a motion control algorithm is in charge of tracking the motion references. The previous scheme can be extended by adding, at the motion control level, an adaptive term, in charge of taking into account modeling uncertainties due to the under-actuation of the system [72]. A gripper on a UAV can serve for object retrieval, sensor installation, courier services, logistics chains, etc. The aerial grasping process can be linked to alignment and grasp, coupled to ground and partial ground contact. However, UAVs have limited positioning accuracy, unstable dynamics, are sensitive to load imbalances and are susceptible to aerodynamic effects. Any disturbance causes drift, and GPS has limited spatial/temporal resolution. It is difficult to have a high-gain position control.

4.4.2.1   Aerial Transportation

The single-lift configuration, where a long rope couples one helicopter and one load, is the only configuration commercially utilized for the transportation of slung loads. However, the manual maneuvering of a helicopter with an attached slung load is very difficult and requires a skillful and experienced pilot. In particular, the active damping of load oscillations is a difficult task, which most pilots avoid. Instead, the pilots stabilize only the helicopter and wait for the load oscillation to die down. The motivation for using two or more small helicopters instead of one with bigger load capacity is as follows:

1. In the case of real manned transport helicopters, the costs for two small helicopters are often less than for one with double load capacity.
2. Independent from the load capacity of the most advanced helicopters, there is always a task, which requires more load capacity than provided by a single helicopter.

In this case, the control software should allow the coupling of the existing helicopters, in order to form a system with sufficient load capacity. The coordinated control of the motion of the UAVs needs to consider the involved forces. Thus, each UAV could be controlled around a common compliance center attached to the transported object. Assuming that each UAV holds the object firmly, the trajectories of all UAVs determine the trajectory of the object. Both centralized and decentralized compliant motion control algorithms have been proposed, including the consideration of nonholonomic constraints [44]. Experimental results with a team of UAVs to manipulate and transport a payload in three dimensions via cables have been recently presented [44]. The transportation of loads using only one UAV is strongly limited by the payload capacity of the UAV itself. Assuming the use of small-sized UAVs, this constraint may prevent the transportation and deployment of loads required for a particular application. The system designed allows the transportation of a single load by means of several helicopters. The number of helicopters is configurable, depending on the capabilities of the helicopters and the load to be transported. The platform comprises self-deployment capabilities using one or several small-sized helicopters. The systems integrated in the platform include UAVs, WSNs, ground-fixed cameras and ground vehicles with actuation capabilities. This load can be a variety of things, such as cargo transportation or sensor deployment.

4.4.2.1.1   Model for the Interaction

The most substantial challenges of this generalized object retrieval and transport of unstructured objects from UAVs can be classified into three categories: **aerial approach and descent**, **object capture** and **UAV stability** after object contact and acquisition. Instability and fragility of hovering UAVs encourages avoiding at all costs contact with surroundings. Landing and take-off generally involve rapidly transitioning through partial contact conditions, with minimal time in intermediate states between static and dynamic stability, where the danger of ground collision is high. The ground effect should also be considered. For grasping and manipulating external objects, operation in these regimes is required, both when grasping objects and in the process of lifting a target clear of the ground. In the case of a UAV interacting with the environment and/or other UAVs, a UAV can perform k different tasks $\Omega = \left\{ \tau^1, \tau^2, \ldots, \tau^k \right\}$ with n logical interactions requiring a change of task in the current plan. Let $E = \left\{ \varepsilon^1, \varepsilon^2, \ldots, \varepsilon^n \right\}$ be a set of discrete events associated with n logical conditions requiring a change of task during the execution. The ith UAV's current task has a discrete dynamics $\delta : \Omega \times E \to \Omega$, $\tau_i^+ = \delta(\tau^i, e^i)$, where $e^i \in E$ is an event requiring a change of task from τ^i to τ_i^+. For example, the following sequence of tasks should be executed in an aerial manipulation task:

1. τ^1: approach from home,
2. τ^2: precise positioning for manipulation,
3. τ^3: grasping and manipulation,
4. τ^4: departure,
5. τ^5: return to home.

The stability of the UAV in coupled and partial contact with ground, and once airborne with payload, must be analyzed and assured. The grasping task can be divided into four phases:

1. Approach band alignment to the target,
2. Grasping hover while coupled to an object resting on the ground,
3. Partial coupling during liftoff,
4. Departure.

Each phase poses specific challenges. Disruptive aerodynamic surface effects make near-ground position-keeping outdoors more difficult than hovering in free air. The wake of the rotor is contained by the surface underneath it, creating a repelling cushion of air referred to as **ground effect**. As a rotor-craft moves laterally through ground effect, the deflected wake is pushed ahead of the UAV and can be entrained and recirculated by the rotor, causing a ground vortex. When the vortex enters the rotor, the thrust decreases rapidly; together, these create an instability that causes the UAV to bounce on descent and then drift and plunge from wake interactions. In practice, UAV hovering is not yet sufficiently precise to enable grasping with a rigid manipulator. When the vehicle is in position and has a secure grasp on the target, its flight dynamics become coupled to the ground through forces transmitted by the gripper. Certain ratios of lateral and angular coupling stiffness can destabilize the UAV. As thrust increases and the surface normal force decreases, this coupling must remain well conditioned. Once the object is lifted clear of the ground, the added load must not destabilize the UAV. The added load changes physical parameters of the system such as the net mass, moment of inertia and location of the center of gravity of the vehicle are altered [250].

4.4.2.1.2 Layers

The interactions between UAVs are not only information exchanges but also physical couplings required to cooperate in the joint transportation of a single load; it requires the consideration of physical interactions between the UAVs. The coordinated control of the motion of the UAVs should consider the involved forces. Thus, each UAV could be controlled around a common compliance center attached to the transported object. Under the assumption that each UAV holds the object firmly, the real trajectories of all of the UAVs are equal to the real trajectory of the object. Both centralized and decentralized compliant motion control algorithms can be proposed. Basically, those vehicles should be able to move to a given location and activate their payload when required. In each UAV, there are two main layers:

1. The **on-board deliberative layer** for high-level distributed decision-making, and
2. The **proprietary executive layer** for the execution of the tasks.

In the interface between both layers, the on-board deliberative layer sends task requests and receives the execution state of each task and the UAV state. For distributed decision-making purposes, interactions among the on-board deliberative layers of different UAVs are required. Finally, the human–machine interface software allows the user to specify the missions and tasks to be executed by the platform, and also to monitor the execution state of the tasks and the status of the different subsystems. The different modules shown in the on-board deliberative layer support the distributed decision-making process involving cooperation and coordination. The vehicles to be integrated in the platform should be able to receive elementary tasks, report their associated execution events and execute them. The task manager module receives the planned tasks from the plan builder module. Those tasks can have pre-conditions and/or post-conditions and the task model assumed is based on elementary events processing, which are expected to occur whenever the states of the tasks and the environment evolve [231].

4.4.2.1.3 Assembling Bar Structure

Initially, low-capacity UAVs incorporated fixed claws under the platform, allowing the system to carry lightweight and small-sized objects. However, the uncertainty in the positioning maneuver during hovering, inherent to rotary-wing platforms and the reduced motion of claws made an autonomous accurate grasping difficult. To face this problem, solutions with either **magnetic devices** or by using poorly articulated claws were proposed. Although this last option had the additional advantage of extending the range of applications of the robotic system, it also implied a significant increase in on-board weight, which was difficult to afford by conventional UAVs. The three basic functions required for an aerial manipulator intended for assembling bar structures are as follows:

1. The capture, including the manipulator approach to the bar and the grasping by the end effector;

2. The transport that involves the displacement of the load from the storage place to the construction site;
3. The assembling of the bar, whose purpose is to install it at the assigned location within the structure.

These functions will be executed in a cyclic manner during the construction process, so that once a bar has been captured, transported and finally assembled, the aerial manipulator returns to the storage place to capture a new one and to repeat the process. The successful achievement of these capabilities is conditioned by the fulfillment of several design requirements. Some of these requirements are generic, typical of any manipulator, and others are specific, due to the particularity of being on-board a rotary-wing UAV. When the rotary-wing UAV hovers, it is not completely stable, but there are small oscillations around the control reference of all degrees of freedom. These oscillations, mostly caused by electromechanical asymmetries and turbulence in the air stream entering the platform rotors, may cause difficulties in the end effector positioning. Manipulator dynamics is usually faster than the aerial platform, and therefore, it is able to compensate quicker this kind of disturbances. However, the manipulator dynamic is strongly coupled with the aerial platform, and it would be necessary to find a compromise between the manipulator speed and the reaction forces at the platform. There are also disturbances caused by displacements of the manipulator center of gravity. When the manipulator configuration changes, the position of its center of gravity varies, generating a reaction torque at the platform. In rotary-wing UAVs, this torque produces an inclination of the propeller's plane, which also induces a displacement of the entire aerial platform. This disturbance is particularly evident in those cases in which combined mass of the manipulator and its load is significant in comparison with the mass of the UAV. On heavy platforms with lightweight manipulators and loads, this disturbance is negligible. The UAV control system can be used to compensate the effect of this perturbation, being not strictly necessary to develop any other additional mechanism [75].

4.4.2.2 Coupled Dynamics

This system is composed of the main components: a helicopter, a manipulator mounted on the fuselage and the object itself. The task of this system is to fly closely enough to the object, to activate the sensor for object tracking, to go closer to the object and to perform the manipulation task. The manipulation task can range from picking up objects to performing force interaction with an object for assembly operations. The control can be realized using one of the following approaches or their combinations [209]:

1. Completely decoupled control,
2. Coupling on the kinematical level,
3. Coupling on the dynamical level.

4.4.2.2.1 Grasping Task

A pick and place or insertion task requires a brief moment of loose coupling with the ground during the grasp or insertion. Gripper compliance reduces coupling forces/torques transmitted to the airframe. When grasping in hover, elastic gripper forces can be added to the helicopter flight dynamics. Another technique can be used such as in [257]. Considering a world-fixed inertial frame $\{O, X, Y, Z\}$ and a body-fixed frame $\{O_b, X_b, Y_b, Z_b\}$ centered at the multi-rotor center of gravity, the dynamic model of a multi-rotor is given by:

$$m\ddot{p}_b + mg_b = \mathbf{R}_b(\eta_b)\left(f_b^b + f_v^b\right)$$
$$\mathbf{I}_b\dot{\omega}_b^b + Sk(\omega_b^b)\mathbf{I}_b\omega_b^b = \tau_b^b + \tau_v^b \qquad (4.86)$$

where $p_b \in \mathbb{R}^3$ is the position of the multi-rotor with respect to the inertial frame, $\mathbf{R}_b(\eta_b) \in SO(3)$ is the rotation matrix, $\eta = (\phi, \theta, \psi)$ Euler angles, m is the multi-rotor mass, $\mathbf{I}_b \in \mathbb{R}^{3 \times 3}$ is its constant diagonal inertia matrix expressed with respect to the body-frame, $\omega_b^b \in \mathbb{R}^3, \dot{\omega}_b^b \in \mathbb{R}^3$ are respectively the angular velocity and acceleration of the multi-rotor expressed in the body frame, $g_b = (0,0,g)^T$ is the gravity vector, $f_b^b \in \mathbb{R}^3$ and $\tau_b^b \in \mathbb{R}^3$ are respectively the external forces and torques input vectors acting on the aerial vehicle and expressed in the body frame (effects of the manipulator on the helicopter, neglected aerodynamic effects, physical interaction of the system with the environment). In the multi-rotor, $f_b^b = (0,0,u)^T$ and $\tau_b^b = (\tau_\phi, \tau_\theta, \tau_\psi)^T$ where u represents the thrust perpendicular to the propellers' plane. Generally, an aerial manipulator consists of three components: a base (fixed to the landing gear of the multi-rotor), a multi-joint arm, and an end effector. The center of gravity of the whole system is as close as possible to the multi-rotor geometric center. The instantaneous center of gravity position of each link i, referred to as the arm fixed axis system $\{O_0, X_0, Y_0, Z_0\}$, is given by:

$$[x_{A_i}^0, y_{A_i}^0, z_{A_i}^0, 1]^T = \mathbf{T}_i^0 [x_{A_i}^i, y_{A_i}^i, z_{A_i}^i, 1]^T \tag{4.87}$$

for $i = 1, \ldots, n$ and $\mathbf{T}_i^0 \in \mathbb{R}^{4 \times 4}$ is the homogeneous transformation matrix of link i. The robotic arm center of gravity position vector $p_A^b \in \mathbb{R}^3$ referred to the body frame is given by:

$$p_A^b = \frac{1}{m_A} \mathbf{E}_3 \mathbf{T}_0^b \left(\sum_{i=1}^n [x_{A_i}^0, y_{A_i}^0, z_{A_i}^0, 1]^T \right) \tag{4.88}$$

where m_i is the ith link mass, $m_A = \sum_{i=11}^n m_i$, $\mathbf{T}_0^b \in SE(3)$ is the homogeneous transformation matrix from arm to body frame, and $\mathbf{E}_0 \in \mathbb{R}^{4 \times 4}$ selects the first three components.

The dynamic model of the system can be written as:

$$\mathbf{M}(\zeta)\ddot{\zeta} + \mathbf{C}(\zeta, \ddot{\zeta}) + g(\mathbf{M}(\zeta) + C(\zeta, \ddot{\zeta})) = u \tag{4.89}$$

where $\zeta = (x_b^T, q^T)^T \in \mathbb{R}^{6+n \times 1}$, and $\mathbf{M}$ represents the symmetric and positive definite inertia matrix of the system:

$$\mathbf{M} = \begin{pmatrix} \mathbf{M}_{pp} & \mathbf{M}_{p\phi} & \mathbf{M}_{pq} \\ \mathbf{M}_{p\phi}^T & \mathbf{M}_{\phi\phi} & \mathbf{M}_{\phi q} \\ \mathbf{M}_{pq}^T & \mathbf{M}_{\phi q}^T & \mathbf{M}_{qq} \end{pmatrix} \tag{4.90}$$

where $\mathbf{M}_{pp}, \mathbf{M}_{p\phi} \in \mathbb{R}^{3 \times 3}$, $\mathbf{M}_{pq}, \mathbf{M}_{\phi q} \in \mathbb{R}^{3 \times n}$. The inertia matrix can be viewed as a block matrix.

Similarly, matrix $\mathbf{C}$ and vector g can be written as:

$$\mathbf{C} = \left(\mathbf{C}_p, \mathbf{C}_\phi, \mathbf{C}_q \right)^T \qquad g = \left(g_p, g_\phi, g_q \right)^T \tag{4.91}$$

with $\mathbf{C}_p \in \mathbb{R}^{3 \times (6+n)}, \mathbf{C}_\phi \in \mathbb{R}^{3 \times (6+n)}, \mathbf{C}_q \in \mathbb{R}^{n \times (6+n)}$ and $g_p \in \mathbb{R}^3, g_\phi \in \mathbb{R}^3, g_q \in \mathbb{R}^n$.

The physical interaction between UAVs and their surrounding environment is currently investigated. The goal is to explore the potentialities of systems that are not only able to fly autonomously, but also to safely interact with remote objects to accomplish tasks such as data acquisition by contact, sample picking, and repairing and assembling objects. This feature affects the design of the control law, particularly because stability has to be preserved even in the presence of disturbances derived from physical interaction. The airframe of a generic quad-rotor helicopter has been adapted to carry a miniature robotic manipulator, which has been specifically designed for aerial inspection of industrial plants [136]. By analyzing the interaction between the quad-rotor, the manipulator and the environment, the dynamical properties of the aerial manipulator are investigated near hovering both during free flight and during a docking maneuver with a vertical surface, i.e., when the end effector is in contact with the environment. Building upon this analysis, an energy-based control strategy is proposed. The main idea is to make the position dynamics of the vehicle passive by relying upon a cascade control strategy in which the attitude is considered a virtual available control input. The closed-loop passive system can then be controlled as a standard robotic manipulator, implementing impedance control strategies, suitable to handle both contact and no-contact cases.

4.4.2.2.2 Valve Turning

Valve, knob and handle turning has been widely studied for use with industrial robots, mobile manipulators and personal assistance robots. A typical requirement involves a grasp and turn of an object that remains fixed to the environment but allowed to rotate. Various techniques such as compliant motion, learning, passive compliance, hybrid position and force control, and impedance control have been implemented. All of these solutions deal with the challenges in the dynamic interaction of the manipulator with its environment. However, the strong coupling required during valve or knob turning greatly influences the dynamics of an aerial manipulator. Rigidity in the manipulator and the propagation of contact forces when interacting with the environment can cause crashes. The direct coupling between the manipulators and valve can cause sudden unexpected changes in the flight dynamics. The aerial manipulator must constantly adjust to compensate for the vehicle movement and further have adequate compliance to prevent a crash, particularly during the manipulator–environment coupling after grasping and while turning. Handles come in different shapes and sizes. The hand-wheel shape is ideally designed for the envisioned scenario: the aerial manipulator grabs onto the valve handle and twists it using its own degrees of freedom. The main approach used for detecting circular shapes of known radius R in 3D environments is by observing their elliptic perspective projection based on projective linear transformation, namely, collineation of a circle, by observing the camera with a pinhole model approach [210].

In order for a 2-arm manipulator A,B to grab the valve and remain symmetric with respect to the UAS ZY plane, certain kinematic constraints on joint movements need to be applied. In order to maintain symmetry, both manipulators need to move in exactly the same way (i.e., $q_1^A = -q_1^B$ and $q_2^A = -q_2^B$). Lengths L_1 and L_2 denote link sizes from Denavit–Hartenberg parameters. Given a desired vertical distance $H(q_1,q_2)$ along the z-axis, and horizontal distance $X(q_1,q_2)$ along the x-axis of the body frame, the following constraints on joint movements can be written [241]:

$$q_1 = \pm \arccos \left[\frac{L_2^2 - H(q_1,q_2)^2 - X(q_1,q_2)^2 - L_1^2}{2L_1 \sqrt{H(q_1,q_2)^2 + X(q_1,q_2)^2}} \right] \tag{4.92}$$

$$q_2 = \pm \arccos \left[\frac{H(q_1,q_2)^2 + X(q_1,q_2)^2 - L_1^2 - L_2^2}{2L_1 H(q_1,q_2)^2 X(q_1,q_2)} \right] \tag{4.93}$$

By keeping the manipulator motion slow with respect to the quad-rotor attitude motion, the dynamic coupling between UAS body and arm torques is minimized. Next, model simplification comes from the fact that the payload limits joint actuator choice, so the manipulators need to be constructed using lightweight servo-motors. Once the UAV-arm system has taken position over the valve and the geometric center of the valve is aligned with the aerial system centroid, the valve is constrained on a plane parallel to the bottom plate of the quad-rotor. In this situation, the quad-rotor center of mass is positioned directly above the pivot of the valve. The valve is balanced; i.e., the pivot of the valve represents the center of mass in the plane on which the valve turns. As the arms are symmetric and articulate equally and opposite of each other, the combined arms' center of mass shifts only along the z-axis of the quad-rotor geometric center.

4.4.2.2.3 Pickup and Delivery Problem

The objects with a priori known masses are located in a bounded 2D space, where the UAV is capable of localizing itself using a state-of-the-art SLAM system. A challenge follows from the requirement to collect a finite number of objects and drop them at a particular spot, both leading to autonomous switchings of the UAV's continuous dynamics. Deterministic versions of this problem can be handled efficiently, e.g., by two-stage optimization or relaxation. Since the UAV has to reach the corresponding locations of the objects or the depot with minimal overall cost, the overall problem also contains an instance of the traveling salesperson problem. Further, optimal exploration of a limited space is a difficult problem by itself. Minimizing the expected time for detecting a target located on a real line can be addressed with a known probability distribution by a searcher that can change its motion direction instantaneously, has a bounded maximal velocity and starts at the origin. Different versions of this problem are a **pursuit-evasion game** or a **coverage problem**, but its solution for a general probability distribution or a general geometry of the region is still an open question [235].

4.4.2.2.4 UAV Tele-Operation

Tele-operation of a group of UAVs in a bilateral way, providing a force feedback to the human operator, is an emerging topic which combines the field of autonomous multi-robot systems and the studies on human–robot interaction. As an alternative to unilateral tele-operation, the use of suitable sensorial feedback has also been proven to improve the tele-presence of the human operator, in particular by exploiting haptic cues via force feedback. It is then interesting to study the possibility of establishing a bilateral tele-operation channel interfacing a human operator with a remote group of UAVs possessing some local autonomy, but still bound to follow the high-level human motion commands [131]. The feedback upon which operators in tele-operation tasks base their control actions differs substantially from the feedback to the pilot of a UAV. There is often a lack of sensory information, and there is an additional status information presented via the visual channel. Haptic feedback can be used to unload the visual channel and to compensate for the lack of feedback in other modalities. For collision avoidance, haptic feedback could provide repulsive forces via the control inceptor [125]. Haptic feedback allows operators to interpret the repulsive forces as impedance to their control deflections when a potential for collision exists. Tele-operation performance and efficiency may be improved by providing additional information. Haptic feedback allows the operator to directly perceive the information about the environment through the sense of touch. Using haptic feedback for a collision avoidance system requires an algorithm to generate artificial forces on the control inceptor in order for the operator to perceive through the haptic channel, information about the environment before actual contact with an obstacle occurs. Both the magnitude of the repulsive forces and the mapping algorithm affect operator performance and workload. Haptic information can be generated from an artificial force field that maps environment constraints to repulsive forces [216]. The parametric risk field allows adjustments of the size, shape and force gradient by means of parameter settings, which determine the sensitivity of the field. Because of its smaller size, the field yields lower repulsive forces, results in less force cancellation effects and allows for larger UAV velocities. This indicates less operator control demand and more effective UAV operations, both expected to lead to lower operator workload, while increasing safety.

The key features of the direct-manipulation interface can be described as follows:

1. **Perception:** The first task of the interface is to help the user perceive the current relationship between the UAV and the world, while not overwhelming the user with unnecessary information.
2. **Comprehension:** The next level of situational awareness is obtained by combining perceptual data to comprehend how these data relate to the overall goal.

3. **Projection:** The highest level of situational awareness is projection: the ability to predict what will happen to the system in the near future.

By using direct manipulation and visual/overlays, the behavior of the UAV can be predictable. The interface maintains constant flight parameters unless it receives a new command, in which case it will seek to match the new parameters.

4.5 CONCLUSION

The aim is to develop a new generation of service robots capable of supporting human beings in all those activities that require the ability to interact actively and safely with environments not constrained on the ground but airborne. Deployment problem is considered first, using location approaches as well as optimal control methods. The application to homogeneous point UAVs and heterogeneous groups of UAVs with various sensori-motor capabilities is considered from different points of view. Patrolling is a very active field of research. UAVs equipped with on-board radar or high-resolution imaging payloads like electro-optic infrared sensors are used to localize and track the targets in perimeter surveillance missions. Foraging combines in fact all the previously presented robotic generic problems. A natural progression is to advance beyond simple motion and observation to interaction with objects and the fixed environment. Of specific interest is grasping and retrieving objects while hovering, combining robotic manipulation capabilities with the range, speed and vertical workspace of UAVs. This last section illustrates how a team of UAVs can cooperatively transport a single payload to distribute and minimize the load on each UAV.

Bibliography

1. Aaron, E.; Krizanc, D.; Meyerson, E.: DMVP: Foremost waypoint coverage of time-varying graphs, *In International Workshop on Graph-Theoretic Concepts in Computer Science*, Vall de Núria, Springer International Publishing, pp. 29–41, 2014.

2. Abbasi, F.; Mesbahi, A.; Mohammadpour, J.: Team-based coverage control of moving sensor networks, *In American Control Conference*, pp. 5691–5696, 2016. DOI: 10.1109/ACC.2016.7526561.

3. Abdollahzadeh, S.; Navimipour, N. J.: Deployment strategies in the wireless sensor network: A comprehensive review, *Computer Communications*, vol. **91**, pp. 1–16, 2016.

4. Acevedo, J. J.; Arrue, B. C.; Maza, I.; Ollero, A.: Distributed approach for coverage and patrolling missions with a team of heterogeneous aerial robots under communication constraints, *International Journal of Advanced Robotic Systems*, vol. **10**, pp. 1–13, 2013.

5. Acevedo, J. J.; Arrue, B. C.; Diaz-Banez, J. M.; Ventura, I.; Maza, I.; Ollero, A.: One-to one coordination algorithm for decentralized area partition in surveillance missions with a team of aerial robots, *Journal of Intelligent and Robotic Systems*, vol. **74**, pp. 269–285, 2014.

6. Acevedo, J. J.; Arrue, B. C.; Maza, I.; Ollero, A.: A distributed algorithm for area partitioning in grid-shape and vector-shape configurations with multiple aerial robots, *Journal of Intelligent and Robotic Systems*, vol. **84**, pp. 543–557, 2015. DOI: 10.1007/s10846-015-0272-5.

7. Adaldo, A.: Event-triggered control of multi-agent systems: Pinning control, cloud coordination and sensor coverage, PhD thesis, Royal Institute of Technology, School of Electrical Engineering, Department of Automatic Control, 2016.

8. Adams, H.; Carlsson, G.: Evasion paths in mobile sensor networks, *The International Journal of Robotics Research*, vol. **34**, pp. 90–104, 2015.

9. Ahuja, R. K.; Magnanti, T. L; Orlin, J. B.: *Network Flows*, Prentice-Hall, Englewood Cliffs, NJ, 1993.

10. Akbas, M.; Solmaz, G.; Turgut, D.: Molecular geometry inspired positioning for aerial networks, *Computer Networks*, vol. **98**, pp. 72–88, 2016.

11. Al-Helal, H.; Sprinkle, J.: UAV search: Maximizing target acquisition, *In 17th IEEE International Conference and Workshops on Engineering of Computer Based Systems*, Oxford, pp. 9–18, 2010.

12. Alejo, D.; Diaz-Banez, J. M.; Cobano, J. A.; Perez-Lantero, P.; Ollero, A.: The velocity assignment problem for conflict resolution with multiple UAV sharing airspace, *Journal of Intelligent Robotics Systems*, 2013. DOI: 10.1007/s10846-012-9768-4.

13. Alejo, D.; Cobano, J. A.; Heredia, G.; Ollero, A.: Collision-free trajectory planning based on Maneuver Selection-Particle Swarm Optimization, *In IEEE International Conference on Unmanned Aircraft Systems*, Denver, CO, pp. 72–81, 2015.

14. Alighanbari, M.; Bertuccelli, L. F.; How, J. P.: A robust approach to the UAV task assignment problem, *IEEE Confernce on Decision and Control*, San Diego, CA, pp. 5935–5940, 2006.

15. Alitappeh, R. J.; Jeddisaravi, K.; Guimaraes, F. G.: Multi-objective multi-robot deployment in a dynamic environment, *Soft Computing*, vol. **21**, pp. 6481–6497, 2017.

16. Althoff, D.; Kuffner, J.; Wollherr, D.; Buss, M.: Safety assessment of robot trajectory for navigation in uncertain and dynamic environment, *Autonomous Robots*, vol. **32**, pp. 285–302, 2010.

17. Altshuler, Y.; Bruckstein, A. M.: Static and expanding grid coverage with ant robots: Complexity results, *Theoretical Computer Science*, vol. **41**, pp. 4661–4674, 2011.

18. Anderson, R. P.; Milutinovic, D.: A stochastic approach to Dubins vehicle tracking problems, *IEEE Transactions on Automatic Control*, vol. **59**, pp. 2801–2806, 2014.

19. Angelov, P.; Filev, D. P.; Kasabov, N.: *Evolving Intelligent Systems*, IEEE Press, Piscataway, NJ, 2010.

20. Angelov, P.: *Sense and Avoid in UAS: Research and Applications*, Wiley Aerospace Series, Hoboken, NJ, 2012.

21. Aragues, R.; Montijano, E.; Sagues, C.: Consistency data association in multi-robot systems with limited communications, In: Matsuoka, Y.; Durrant-White, H.; Neira, J. (eds) *Robotics, Science and Systems*, The MIT Press, Cambridge, MA, pp. 97–104, 2010.

22. Arezoumand, R.; Mashohor, S.; Marhaban, M. H.: Efficient terrain coverage for deploying wireless sensor nodes on multi-robot system, *Intelligent Service Robotics*, vol. **9**, pp. 163–175, 2016.

23. Arsie, A.; Frazzoli, E. Motion planning for a quantized control system on SO(3), *In 46th IEEE Conference on Decision and Control*, New Orleans, LA, pp. 2265–2270, 2007.

24. Arslan, O.; Koditschek, D. E.: Voronoi-based coverage control of heterogeneous disk-shaped robots, *In IEEE International Conference on Robotics and Automation (ICRA)*, pp. 4259–4266, 2016. DOI: 10.1109/ICRA.2016.7487622.

25. Arslan, O.; Koditschek, D. E.: Voronoi-based coverage control of heterogeneous disk-shaped robots, *In IEEE International Conference on Robotics and Automation*, Singapore, pp. 4259–4266, 2016.

26. Asmare, E.; Gopalan, A.; Sloman, M.; Dulay, N.; Lupu, E.: Self-management framework for mobile autonomous systems, *Journal of Networked systems management*, vol. **20**, pp. 244–275, 2012.

27. Atkins, E.; Moylan, G.; Hoskins, A.: Space based assembly with symbolic and continuous planning experts, *IEEE Aerospace Conference*, Big Sky, MT, 2006.

28. Aurenhammer, F.: Voronoi diagrams, a survey of fundamental geometric data structure, *ACM Computing Surveys*, vol. **23**, pp. 345–405, 1991.

29. Avanzini, G.: Frenet based algorithm for trajectory prediction, *AIAA Journal of Guidance, Control and Dynamics*, vol. **27**, pp. 127–135, 2004.

30. Avis, D.; Hertz, A.; Marcotte, O.: *Graph Theory and Combinatorial Optimization*, Springer Verlag, Berlin, Heidelberg, 2005.

31. Ayanian, N.; Kallem, V.; Kumar, V.: Synthesis of feedback controllers for multiple aerial robots with geometric constraints, *IEEE/RSJ International Conference on Intelligent Robots and Systems*, San Francisco, CA, pp. 3126–3131, 2011.

32. Babel, L.: Three dimensional route planning for unmanned aerial vehicles in a risk environment, *Jouranl of Intelligent and Robotic Systems*, 2012. DOI: 10.1007/s10846-012-9773-7.

33. Bakolas, E.; Tsiotras, P.: The Zermelo–Voronoi diagram, a dynamic partition problem, *Automatica*, vol. **46**, pp. 2059–2067, 2012.

34. Bakolas, E.; Tsiotras, P.: Feedback navigation in an uncertain flowfield and connections with pursuit strategies, *AIAA Journal of Guidance, Control and Dynamics*, vol. **35**, pp. 1268–1279, 2012.

35. Barbato, M.; Grappe, R.; Lacroix, M.; Calvo, R. W.: A set covering approach for the double traveling salesman problem with multiple stacks, *In International Symposium on Combinatorial Optimization*, pp. 260–272, 2016. DOI: 10.1007/978-3-319-45587-7-23.

36. Baron, O.; Berman, O.; Krass, D.; Wang, Q.: The equitable location problem on the plane, *European Journal of Operational Research*, vol. **183**, pp. 578–590, 2007.

37. Basilico, N.; Amigoni, F.: Exploration strategies based on multi criteria decision making for searching environments in rescue operations, *Autonomous Robots*, vol. **31**, pp. 401–417, 2011.

38. Bayat, B.; Crasta, N.; Crespi, A.; Pascoal, A. M.; Ijspeert, A.: Environmental monitoring using autonomous vehicles: A survey of recent searching techniques, *Current Opinion in Biotechnology*, vol. **45**, pp. 76–84, 2017.

39. Bektas, T.: The multiple traveling salesman problem: An overview of formulations and solution procedures, *Omega*, vol. **34**, pp. 209–219, 2006.

40. Belkhouche, F.; Vadhva, S.; Vaziri, M.: Modeling and controlling 3D formations and flocking behavior of UAV, *IEEE Information Reuse and Integration Conference*, pp. 449–454, 2011. DOI: 10.1109/IRI.2011.6009590.

41. Bennet, D.; McInnes, C.; Suzuki, Uchiyama, K.: Autonomous three-dimensional formation flight for a swarm of unmanned aerial vehicles, *AIAA Journal of Guidance, Control and Dynamics*, vol. **34**, pp. 1899–1908, 2011.

42. Berger, J.; Boukhtouta, A.; Benmoussa, A.; Kettani, O.: A new mixed integer linear programming model for rescue path planning in uncertain adversarial environments, *Computers and Operations Research*, vol. **39**, pp. 3420–3430, 2012.

43. Berger, J.; Lo, N.: An innovative multi-agent search-and-rescue path planning approach, *Computers and Operations Research*, vol. **53**, pp. 24–31, 2015.

44. Bernard, M.; Kondak, K.; Maza, I.; Ollero, A.: Autonomous transportation and deployment with aerial robots for search and rescue missions, *Journal of Field Robotics*, vol. **28**, pp. 914–931, 2011.

45. Bernardini, S.; Fox, M.; Long, D.: Combining temporal planning with probabilistic reasoning for autonomous surveillance missions, *Autonomous Robots*, pp. 1–23, 2015. DOI: 10.1007/s10514-015-9534-0.

46. Bertsimas, D.; VanRyzin, G.: The dynamic traveling repairman problem, MIT Sloan paper 3036-89-MS, 2011.

47. Bertuccelli, L. F.; Pellegrino, N.; Cummings, M. L.: Choice modeling of relook tasks for UAV search mission, *American Control Conference*, Baltimore, MD, pp. 2410–2415, 2010.

48. Bertsekas, D. P.: *Dynamic Programming and Optimal Control*, Athena Scientific, Nashua, NH, 1995.

49. Besada-Portas, E.; De La Torre, L.; de la Cruz, J. M.; de Andrs-Toro, B.: Evolutionary trajectory planner for multiple UAVs in realistic scenarios, *IEEE Transactions on Robotics*, vol. **26**, pp. 619–634, 2010.

50. Bestaoui, Y.; Dahmani, H.; Belharet, K.: Geometry of translational trajectories for an autonomous aerospace vehicle with wind effect, *In 47th AIAA Aerospace Sciences Meeting*, Orlando, FL, paper AIAA-1352, 2009.

51. Bestaoui, Y.: Geometric properties of aircraft equilibrium and non equilibrium trajectory arcs, In: Kozlowski, K. (ed.) *Lectures Notes in Control and Information Sciences*, Springer, Berlin, Heidelberg, pp. 1297–1307, 2009.

52. Bestaoui, Y.; Lakhlef, F.: Flight plan for an autonomous aircraft in a windy environment, In: Lozano, R. (ed.) *Unmanned Aerial Vehicles Embedded Control*, Wiley, Hoboken, NJ, 301–325, 2010. ISBN: 13-9781848211278.

53. Bestaoui, Y.: Collision avoidance space debris for a space vehicle, *IAASS Conference*, Versailles, France, In ESA Special Publication, vol. **699**, pp. 74–79, 2011.

54. Bestaoui, Y.: 3D flyable curve for an autonomous aircraft, *ICNPAA World Congress on Mathematical Problems in Engineering, Sciences and Aerospace*, AIP, Vienna, pp. 132–139, 2012.

55. Bestaoui, Y.; Kahale, E.: Time optimal trajectories of a lighter than air robot with second order constraints and a piecewise constant velocity wind, *AIAA Journal of Information Systems*, vol. **10**, pp. 155–171, 2013. DOI: 10.2514/1.55643.

56. Bestaoui Sebbane, Y.: *Planning and Decision Making for Aerial Robots*, Springer, Switzeland, 2014.

57. Bircher, A.; Kamel, M.; Alexis, K.; Burri, M.; Oettershagen, P.; Omari, S.; Mantel, T.; Siegwart, R.: Three dimensional coverage path planning via viewpoint resampling and tour optimization for aerial robots, *Autonomous Robots*, vol. **40**, pp. 1059–1078, 2016.

58. Bhatia, A.; Maly, M.; Kavraki, L.; Vardi, M.: Motion planing with complex goals, *IEEE Robotics and Automation Magazine*, vol. **18**, pp. 55–64, 2011.

59. Bicho, E.; Moreira, A.; Carvalheira, M.; Erlhagen, W.: Autonomous flight trajectory generator via attractor dynamics, *Proceedings of IEEE/RSJ Intelligents Robots and Systems*, vol. **2**, pp. 1379–1385, 2005.

60. Bijlsma, S. J.: Optimal aircraft routing in general wind fields, *AIAA Journal of Guidance, Control, and Dynamics*, vol. **32**, pp. 1025–1029, 2009.

61. Bloch, A. M.: *Non Holonomics Mechanics and Control*, Springer-Verlag, Berlin, 2003.

62. Bolanos, R.; Echeverry, M.; Escobar, J.: A multi-objective non-dominated sorting genetic algorithm (NSGA-II) for the multiple traveling salesman problem, *Decision Science Letters*, vol. **4**, pp. 559–568, 2015.

63. Boizot, N.; Gauthier, J. P.: Motion planning for kinematic systems, *IEEE Transactions on Automatic Control*, vol. **58**, pp. 1430–1442, 2013.

64. Boukraa, D.; Bestaoui, Y.; Azouz, N.: Three dimensional trajectory generation for an autonomous plane, *The International Review of Aerospace Engineering*, vol. **4**, pp. 355–365, 2013.

65. Braunl, T.: *Embedded Robotics*, Springer, Berlin, 2013.

66. Breitenmoser, A.; Schwager, M.; Metzger, J. C.; Siegwart, R.; Rus, D.: Voronoi coverage of non-convex environments with a group of networked robots, *In IEEE International Conference on Robotics and Automation*, Anchorage, AK, pp. 4982–4989, 2010.

67. Brooks, A.; Makarenko, A.; Williams, S.; Durrant-Whyte, H.: Parametric POMDP for planning in continuous state spaces, *Robotics and Autonomous Systems*, vol. **54**, pp. 887–897, 2006.

68. Brown Kramer, J.; Sabalka, L.: Multidimensional online robot motion, 2009. arXiv preprint arXiv:0903.4696.

69. Bryngelsson, E.: Multi-robot distributed coverage in realistic environments, MS thesis in Computer science, Chalmers University of Technology, Gothenburg, Sweeden, pp. 1–46, 2008.

70. Budiyono, A.; Riyanto, B.; Joelianto, E.: *Intelligent Unmanned Systems: Theory and Applications*, Springer, Berlin, 2013.

71. Bullo, F.; Carli, R.; Frasca, P.: Gossip coverage control for robotic networks: Dynamical systems on the space of partitions, *SIAM Journal on Control and Optimization*, vol. **50**, pp. 419–447, 2012.

72. Caccavale, F.; Giglio, G.; Muscio, G.; Pierri, F.: Adaptive control for UAVs equipped with a robotic arm, *In 19th World Congress of the the International Federation of Automatic Control*, Cape Town, South Africa, pp. 11049–11054, 2014.

73. Calvo, O.; Sousa, A.; Rozenfeld, A.; Acosta, G.: Smooth path planning for autonomous pipeline inspections, *IEEE Multi-Conference on Systems, Signals and Devices, SSD'09*, IEEE, pp. 1–9, 2013. DOI: 978-1-4244-4346-8/09/.

74. Campbell, S.; Naeem, W.; Irwin, G. W.: A review on improving the autonomy of unmanned surface vehicles through intelligent collision avoidance maneuvers, *Annual Reviews in Control*, vol. **36**, pp. 267–283, 2013.

75. Cano, R.; Perez, C.; Pruano, F.; Ollero, A.; Heredia, G.: Mechanical design of a 6-DOF aerial manipulator for assembling bar structures using UAVs, *In 2nd RED-UAS Workshop on Research, Education and Development of Unmanned Aerial Systems*, Compiegne, France, 2013.

76. Cao, Y.; Ren, W.: Multi vehicle coordination for double integrator dynamics under fixed undirected/directed interaction with a sampled data setting, *International Journal of Robust and Nonlinear control*, vol. **20**, pp. 987–1000, 2010.

77. Cao, Y.; Yu, W.; Ren, W.; Chen G.: An overview of recent progress in the study of distributed multi-agent coordination, *IEEE Transactions on Industrial Informatics*, vol. **9**, pp. 427–438, 2013.

78. Carlsson, G.; De Silva, V.; Morozov, D.: Zigzag persistent homology and real-valued functions, *Proceedings of the Annual Symposium on Computational Geometry*, Aarhus, Denmark, pp. 247–256, 2009.

79. Chakravarthy, A.; Ghose, D.: Generalization of the collision cone approach for motion safety in 3D environments, *Autonomous Robots*, vol. **32**, pp. 243–266, 2013.

80. Chakravorty, S.; Junkins, J.: A methodology for intelligent path planning, *IEEE International Symposium on Mediterranean Control*, Cyprus, pp. 592–597, 2005. DOI: 10.1109/.2005.1467081.

81. Chang D. E.: A simple proof of the Pontryaguin maximum principle on manifolds, *Automatica*, vol. **47**, pp. 630–633, 2013.

82. Chaumont, N.; Adami, C.: Evolution of sustained foraging in three-dimensional environments with physics, *Genetic Programming and Evolvable Machines*, vol. **17**, pp. 359–390, 2016.

83. Chavel, I. (ed.): *Eigenvalues in Riemannian Geometry*, Academic Press, Cambridge, MA, 2013.

84. Cheng, C. F.; Tsai, K. T.: Encircled belt-barrier coverage in wireless visual sensor networks, *Pervasive and Mobile Computing*, 2017. DOI: 10.1016/j.pmcj.2016.08.005.

85. Choset, H.: Coverage of known spaces: The Boustrophedon cellular decomposition, *Autonomous Robots*, vol. **9**, pp. 247–253, 2000.

86. Choset, H.; Lynch, K.; Hutchinson, S.; Kantor, G.; Burgard, W.; Kavraki, L.; Thrum, S.: *Principles of Robot Motion, Theory, Algorithms and Implementation*, The MIT Press, Cambridge, MA, 2013.

87. Chryssanthacopoulos, J.; Kochender, M. J.: Decomposition methods for optimized collision avoidance with multiple threats, *AIAA Journal of Guidance, Control and Dynamics*, vol. **35**, pp. 368–405, 2013.

88. Clelland, J. N.; Moseley, C.; Wilkens, G.: Geometry of control affine systems, *Symmetry, Integrability and Geometry Methods and Applications (SIGMA5)*, vol. **5**, pp. 28–45, 2013.

89. Cook, W. J.: *In Pursuit of the Traveling Salesman: Mathematics at the Limits of Computation*, Princeton University Press, Princeton, NJ, 2013.

90. Coradeschi, S.; Saffiotti, A.: Anchoring symbols to sensor data: Preliminary report, *In AAAI/IAAI American Association for Artificial Intelligence*, Austin, TX, pp. 129–135, 2000.

91. Coradeschi, S.; Saffiotti, A.: An introduction to the anchoring problem, *Robotics and Autonomous Systems*, vol. **43**, pp. 85–96, 2003.

92. Cormen, T. H.: *Introduction to Algorithms*, The MIT Press, Cambridge, MA, 2013.

93. Cotta, C.; Van Hemert, I.: *Recent Advances in Evolutionary Computation for Combinatorial Optimization*, Springer, Berlin, 2008.

94. Couceiro, M. S.; Figueiredo, C. M.; Rocha, R. P.; Ferreira, N. M.: Darwinian swarm exploration under communication constraints: Initial deployment and fault-tolerance assessment, *Robotics and Autonomous Systems*, vol. **62**, pp. 528–544, 2014.

95. Cruz G. C. S.; Encarnacao P. M.: Obstacle avoidance for unmanned aerial vehicles, *Journal of Intelligent and Robotics Systems*, vol. **65**, pp. 203–217, 2013.

96. Czyzowicz, J.; Kranakis, E.; Krizanc, D.; Narayanan, L.; Opatrny, J.: Optimal online and offline algorithms for robot-assisted restoration of barrier coverage, *In International Workshop on Approximation and Online Algorithms*, Wrocław, Poland, Springer International Publishing, pp. 119–131, 2014.

97. Dai, R.; Cochran, J. E.: Path planning and state estimation for unmanned aerial vehicles in hostile environments, *AIAA Journal of Guidance, Control and Dynamics*, vol. **33**, pp. 595–601, 2013.

98. Dai, X.; Jiang, L.; Zhao, Y.: Cooperative exploration based on supervisory control of multi-robot systems, *Applied Intelligence*, vol. **45**, pp. 1–12, 2016.

99. Dantzig, G.; Fulkerson, R.; Johnson, S.: Solution of a large-scale traveling-salesman problem, *Journal of the Operations Research Society of America*, vol. **2**, pp. 393–410, 2013.

100. Dantzig, G. B.; Ramser, J. H.: The truck dispatching problem, *Management Science*, vol. **6**, pp. 80–91, 2013.

101. Dasgupta, P.: Multi-robot task allocation for performing cooperative foraging tasks in an initially unknown environment, In: Jain, L.; Aidman, E.; Abeynayake, C. (eds) *In Innovations in Defence Support Systems*, Springer, Berlin, Heidelberg, pp. 5–20, 2011.

102. De Filippis, L.; Guglieri, G.: Path planning strategies for UAV in 3D environments, *Journal of Intelligent and Robotics Systems*, vol. **65**, pp. 247–264, 2013.

103. Delahaye, D.; Puechmurel, S.; Tsiotras, P.; Feron, E.: Mathematical models for aircraft design: A survey, In: *In Air Traffic Management and Systems*, Springer, Berlin, Heidelberg, pp. 205–247, 2013.

104. Deng, Q.; Yu, J.; Mei, Y.: Deadlock free consecutive task assignment of multiple heterogeneous unmanned aerial vehicles, *AIAA Journal of Aircraft*, vol. **51**, pp. 596–605, 2013.

105. Devasia, S.: Nonlinear minimum-time control with pre- and post-actuation, *Automatica*, vol. **47**, pp. 1379–1387, 2013.

106. Dicheva, S.; Bestaoui, Y.: Route finding for an autonomous aircraft, *AIAA Aerospace Sciences Meeting*, Orlando, FL, vol. **10**, pp. 6–2011, 2013.

107. Dille, M.; Singh, S.: Efficient aerial coverage search in road networks, *In AIAA Guidance, Navigation, and Control (GNC) Conference*, Boston, MA, pp. 5094–5109, 2013.

108. Dimarogonas, D.; Loizon, S. J.; Kyriakopoulos, K.; Zavlanos, M.: A feedback stabilization and collision avoidance scheme for multiple independent non point agents, *Automatica*, vol. **42**, pp. 229–243, 2006.

109. Ding, H.; Castanon, D.: Fast algorithms for UAV tasking and routing, *In IEEE Conference on Control Applications (CCA)*, Buenos Aires, Argentina, pp. 368–373, 2016.

110. Dubins, L. E.: On curves of minimal length with a constraint on average curvature and with prescribed initial and terminal positions and tangents, *American Journal of Mathematics*, vol. **79**, pp. 497–517, 2013.

111. Duan, H.; Zhang, X.; Wu, J.; Ma, G.: Max-min adaptive ant colony optimization approach to multi UAV coordinates trajectory replanning in dynamic and uncertain environments, *Journal of Bionic Engineering*, vol. **6**, pp. 161–173, 2009.

112. Durham, J. W.; Carli, R.; Bullo, F.: Pairwise optimal discrete coverage control for gossiping robots, *In 49th IEEE Conference on Decision and Control*, Atlanta, pp. 7286–7291, 2010.

113. Durham, J. W.; Carli, R.; Frasca, P.; Bullo, F.: Discrete partitioning and coverage control for gossiping robots, *IEEE Transactions on Robotics*, vol. **28**, pp. 364–378, 2012.

114. Edison, E.; Shima, T.: Integrating task assignment and path optimization for cooperating UAV using genetic algorithms, *Computers and Operations Research*, vol. **38**, pp. 340–356, 2011.

115. Eele, A.; Richards, A.: Path planning with avoidance using nonlinear branch and bound optimization, *AIAA Journal of Guidance, Control and Dynamics*, vol. **32**, pp. 384–394, 2013.

116. El Ferik, S.; Thompson, O. R.: Biologically inspired control of a fleet of UAVs with threat evasion strategy, *Asian Journal of Control*, vol. **18**, pp. 2283–2300, 2016.

117. Enes, A.; Book, W.: Blended shared control of Zermelo's navigation problem, *American Control Conference*, Baltimore, MD, pp. 4307–4312, 2013.

118. Enright, J. J.; Savla, K.; Frazzoli, E.; Bullo, F.: Stochastic routing problems for multiple UAV, *AIAA Journal of Guidance, Control and Dynamics*, vol. **32**, pp. 1152–116, 2009.

119. Ergezer, H.; Leblebiciolu, K.: 3-D path planning for multiple UAVs for maximum information collection, *Journal of Intelligent and Robotic Systems*, vol. **73**, pp. 737–762, 2014.

120. Evers, L.; Dollevoet, T.; Barros, A. I.; Monsuur, H.: Robust UAV mission planning, *Annals of Operations Research*, vol. **222**, pp. 293–315, 2014.

121. Evers, L.; Barros, A. I.; Monsuur, H.; Wagelmans, A.: Online stochastic UAV mission planning with time windows and time-sensitive targets, *European Journal of Operational Research*, vol. **238**, pp. 348–362, 2014.

122. Faied, M.; Mostafa, A.; Girard, A.: Dynamic optimal control of multiple depot routing problem with metric temporal logic, *IEEE American Control Conference*, Saint Louis, MO, pp. 3268–3273, 2009.

123. Farault, J.: *Analysis on Lie Groups: An Introduction (Cambridge Studies in Advanced Mathematics)*, Cambridge University Press, Cambridge, MA, 2013.

124. Farouki, R. T.: *Pythagorean Hodograph Curves*, Springer, Berlin, 2013.

125. Field, E.; Harris, D.: A comparative survey of the utility of cross-cockpit linkages and autoflight systems' backfeed to the control inceptors of commercial aircraft, *Ergonomics*, vol. **41**, pp. 1462–1477, 1998.

126. Foka, A.; Trahanias, P.: Real time hierarchical POMDP for autonomous robot navigation, *Robotics and Autonomous Systems*, vol. **55**, pp. 561–571, 2013.

127. Foo, J.; Knutzon, J.; Kalivarapu, V.; Oliver, J.; Winer, E.: Path planning of UAV using B-splines and particles swarm optimization, *AIAA Journl of Aerospace Computing, Information and Communication*, vol. **6**, pp. 271–290.

128. Forstenhaeusler, M.; Funada, R.; Hatanaka, T.; Fujita, M.: Experimental study of gradient-based visual coverage control on SO (3) toward moving object/human monitoring, *In American Control Conference*, pp. 2125–2130, 2015. DOI: 10.1109/ACC.2015.7171047.

129. Fraccone, G. C.; Valenzuela-Vega, R.; Siddique, S.; Volovoi, V.: Nested modeling of hazards in the national air space system, *AIAA Journal of Aircraft*, 2013. DOI: 10.2514/1.C031690.

130. Fraichard, T.; Scheuer, A.: From Reeds and Shepp's to continuous curvature paths, *IEEE Transactions on Robotics*, vol. **20**, pp. 1025–10355, 2013.

131. Franchi, A.; Bulthoff, H. H.; Giordano, P. R.: Distributed online leader selection in the bilateral tele-operation of multiple UAVs, *In 50th IEEE Conference on Decision and Control and European Control Conference*, pp. 3559–3565, 2011. DOI: 10.1109/CDC.2011.6160944.

132. Franchi, A.; Stegagno, P.; Oriolo, G.: Decentralized multi-robot target encirclement in 3D space, 2013. arXiv preprint arXiv:1307.7170.

133. Fraser, C.; Bertucelli, L.; Choi, H.; How, J.: A hyperparameter consensus method for agreement under uncertainty, *Automatica*, vol. **48**, pp. 374–380, 2012.

134. Frederickson, G.; Wittman, B.: Speedup in the traveling repairman problem with unit time window, 2009. arXiv:0907.5372 [cs.DS].

135. Frost, J. R.; Stone, L. D.: Review of search theory: Advances and application to search and rescue decision support, U.S. Coast Guard Research and Development Center, report CG-D-15-01, 2001.

136. Fumagalli, M.; Naldi, R.; Macchelli, A.; Forte, F.; Keemink, A. Q.; Stramigioli, S.; Carloni, R.; Marconi, L.: Developing an aerial manipulator prototype: Physical interaction with the environment, *IEEE Robotics and Automation Magazine*, vol. **21**, pp. 41–50, 2014.

137. Funabiki, K.; Ijima, T.; Nojima, T.: Method of trajectory generation for perspective flight path display in estimated wind condition, *AIAA Journal of Aerospace Information Systems*, vol. **10**, pp. 240–249, 2009. DOI: 10.2514/1.37527.

138. Furini, F.; Ljubic, I.; Sinnl, M.: An effective dynamic programming algorithm for the minimum-cost maximal knapsack packing problem, *European Journal of Operational Research*, 2017. DOI: 10.1016/j.ejor.2017.03.061.

139. Galceran, E.; Carreras, M.: A survey on coverage path planning for robotics, *Robotics and Autonomous Systems*, vol. **61**, pp. 1258–1276, 2013.

140. Galceran, E.; Campos, R.; Palomeras, N.; Ribas, D.; Carreras, M.; Ridao, P.: Coverage path planning with real time replanning and surface reconstruction for inspection of three dimensional underwater structures using autonomous underwater vehicles, *Journal of Field Robotics*, vol. **32**, pp. 952–983, 2015.

141. Gao, C.; Zhen, Z.; Gong, H.: A self-organized search and attack algorithm for multiple unmanned aerial vehicles, *Aerospace Science and Technology*, vol. **54**, pp. 229–240, 2016.

142. Garcia, E.; Casbeer, D. W.: Cooperative task allocation for unmanned vehicles with communication delays and conflict resolution, *AIAA Journal of Aerospace Information Systems*, vol. **13**, pp. 1–13, 2016.

143. Garone, E.; Determe, J. F.; Naldi, R.: Generalized traveling salesman problem for carrier-vehicle system, *AIAA Journal of Guidance, Control and Dynamics*, vol. **37**, pp. 766–774, 2009.

144. Garzon, M.; Valente, J.; Roldan, J. J.; Cancar, L.; Barrientos, A.; Del Cerro, J.: A multirobot system for distributed area coverage and signal searching in large outdoor scenarios, *Journal of Field Robotics*, vol. **33**, pp. 1096–1106, 2016.

145. Gattani, A.; Benhardsson, B.; Rantzer, A.: Robust team decision theory, *IEEE Transactions on Automatic Control*, vol. **57**, pp. 794–798, 2012.

146. Gaynor, P.; Coore, D.: Towards distributed wilderness search using a reliable distributed storage device built from a swarm of miniature UAVs, *In International Conference on Unmanned Aircraft Systems (ICUAS)*, pp. 596–601, 2014. DOI: 10.1109/ICUAS.2014.6842302.

147. Gazi, V.; Fidan, B.: Coordination and control of multi-agent dynamic systems: Modes and apporaches, In: Sahin, E. (ed) *Swarm Robotics*, LNCS 4433, Springer, Berlin, pp. 71–102, 2007.

148. Geramifard, A.; Redding, J.; Joseph, J.; Roy, N.; How, J.: Model estimation within planning and learning, *American Control Conference*, Montreal, pp. 793–799, 2012.

149. Giardinu, G.; Kalman-Nagy, T.: Genetic algorithms for multi agent space exploration, *AIAA Infotech@Aerospace Conference*, Honolulu, HI, paper AIAA2007-2824, 2007.

150. Glad, A.; Buffet, O.; Simonin, O.; Charpillet, F.: Self-organization of patrolling-ant algorithms, *IEEE 7th International Conference on Self-Adaptive and Self-Organizing Systems*, San Francisco, CA, pp. 61–70, 2009.

151. Goel, A.; Gruhn, V.: A general vehicle routing problem, *European Journal of Operational Research*, vol. **191**, pp. 650–660, 2008.

152. Goerzen, C.; Kong, Z.; Mettler, B.: A survey of motion planning algorithms from the perspective of autonomous UAV guidance, *Journal of Intelligent Robot Systems*, vol. **20**, pp. 65–100, 2010.

153. Gorain, B.; Mandal, P. S.: Solving energy issues for sweep coverage in wireless sensor networks, *Discrete Applied Mathematics*, 2016. DOI: 10.1016/j.dam.2016.09.028.

154. Grace, J.; Baillieul, J.: Stochastic strategies for autonomous robotic surveillance, *IEEE Conference on Decision and Control*, Seville, Spain, pp. 2200–2205, 2005.

155. Greengard, C.; Ruszczynski, R.: *Decision Making under Uncertainty: Energy and Power*, Springer, Berlin, 2002.

156. Guerrero, J. A.; Bestaoui, Y.: UAV path planning for structure inspection in windy environments, *Journal of Intelligent and Robotics Systems*, vol. **69**, pp. 297–311, 2013.

157. Guha, S.; Munagala, K.; Shi, P.: Approximation algorithms for restless bandit problems, *Journal of the ACM*, vol. **58**, 2010. DOI: 10.1145/1870103.1870106.

158. Gusrialdi, A.; Hirche, S.; Hatanaka, T.; Fujita, M.: Voronoi based coverage control with anisotropic sensors, *In 53rd Proceedings of American Control Conference*, Seattle, WA, pp. 736–741, 2008.

159. Habib, Z.; Sakai, M.: Family of G^2 cubic transition curves, *IEEE International Conference on Geometric Modeling and Graphics*, London, pp. 117–122, 2003.

160. Hameed, T. A.: Intelligent coverage path planning for agricultural robots and autonomous machines on three dimensional terrain, *Journal of Intelligent Robot Systems*, vol. **74**, pp. 965–983, 2014.

161. Hansen, E. A.; Bernstein, D. S.; Zilberstein, S.: Dynamic programming for partially observable stochastic games, *In AAAI Conference on Artifical Intelligence*, vol. **4**, pp. 709–715, 2004.

162. Hatanaka, T.; Funada, R.; Fujita, M.: 3-D visual coverage based on gradient descent algorithm on matrix manifolds and its application to moving objects monitoring, *In 2014 American Control Conference*, pp. 110–116, 2014. DOI: 10.1109/ACC.2014.6858663.

163. Hazon, N.; Gonen, M.; Kleb, M.: Approximation and heuristic algorithms for probabilistic physical search on general graphs, 2015. arXiv preprint arXiv:1509.08088.

164. Heintz, F.; Kvarnstrom, J.; Doherty, P.: Stream-based hierarchical anchoring, *KI-Kunstliche Intelligenz*, vol. **27**, pp. 119–128, 2013.

165. Holdsworth, R.: Autonomous in flight path planning to replace pure collision avoidance for free flight aircraft using automatic dependent surveillance broadcast, PhD thesis, Swinburne University, 2003.

166. Hong, Y.; Kim, D.; Li, D.; Xu, B.; Chen, W.; Tokuta, A. O.: Maximum lifetime effective-sensing partial target-coverage in camera sensor networks, *In 11th IEEE International Symposium on Modeling and Optimization in Mobile, Ad Hoc and Wireless Networks (WiOpt)*, Tsukuba Science City, Japan, pp. 619–626, 2013.

167. Holt, J.; Biaz, S.; Aj, C. A.: Comparison of unmanned aerial system collision avoidance algorithm in a simulated environment, *AIAA Journal of Guidance, Control, and Dynamics*, vol. **36**, pp. 881–883, 2013.

168. Holzapfel, F.; Theil, S. (eds): *Advances in Aerospace Guidance, Navigation and Control*, Springer, Berlin, 2011.

169. Hota, S.; Ghose, D.: Optimal trajectory planning for unmanned aerial vehicles in three-dimensional space, *AIAA Journal of Aircraft*, vol. **51**, pp. 681–687, 2014.

170. Hota, S.; Ghose, D.: Time optimal convergence to a rectilinear path in the presence of wind, *Journal of Intelligent and Robotic Systems*, vol. **74**, pp. 791–815, 2014.

171. Howlett, J. K.; McLain, T.; Goodrich, M. A.: Learning real-time A* path planner for unmanned air vehicle target sensing, *AIAA Journal of Aerospace Computing, Information and Communication*, vol. **23**, pp. 108–122, 2006.

172. Hsu, D.; Isler, V.; Latombe, J. C.; Lin, M. C.: *Algorithmic Foundations of Robotic*, Springer, Berlin, 2010.

173. Huang, Z.; Zheng, Q. P.; Pasiliao, E. L.; Simmons, D.: Exact algorithms on reliable routing problems under uncertain topology using aggregation techniques for exponentially many scenarios, *Annals of Operations Research*, vol. 249, pp. 141–162, 2017.

174. Hunt, S.; Meng, Q.; Hinde, C.; Huang, T.: A consensus-based grouping algorithm for multi-agent cooperative task allocation with complex requirements, *Cognitive Computation*, vol. **6**, pp. 338–350, 2014.

175. Hutter, M.: *Universal Artificial Intelligence, Sequential Decisions Based on Algorithmic Probability*, Springer, Berlin, 2005.

176. Huynh, U.; Fulton, N.: Aircraft proximity termination conditions for 3D turn centric modes, *Applied Mathematical Modeling*, vol. **36**, pp. 521–544, 2012.

177. Ibarguren, A.; Molina, J.; Susperregi, L.; Maurtua, I.: Thermal tracking in mobile robots for leak inspection activities, *Sensors*, vol. **13**, pp. 13560–13574, 2013.

178. Ibe, O.; Bognor, R.: *Fundamentals of Stochastic Networks*, Wiley, Hoboken, NJ, 2011.

179. Igarashi, H.; Loi. K: Path-planning and navigation of a mobile robot as a discrete optimisation problems, *Art Life and Robotics*, vol. **5**, pp. 72–76, 2001.

180. Innocenti, M.; Pollini, L.; Turra, D.: Guidance of unmanned air vehicles based on fuzzy sets and fixed way-points, *AIAA Journal on Guidance, Control and Dynamics*, vol. **27**, pp. 715–720, 2002.

181. Itani, S.; Frazzoli, E.; Dahleh, M. A.: Dynamic traveling repair-person problem for dynamic systems, *In IEEE Conference on Decision and Control*, Cancun, Mexico, pp. 465–470, 2008.

182. Itoh, H.; Nakamura, K.: Partially observable Markov decision processes with imprecise parameters, *Artificial Intelligence*, vol. **171**, pp. 453–490, 2007.

183. Jaklic, G.; Kozak, J.; Krajnc, M.; Vitrih, V.; Zagar, E.: Geometric lagrange interpolation by planar cubic pythagorean hodograph curves, *Computer Aided Design*, vol **25**, pp. 720–728, 2008.

184. Jardin, M. R.; Bryson, A. E.: Neighboring optimal aircraft guidance in winds, *AIAA Journal of Guidance, Control and Dynamics*, vol. **24**, pp. 710–715, 2001.

185. Jardin, M. R.; Bryson, A. E.: Methods for computing minimum time paths in strong winds, *AIAA Journal of Guidance, Control and Dynamics*, vol. **35**, pp. 165–171, 2012.

186. Jennings, A. L.; Ordonez, R.; Ceccarelli, N.: Dynamic programming applied to UAV way point path planning in wind, *IEEE International Symposium on Computer-Aided Control System Design*, San Antonio, TX, pp. 215–220, 2008.

187. Jiang, Z.; Ordonez, R.: Robust approach and landing trajectory generation for reusable launch vehicles in winds, *In 17th IEEE International Conference on Control Applications*, San Antonio, TX, pp. 930–935, 2008.

188. Johnson, B.; Lind, R.: Three dimensional tree-based trajectory planning with highly maneuverable vehicles, *In 49th AIAA Aerospace Sciences Meeting*, Orlando, FL, paper AIAA2011-1286, 2011.

189. Jung, S.; Ariyur, K. B.: Enabling operational autonomy for unmanned aerial vehicles with scalability, *AIAA Journal of Aerospace Information Systems*, vol. **10**, pp. 517–529, 2013.

190. Kampke, T.; Elfes, A.: Optimal aerobot trajectory planning for wind based opportunity flight control, *IEEE/RSJ International Conference on Intelligent Robots and Systems*, Las Vegas, NV, pp. 67–74, 2003.

191. Kang, Y.; Caveney, D. S.; Hedrick, J. S.: Real time obstacle map building with target tracking, *AIAA Journal of Aerospace Computing, Information and Communication*, vol. **5**, pp. 120–134, 2008.

192. Kaelbling, L.; Littman, M.; Cassandra, A.: Planning and acting in partially observable stochastic domains, *Artificial Intelligence*, vol. **101**, pp. 99–134, 1998.

193. Kalyanam, K.; Chandler, P.; Pachter, M.; Darbha, S.: Optimization of perimeter patrol operations using UAV, *AIAA Journal of Guidance, Control and Dynamics*, vol. **35**, pp. 434–441, 2012.

194. Kalyanam, K.; Park, M.; Darbha, S.; Casbeer, D.; Chandler, P.; Pachter, M.: Lower bounding linear program for the perimeter patrol optimization, *AIAA Journal of Guidance, Control and Dynamics*, vol. **37**, pp. 558–565, 2014.

195. Khatib, O.: Real time obstacle avoidance for manipulators and mobile robots, *IEEE Interernational Conference on Robotics and Automation*, Saint Louis, MO, pp. 500–505, 1985.

196. Kothari, M.; Postlethwaite, I.; Gu, D. W.: UAV path following in windy urban environments, *Journal of Intelligent and Robotic Systems*, vol. **74**, pp. 1013–1028, 2014.

197. Kluever, C. A.: Terminal guidance for an unpowered reusable launch vehicle with bank constraints, *AIAA Journal of Guidance, Control, and Dynamics*, vol. **30**, no. 1, pp. 162–168, 2007.

198. Kuwata, Y.; Schouwenaars, T.; Richards, A.; How, J.: Robust constrained receding horizon control for trajectory planning, *AIAA Conference on Guidance, Navigation and Control*, San Francisco, CA, 2005.

199. Kalra, N.; Martinoli, A.: Optimal multiplicative Bayesian search for a lost target, In: Asama, H.; Fukuda, T.; Arai, T.; Endo, I. (eds) *In Distributed Autonomous Robotic Systems*, Springer, Japan, pp. 91–101, 2006.

200. Kalyanam, K.; Chandler, P.; Pachter, M.; Darbha, S.: Optimization of perimeter patrol operations using unmanned aerial vehicles, *AIAA Journal of Guidance, Control, and Dynamics*, vol. **35**, pp. 434–441, 2012.

201. Kalyanam, K.; Park, M.; Darbha, S.; Casbeer, D.; Chandler, P.; Pachter, M.: Lower bounding linear program for the perimeter patrol optimization problem, *AIAA Journal of Guidance, Control, and Dynamics*, vol. **37**, pp. 558–565, 2014.

202. Kantaros, Y.; Thanou, M.; Tzes, A.: Distributed coverage control for concave areas by a heterogeneous robot swarm with visibility sensing constraints, *Automatica*, vol. **53**, pp. 195–207, 2015.

203. Ke, L.; Zhai, L.; Li, J.; Chan, F. T.: Pareto mimic algorithm: An approach to the team orienteering problem, *Omega*, vol. **61**, pp. 155–166, 2016.

204. Kerkkamp, R. B. O.; Aardal, K.: A constructive proof of swap local search worst-case instances for the maximum coverage problem, *Operations Research Letters*, vol. **44**, pp. 329–335, 2016.

205. Khan, A.; Noreen, I.; Ryu, H.; Doh, N. L.; Habib, Z.: Online complete coverage path planning using two-way proximity search, *Intelligent Service Robotics*, pp. 1–12, 2017. DOI: 10.1007/s11370-017-0223-z.

206. Kim, S.; Oh, H.; Suk, J.; Tsourdos, A.: Coordinated trajectory planning for efficient communication relay using multiple UAVs, *Control Engineering Practice*, vol. **29**, pp. 42–49, 2014.

207. Kiraly, A.; Christidou, M.; Chovan, T.; Karlopoulos, E.; Abonyi, J.: Minimization of off-grade production in multi-site multi-product plants by solving multiple traveling salesman problem, *Journal of Cleaner Production*, vol. **111**, pp. 253–261, 2016.

208. Klein, R.; Kriesel, D.; Langetepe, E.: A local strategy for cleaning expanding cellular domains by simple robots, *Theoretical Computer Science*, vol. **605**, pp. 80–94, 2015.

209. Kondak, K.; Ollero, A.; Maza, I.; Krieger, K.; Albu-Schaeffer, A.; Schwarzbach, M.; Laiacker, M.: Unmanned aerial systems physically interacting with the environment: Load transportation, deployment, and aerial manipulation, In: Valavanis, K. P.; Vachtsevanos, G. J. (eds) *In Handbook of Unmanned Aerial Vehicles*, Springer, Netherlands, pp. 2755–2785, 2015.

210. Korpela, C.; Orsag, M.; Oh, P.: Towards valve turning using a dual-arm aerial manipulator, *In IEEE/RSJ International Conference on Intelligent Robots and Systems*, pp. 3411–3416, 2014. DOI: 10.1109/IROS.2014.6943037.

211. Korsah, G. A.; Stentz, A.; Dias, M. B.: A comprehensive taxonomy for multi-robot task allocation, *The International Journal of Robotics Research*, vol. **32**, pp. 1495–1512, 2013.

212. Koubaa, A.; Cheikhrouhou, O.; Bennaceur, H.; Sriti, M. F.; Javed, Y.; Ammar, A.: Move and mmprove: A market-based mechanism for the multiple depot multiple travelling salesmen problem, *Journal of Intelligent and Robotic Systems*, vol. **85**, pp. 307–330, 2017.

213. Kriheli, B.; Levner, E.: Optimal search and detection of clustered hidden targets under imperfect inspections, *IFAC Proceedings Volumes*, vol. **46**, pp. 1656–1661, 2016.

214. Kumar, G. P.; Berman, S.: The probabilistic analysis of the network created by dynamic boundary coverage, 2016, arXiv preprint arXiv:1604.01452.

215. Lalish, E.; Morgansen, K. A.: Distributed reactive collision avoidance, *Autonomous Robots*, vol. **32**, pp. 207–226, 2012.

216. Lam, T. M.; Boschloo, H. W.; Mulder, M.; Van Paassen, M. M.: Artificial force field for haptic feedback in UAV teleoperation, *IEEE Transactions on Systems, Man and Cybernetics, Part A: Systems and Humans*, vol. **39**, pp. 1316–1330, 2009.

217. Lee, S. K.; Becker, A.; Fekete, S. P.; Kroller, A.; McLurkin, J.: Exploration via structured triangulation by a multi-robot system with bearing-only low-resolution sensors, *In IEEE International Conference on Robotics and Automation*, Hong Kong, pp. 2150–2157, 2014.

218. Levine, D.; Luders, B.; How, J. P.: Information-theoretic motion planning for constrained sensor networks, *AIAA Journal of Aerospace Information Systems*, vol. **10**, pp. 476–496, 2013.

219. Li, W.; Wu, Y.: Tree-based coverage hole detection and healing method in wireless sensor networks, *Computer Networks*, vol. **103**, pp. 33–43, 2016.

220. Liemhetcharat, S.; Yan, R.; Tee, K. P.: Continuous foraging and information gathering in a multi-agent team, *In Proceedings of the International Conference on Autonomous Agents and Multiagent Systems*, Stanbul, Turkey, pp. 1325–1333, 2015.

221. Lee, S. G.; Diaz-Mercado, Y.; Egerstedt, M.: Multirobot control using time-varying density functions, *IEEE Transactions on Robotics*, vol. **31**, pp. 489–493, 2015.

222. Le Ny, J.; Pappas, G. J.: Adaptive algorithms for coverage control and space partitioning in mobile robotic networks, 2010, arXiv preprint arXiv:1011.0520.

223. Le Ny, J.; Feron, E.; Frazzoli, E.: On the Dubins traveling salesman problem, *IEEE Transactions on Automatic Control*, vol. **57**, pp. 265–270, 2012.

224. Leonard, N. E.; Olshevsky, A.: Nonuniform coverage control on the line, *IEEE Transactions on Automatic Control*, vol. **58**, pp. 2743–2755, 2013.

225. Liu, L.; Zlatanova, S.: An approach for indoor path computation among obstacles that considers user dimension, *ISPRS International Journal of Geo-Information*, vol. **4**, pp. 2821–2841, 2015.

226. Lu, X.: Dynamic and stochastic routing optimization: Algorithmic development and analysis, PhD thesis, University of California, Irvine, CA, 2001.

227. Maftuleac, D.; Lee, S. K.; Fekete, S. P.; Akash, A. K.; Lopez-Ortiz, A.; McLurkin, J.: Local policies for efficiently patrolling a triangulated region by a robot swarm, *In IEEE International Conference on Robotics and Automation (ICRA)*, Seattle, WA, pp. 1809–1815, 2015.

228. Maravall, D.; De Lope, J.; Martin, J. A.: Hybridizing evolutionary computation and reinforcement learning for the design of almost universal controllers for autonomous robots, *Neurocomputing*, vol. **72**, pp. 887–894, 2009.

229. Marier, J. S.; Besse, C.; Chaib-Draa, B.: A Markov model for multiagent patrolling in continuous time, In *Neural Information Processing: 16th International Conference, ICONIP 2009*, Bangkok, Thailand, pp. 648–656, 2009.

230. Matveev, A. S.; Teimoori, H.; Savkin, A.: Navigation of a uni-cycle like mobile robot for environmental extremum seeking, *Automatica*, vol. **47**, pp. 85–91, 2011.

231. Maza, I.; Kondak, K.; Bernard, M.; Ollero, A.: Multi-UAV cooperation and control for load transportation and deployment, *Journal of Intelligent and Robotic Systems*, vol. **57**, pp. 417–449, 2010.

232. Miller, L. M.; Silverman, Y.; MacIver, M. A.; Murphey, T. D.: Ergodic exploration of distributed information, *IEEE Transactions on Robotics*, vol. **32**, pp. 36–52, 2016.

233. Mladenovic, N.; Brimberg, J.; Hansen, P.; Moreno-Perez, J. A.: The p-median problem: A survey of metaheuristic approaches, *European Journal of Operational Research*, vol. **179**, pp. 927–939, 2007.

234. Mufalli, F.; Batta, R.; Nagi, R.: Simultaneous sensor selection and routing of UAV for complex mission plans, *Computers and Operations Research*, vol. **39**, pp. 2787–2799, 2012.

235. Nenchev, V.; Cassandras, C. G.; Raisch, J.: Optimal control for a robotic exploration, pick-up and delivery problem, 2016. arXiv preprint arXiv:1607.01202.

236. Nestmeyer, T.; Giordano, P. R.; Bulthoff, H. H.; Franchi, A.: Decentralized simultaneous multi-target exploration using a connected network of multiple robots, *Autonomous Robots*, pp. 1–23, 2016. DOI: 10.1007/s10514-016-9578-9.

237. Nourani-Vatani, N.: Coverage algorithms for under-actuated car-like vehicle in an uncertain environment, PhD thesis, Technical University of Denmark, Lyngby, 2006.

238. Obermeyer, K. J.; Ganguli, A.; Bullo, F.: Multi-agent deployment for visibility coverage in polygonal environments with holes, *International Journal of Robust and Nonlinear Control*, vol. **21**, pp. 1467–1492, 2011.

239. Obermeyer, K.; Oberlin, P.; Darbha, S.: Sampling based path planning for a visual reconnaissance UAV, *AIAA Journal of Guidance, Control and Dynamics*, vol. **35**, pp. 619–631, 2012.

240. Okabe, A.; Boots, B.; Sugihara, K.; Chiu, S. N.: *Spatial tessellations: Concepts and applications of Voronoi diagrams*, John Wiley, Hoboken, NJ, 2009.

241. Orsag, M.; Korpela, C.; Bogdan, S.; Oh, P.: Valve turning using a dual-arm aerial manipulator, *In IEEE International Conference on Unmanned Aircraft Systems*, pp. 836–841, 2014. DOI: 10.1109/ICUAS.2014.6842330.

242. Ozbaygin, G.; Yaman, H.; Karasan, O. E.: Time constrained maximal covering salesman problem with weighted demands and partial coverage, *Computers and Operations Research*, vol. **76**, pp. 226–237, 2016.

243. Palacios-Gass, J. M.; Montijano, E.; Sags, C.; Llorente, S.: Distributed coverage estimation and control for multirobot persistent tasks, *IEEE Transactions on Robotics*, vol. **32**, pp. 1444–1460, 2016.

244. Pastor, E.; Royo, P.; Santamaria, E.; Prats, X.: In flight contingency management for unmanned aircraft systems, *Journal of Aerospace Computing Information and Communication*, vol. **9**, pp. 144–160, 2012.

245. Penicka, R.; Faigl, J.; Vana, P.; Saska, M.: Dubins orienteering problem, *IEEE Robotics and Automation Letters*, vol. **2**, pp. 1210–1217, 2017.

246. Pierson, A.; Schwager, M.: Adaptive inter-robot trust for robust multi-robot sensor coverage, In: Inaba, M.; Corke, P. (eds) *Robotics Research*, Springer, Berlin, pp. 167–183, 2016.

247. Pita, J.; Jain, M.; Tambe, M.; Ordonez, F.; Kraus, S.: Robust solutions to stackelberg games: Addressing bounded rationality and limited observations in human cognition, *Artificial Intelligence*, vol. **174**, pp. 1142–1171, 2010.

248. Poduri, S.; Sukhatme, G. S.: Constrained coverage for mobile sensor networks, *In IEEE International Conference on Robotics and Automation*, vol. **1**, pp. 165–171, 2004.

249. Portugal, D.; Rocha, R. P.: Cooperative multi-robot patrol with Bayesian learning, *Autonomous Robots*, **40**, pp. 929–953, 2016.

250. Pounds, P. E.; Bersak, D. R.; Dollar, A. M.: Grasping from the air: Hovering capture and load stability, *In IEEE International Conference on Robotics and Automation*, Shanghai, China, pp. 2491–2498, 2011.

251. Regev, E.; Altshuler, Y.; Bruckstein, A. M.: The cooperative cleaners problem in stochastic dynamic environments, 2012. arXiv preprint arXiv:1201.6322.

252. Renzaglia, A.; Doitsidis, L.; Martinelli, A.; Kosmatopoulos, E.: Multi-robot three dimensional coverage of unknown areas, *International Journal of Robotics Research*, vol. **31**, pp. 738–752, 2012.

253. Riera-Ledesma, J.; Salazar-Gonzalez, J. J.: Solving the team orienteering arc routing problem with a column generation approach, *European Journal of Operational Research*, 2017. doi: 10.1016/j.ejor.2017.03.027.

254. Rojas, I. Y.: Optimized photogrammetric network design with flight path planner for UAV-based terrain surveillance, MS thesis, Brigham Young university, 2014.

255. Rout, M.; Roy, R.: Dynamic deployment of randomly deployed mobile sensor nodes in the presence of obstacles, *Ad Hoc Networks*, vol. **46**, pp. 12–22, 2016.

256. Roy, A.; Mitra, D.: Unscented Kalman filter based multi-target tracking algorithms for airborne surveillance applications, *AIAA Journal of Guidance, Control and Dynamics*, vol. **39**, pp. 1949–1966, 2016.

257. Ruggiero, F.; Trujillo, M. A.; Cano, R.; Ascorbe, H.; Viguria, A.; Perez, C.; Lippiello, V.; Ollero, A.; Siciliano, B.: A multilayer control for multirotor UAVs equipped with a servo robot arm, *In IEEE International Conference on Robotics and Automation (ICRA)*, Seattle, WA, pp. 4014–4020, 2015.

258. Sadovsky, A. V.; Davis, D.; Isaacson, D. R.: Efficient computation of separation-compliant speed advisories for air traffic arriving in terminal airspace, *ASME Journal of Dynamic Systems, Measurement, and Control*, vol. **136**, pp. 536–547, 2014.

259. Savla, K.; Frazzoli, E.; Bullo, F.: Traveling salesperson problems for the Dubbins vehicle, *IEEE Transactions on Automatic Control*, vol. **53**, pp. 1378–1391, 2008.

260. Schouwenaars, T.; Mettler, B.; Feron, E.: Hybrid model for trajectory planning of agile autonomous vehicles, *AIAA Journal on Aerospace Computing, Inforation and Communication*, vol. **12**, pp. 629–651, 2004.

261. Schwertfeger, S.; Birk, A.: Map evaluation using matched topology graphs, *Autonomous Robots*, vol. **40**, pp. 761–787, 2016.

262. Sharifi, F.; Chamseddine, A.; Mahboubi, H.; Zhang, Y.; Aghdam, A. G.: A distributed deployment strategy for a network of cooperative autonomous vehicles, *IEEE Transactions on Control Systems Technology*, vol. **23**, pp. 737–745, 2015.

263. Sharma, V.; Patel, R. B.; Bhadauria, H. S.; Prasad, D.: Deployment schemes in wireless sensor network to achieve blanket coverage in large-scale open area: A review, *Egyptian Informatics Journal*, vol. **17**, pp. 45–56, 2016.

264. Sharma, V.; Srinivasan, K.; Chao, H. C.; Hua, K. L.: Intelligent deployment of UAVs in 5G heterogeneous communication environment for improved coverage, *Journal of Network and Computer Applications*, 2017. DOI: 10.1016/j.jnca.2016.12.012.

265. Singh, M. K.: Evaluating levy flight parameters for random searches in a 2D space, Doctoral dissertation, Massachusetts Institute of Technology, 2013.

266. Sivaram Kumar, M. P.; Rajasekaran, S.: Path planning algorithm for extinguishing forest fires, *Journal of Computing*, vol. **4**, pp. 108–113, 2012.

267. Smith, S.; Tumova, J.; Belta, C.; Rus, D.: Optimal path planning for surveillance with temporal logic constraints, *International Journal of Robotics Research*, vol. **30**, pp. 1695–1708, 2011.

268. Song, C.; Liu, L.; Feng, G.; Xu, S.: Optimal control for multi-agent persistent monitoring, *Automatica*, vol. **50**, pp. 1663–1668, 2014.

269. Song, C.; Liu, L.; Feng, G.; Xu, S.: Coverage control for heterogeneous mobile sensor networks on a circle, *Automatica*, vol. **63**, pp. 349–358, 2016.

270. Sposato, M.: Multiagent cooperative coverage control, MS thesis, KTH, Royal Institute of Technology, Stockholm, Sweden, p. 67, 2016.

271. Srivastava, V.; Reverdy, P.; Leonard, N. E.: On optimal foraging and multi-armed bandits, *In 51st IEEE Annual Allerton Conference on Communication, Control, and Computing (Allerton)*, Monticello, IL, pp. 494–499, 2013.

272. Stranders, R.; Munoz, E.; Rogers, A.; Jenning N. R.: Near-optimal continuous patrolling with teams of mobile information gathering agents, *Artificial Intelligence*, vol. **195.**, pp. 63–105, 2013.

273. Sun, S.; Sun, L.; Chen, S.: Research on the target coverage algorithms for 3D curved surface, *Chaos, Solitons and Fractals*, vol. **89**, pp. 397–404, 2016.

274. Sutantyo, D. K.; Kernbach, S.; Nepomnyashchikh, V. A.; Levi, P.: Multi-robot searching algorithm using Levy flight and artificial potential field, *IEEE International Workshop on Safety Security and Rescue Robotics (SSRR)*, Bremem, Germany, 2010.

275. Sydney, N.; Paley, D. A.: Multiple coverage control for a non stationary spatio-temporal field, *Automatica*, vol. **50**, pp. 1381–1390, 2014.

276. Tambe, M.: *Security and Game Theory: Algorithms, Deployed Systems, Lessons Learned*, Cambridge University Press, Cambridge, 2012.

277. Tuna, G.; Gulez, K.; Gungor, V. C.: The effects of exploration strategies and communication models on the performance of cooperative exploration, *Ad Hoc Networks*, vol. **11**, pp. 1931–1941, 2013.

278. Wang, Y. P.: Regret-based automated decision-making aids for domain search tasks using human-agent collaborative teams, *IEEE Transactions on Control Systems Technology*, vol. **24**, pp. 1680–1695, 2016.

279. Winfield, A. F.: Towards an engineering science of robot foraging, In: Asama, H.; Fukuda, T.; Arai, T.; Endo, I. (eds) *In Distributed Autonomous Robotic Systems*, Springer, Berlin, Heidelberg, vol. **8**, pp. 185–192, 2009.

280. Wilkins, D. E.; Smith, S. F.; Kramer, L. A.; Lee, T.; Rauenbusch, T.: Airlift mission monitoring and dynamic rescheduling, *Engineering Application of Artificial Intelligence*, vol. **21**, pp. 141–155, 2008.

281. Xue, M. H.; Wang, T. Z.; Mao, S.: Double evolutsional artificial bee colony algorithm for multiple traveling salesman problem, *In MATEC Web of Conferences*, vol. **44**, EDP Sciences, 2016, DOI:10.1051/matecconf/20164402025.

282. Yakici, E.: *Solving location and routing problem for UAVs*, Computers and Industrial Engineering, vol. **102**, pp. 294–301, 2016.

283. Yang, R.; Kiekintvled, C.; Ordonez, R.; Tambe, M.; John, R.: Improving resource allocation strategies against human adversaries in security games: An extended study, *Artificial Intelligence Journal*, vol. **195**, pp. 440–469, 2013.

284. Yin, Z.; Xin Jiang, A.; Tambe, M.; Kiekintveld, C.; Leyton-Brown, K.; Sandholm, T.; Sullivan, J. P.: Trusts: Scheduling randomized patrols for fare inspection in transit systems using game theory, *AI Magazine*, vol. **33**, pp. 59–72, 2012.

285. Younis, M.; Akkaya, K.: Strategies and techniques for node placement in wireless sensor networks: A survey, *Ad Hoc Networks*, vol. **6**, pp. 621–655, 2008.

286. Zannat, H.; Akter, T.; Tasnim, M.; Rahman, A.: The coverage problem in visual sensor networks: A target oriented approach, *Journal of Network and Computer Applications*, vol. **75**, pp. 1–15, 2016.

287. Zhao, W.; Meng, Q.; Chung, P. W.: A Heuristic distributed task allocation method for multi-vehicle multi-task problems and its application to search and rescue scenario, *IEEE Transactions on Cybernetics*, vol. **46**, pp. 902–915, 2016.

288. Zhu, C.; Zheng, C.; Shu, L.; Han, G.: A survey on coverage and connectivity issues in wireless sensor networks, *Journal of Network and Computer Applications*, vol. **35** pp. 619–632, 2012.

289. Zorbas, D.; Pugliese, L. D. P.; Razafindralambo, T.; Guerriero, F.: Optimal drone placement and cost-efficient target coverage, *Journal of Network and Computer Applications*, vol. **75**, pp. 16–31, 2016.

290. Tang, J.; Alam, S.; Lokan, C.; Abbass, H. A.: A multi-objective approach for dynamic airspace sectorization using agent based and geometric models, *Transportation Research Part C*, vol. **21**, pp. 89–121, 2012.

291. Tapia-Tarifa, S. L.: The cooperative cleaners case study: Modelling and analysis in real-time ABS, MS thesis, Department of Informatics, University of Oslo, 116 p., 2013.

292. Temizer, S.: Planning under uncertainty for dynamic collision avoidance, PhD thesis, MIT, Cambridge, MA, 2011.

293. Thanou, M.; Stergiopoulos, Y.; Tzes, A.: Distributed coverage using geodesic metric for non-convex environments, *In IEEE International Conference on Robotics and Automation*, Karlsruhe, Germany, pp. 933–938, 2013.

294. Tian, J.; Liang, X.; Wang, G.: Deployment and reallocation in mobile survivability heterogeneous wireless sensor networks for barrier coverage, *Ad Hoc Networks*, vol. **36**, pp. 321–331, 2016.

295. Tilk, C.; Rothenbacher, A. K.; Gschwind, T.; Irnich, S.: Asymmetry matters: Dynamic half-way points in bidirectional labeling for solving shortest path problems with resource constraints faster, *European Journal of Operational Research*, vol. **261**, pp. 530–539, 2017. DOI: 10.1016/j.ejor.2017.03.017.

296. Torres, M.; Pelta, D. A.; Verdegay, J. L.; Torres, J. C.: Coverage path planning with unmanned aerial vehicles for 3D terrain reconstruction, *Expert Systems with Applications*, vol. **55**, pp. 441–451, 2016.

297. Toth, P.; Vigo, D.: *The Vehicle Routing Problem*, SIAM, Philadelphia, PA, 2002.

298. Troiani, C.; Martinelli, A.; Laugier, C.; Scaramuzza, D.: Low computational-complexity algorithms for vision-aided inertial navigation of micro aerial vehicles, *Robotics and Autonomous Systems*, vol. **69**, pp. 80–97, 2015.

299. Tseng, K. S.; Mettler, B.: Near-optimal probabilistic search via submodularity and sparse regression, *Autonomous Robots*, vol. **41**, pp. 205–229, 2017.

300. Tuna, G.; Gungor, V. C.; Gulez, K.: An autonomous wireless sensor network deployment system using mobile robots for human existence detection in case of disasters, *Ad Hoc Networks*, vol. **13**, pp. 54–68, 2014.

301. Valente, J.; Del Cerro, J.; Barrientos, A.; Sanz, D.: Aerial coverage optimization in precision agriculture management: A musical harmony inspired approach, *Computers and Electronics in Agriculture*, vol. **99**, pp. 153–159, 2013.

302. Valente, J.: Aerial coverage path planning applied to mapping, PhD thesis, Universidad Politecnica de Madrid, 2014.

303. VanderBerg, J. P.; Patil, S.; Alterovitz, R.: Motion planning under uncertainty using differential dynamic programming in Belief space, *International Symposium of Robotics Research*, Flagstaff, AZ, 2011.

304. Verbeeck, C.; Vansteenwegen, P.; Aghezzaf, E. H.: Solving the stochastic time-dependent orienteering problem with time windows, *European Journal of Operational Research*, vol. **255**, pp. 699–718, 2016.

305. Vieira, L. F. M.; Almiron, M. G.; Loureiro, A. A.: Link probability, node degree and coverage in three-dimensional networks, *Ad Hoc Networks*, vol. **37**, pp. 153–159, 2016.

306. Viet, H. H.; Dang, V. H.; Laskar, M. N.; Chung T. C.: BA*: An online complete coverage algorithm for cleaning robots, *Applied Intelligence*, vol. **39**, pp. 217–237, 2013.

307. Wang, H. F.; Wen, Y. P.: Time-constrained Chinese postman problems, *Computers and Mathematics with Applications*, vol. **44**, pp. 375–387, 2002.

308. Wang, Z.; Guo, J.; Zheng, M.; Wang, Y.: Uncertain multiobjective traveling salesman problem, *European Journal of Operational Research*, vol. **241**, pp. 478–489, 2015.

309. Waqqas, A.: Distributed navigation of multi-robot systems for sensing coverage, PhD thesis, School of Electrical Engineering and Telecommunications, The University of New South Wales, Australia, 224 p., 2016.

310. Xu, A.; Viriyasuthee, C.; Rekleitis, I.: Optimal complete terrain coverage using an unmanned aerial vehicle, *In IEEE International Conference on Robotics and Automation*, Shanghai, China, pp. 2513–2519, 2011.

311. Xu, A.; Viriyasuthee, C.; Rekleitis, I.: Efficient complete coverage of a known arbitrary environment with applications to aerial operations, *Autonomous Robots*, vol. **36**, pp. 365–381, 2014.

312. Yazicioglu, A. Y.; Egerstedt, M.; Shamma, J. S.: Communication-free distributed coverage for networked systems, *IEEE Transactions on Control of Network Systems*, 2016. DOI: 10.1109/TCNS.2016.2518083.

313. Yu, J.; Schwager, M.; Rus, D.: Correlated orienteering problem and its application to persistent monitoring tasks, *IEEE Transactions on Robotics*, vol. **32**, pp. 1106–1118, 2016.

314. Zhong, M.; Cassandras, C. G.: Distributed coverage control in sensor network environments with polygonal obstacles, *IFAC Proceedings Volumes*, vol. **41**, pp. 4162–4167, 2008.

315. Laumond, J. P.: *Robot Motion Planning and Control*, Springer, Berlin, 1998.

316. Laugier, C.; Chatila, R. (eds): *Autonomous Navigation in Dynamic Environments*, Springer, Berlin, 2007.

317. Lavalle, S. M.: *Planning Algorithms*. Cambridge University Press, Cambridge, 2006.

318. Lavalle, S. M.: Motion Planning, *IEEE Robotics and Automation Magazine*, vol. **18**, pp. 108–118, 2011.

319. Lawler, E. L.; Lenstra, J. K.; Rinnoy Kan, A. H. G; Shmoys, D.B.: *A Guided Tour of Combinatorial Optimization*, Wiley. Hoboken, NJ, 1995.

320. Lee, J.; Kwon, O.; Zhang, L.; Yoon S.: A selective retraction based RRT planner for various environments, *IEEE Transactions on Robotics*, 2014. DOI: 10.1109/TRO.2014.2309836.

321. Lekkas A. M.; Fossen T. I.: Integral LOS path following for curved paths based on a monotone cubic Hermite spline parametrization, *IEEE Transactions on Control System Technology*, 2014. DOI: 10.1109/TCST.2014.2306774.

322. Le Ny, J.; Feron, E.; Frazzoli, E.: On the Dubins traveling salesman problem, *IEEE Transactions on Automatic Control*, vol. **57**, pp. 265–270, 2012.

323. Li, Z.; Canny, J. F.: *Non Holonomic Motion Planning*. Kluwer Academic Press, Berlin, 1992.

324. Li, B.; Xu, C.; Teo, K. L.; Chu, J.: Time optimal Zermelo's navigation problem with moving and fixed obstacles, *Applied Mathematics and Computation*, vol. **224**, pp. 866–875, 2013.

325. Lin, L.; Goodrich, M. A.: Hierarchical heuristic search using a Gaussian mixture model for UAV coverage planning, *IEEE Transactions on Cybernetics*, 2014. DOI: 10.1109/TCYB.2014.2309898.

326. Liu, Y.; Saripelli, S.: Path planning using 3D Dubins curve for unmanned aerial vehicles, *International Conference on Unmanned Aircraft System*, pp. 296–304, 2014. DOI: 10.1109/icuas.2014.6842268.

327. Littman, M.: A tutorial on partially observable Markov decision process, *Journal of Mathematical Psychology*, vol. **53**, pp. 119–125, 2009.

328. Macharet, D.; Neto, A. A.; Campos, M.: On the generation of feasible paths for aerial robots in environments with obstacles, *IEEE/RSJ International Conference on Intelligent Robots and Systems*, St. Louis, MO, pp. 3380–3385, 2009.

329. Maggiar, A.; Dolinskaya, I.S.: Construction of fastest curvature constrained path in direction dependent media, *AIAA Journal of Guidance, Control and Dynamics*, vol. **37**, pp. 813–827, 2014.

330. Marigo, A.; Bichi, A.: Steering driftless non-holonomic systems by control quanta, *IEEE Interernational Conference on Decision and Control*, vol. **4**, pp. 466–478, 1998.

331. Masoud, A. A.: A harmonic potential approach for simultaneous planning and control of a generic UAV platform, *Journal of Intelligent and Robotics Systems*, vol. **65**, pp. 153–173, 2012.

332. Matsuoka, Y.; Durrant-Whyte, H.; Neira, J.: *Robotics, Science and Systems*, The MIT Press, Cambridge, MA, 2010.

333. Mattei, M.; Blasi, L.: Smooth flight trajectory planning in the presence of no-fly zones and obstacles, *AIAA Journal of Guidance, Control and Dynamics*, vol. **33**, No. 2, pp. 454–462, 2010.

334. Matveev, A. S.; Teimoori, H.; Savkin, A.: Navigation of a uni-cycle like mobile robot for environmental extremum seeking, *Automatica*, vol. **47**, pp. 85–91, 2011.

335. McGee, T.; Hedrick, J. K.: Optimal path planning with a kinematic airplane model, *AIAA Journal of Guidance, Control and Dynamics*, vol. **30**, pp. 1123–1133, 2007.

336. Miele, A.; Wang, T.; Melvin, W.: Optimal take-off trajectories in the presence of windshear, *Journal of Optimization, Theory and Applications*, vol. **49**, pp. 1–45, 1986.

337. Miele, A.; Wang, T.; Melvin, W.: Penetration landing guidance trajectories in the presence of windshear, *AIAA Journal of Guidance*, vol. **12**, pp. 806–814, 1989.

338. Missiuro, P.; Roy, N.: Adaptive probabilistic roadmaps to handle uncertain maps, *IEEE International Conference on Robotics and Automation*, Orlando, FL, pp. 1261–1267, 2006.

339. Mufalli, F.; Batta, R.; Nagi, R.: Simultaneous sensor selection and routing of unmanned aerial vehicles for complex mission plans, *Computers and Operations Research*, vol. **39**, pp. 2787–2799, 2012.

340. Mujumda, A.; Padhi, R.: Evolving philosophies on autonomous obstacles/collision avoidance of unmanned aerial vehicles, *AIAA Journal of Aerospace Computing, Information and Communication*, vol. **8**, pp. 17–41, 2011.

341. Naldi, R.; Marconi, L.: Optimal transition maneuvers for a class of V/STOL aircraft, *Automatica*, vol. **47**, pp. 870–879, 2011.

342. Ng, H. K.; Sridhar, B.; Grabbe, S.: Optimizing trajectories with multiple cruise altitudes in the presence of winds, *AIAA Journal of Aerospace Information Systems*, vol. **11**, pp. 35–46, 2014.

343. Oberlin, P.; Rathinam, S.; Darbha, S.: Todays traveling salesman problem, *IEEE Robotics and Automation Magazine*, vol. **17**, pp. 70–77, 2010.

344. Obermeyer, K.; Oberlin, P.; Darbha, S.: Sampling based path planning for a visual reconnaissance unmanned air vehicle, *AIAA Journal of Guidance, Control and Dynamics*, vol. **35**, pp. 619–631, 2012.

345. Oikonomopoulos, A. S.; Kyriakopoulos, K. J.; Loizou, S. G.: Modeling and control of heterogeneous nonholonomic input-constrained multi-agent systems, *In 49th IEEE Conference on Decision and Control*, Atlanta, GA, pp. 4204–4209, 2010.

346. Patterson, T.; McClean, S.; Morrow, P.; Parr, G.: Modeling safe landing zone detection options to assist in safety critical UAV decision making, *Procedia Computer science*, vol. **10**, pp. 1146–1151, 2012.

347. Pavone, M.; Frazzoli, E.; Bullo, F.: Adaptive and distributive algorithms for Vehicle routing in a stochastic and dynamic environment, *IEEE Transactions on Automatic Control*, vol. **56**, pp. 1259–1274, 2011.

348. Peng, R.; Wang, H.; Wang, Z.; Lin, Y.: Decision making of aircraft optimum configuration utilizing multi dimensional game theory, *Chinese Journal of Aeronautics*, vol. **23**, pp. 194–197, 2010.

349. Persiani, F.; De Crescenzio, F.; Miranda, G.; Bombardi, T.; Fabbri, M.; Boscolo, F.: Three dimensional obstacle avoidance strategies for uninhabited aerial systems mission planning and replanning, *AIAA Journal of Aircraft*, vol. **46**, pp. 832–846, 2009.

350. Pettersson, P. O.; Doherty, P.: Probabilistic road map based path planning for an autonomous unmanned aerial vehicle, *Workshop on Connecting Planning Theory with Practice*, 2004.

351. Petres, C.; Pailhas, Y.; Pation, P.; Petillot, Y.; Evans, J.; Lame, D.: Path planning for autonomous underwater vehicles, *IEEE Transactions on Robotics*, vol. **23**, pp. 331–341, 2007.

352. Phillips, J. M.; Bedrossian, N.; Kavraki, L. E.: Guided expansive space trees: A search strategy for motion and cost constrained state spaces, *IEEE International Conference on Robotics and Automation*, vol. **5**, pp. 3968–3973, 2004.

353. Piazzi, A.; Guarino Lo Bianco, C.; Romano, M.: η^3 Splines for the Smooth Path Generation of Wheeled Mobile Robot, *IEEE Transactions on Robotics*, vol. **5**, pp. 1089–1095, 2007.

354. Plaku, E.; Hager, G. D.: Sampling based motion and symbolic action planning with geometric and differential constraints, *IEEE International Conferennce on Robotics and Automation*, Anchorage, AK, pp. 5002–5008, 2010.

355. Poggiolini, L.; Stefani, G.: Minimum time optimality for a bang-singular arc: Second order sufficient conditions, *IEEE 44th Conference on Decision and Control*, Seville, Spain, pp. 1433–1438, 2005.

356. Powell, W. B.: *Approximate Dynamic Programming: Solving the Curse of Dimensionality*, Halsted Press, New York, 2011.

357. Prasanth, R. K.; Boskovic, J. D.; Li, S. M.; Mehra, R.: Initial study of autonomous trajectory generation for UAV, *IEEE International Conference on Decision and Control*, Orlando, FL, pp. 640–645, 2001.

358. Prats, X.; Puig, V.; Quevedo, J.; Nejjari, F.: Lexicographic optimization for optimal departure aircraft trajectories, *Aerospace Science and Technology*, vol. **14**, pp. 26–37, 2010.

359. Puterman, M. L.: *Markov Decision Processes Discrete Stochastic Dynamic Programming*, Wiley, Hoboken, NJ, 2005.

360. Qu, Y.; Zhang, Y.; Zhang, Y.: Optimal flight planning for UAV in 3D threat environment, *International Conference on Unmanned Aerial Systems*, 2014. DOI: 10.1109/ICUAS.2014.6842274.

361. Rabier, P. J.; Rheinboldt, W. C.: *Nonholonomic Motion of Rigid Mechanical Systems from a DAE Viewpoint*, SIAM Press, Philadelphia, PA, 2000.

362. Richards, A.; Schouwenaars, T.; How, J.; Feron, E.: Spacecraft trajectory planning with avoidance constraints using Mixed-Integer Linear Programming, *AIAA Journal of Guidance, Control and Dynamics*, vol. **25**, pp. 755–764, 2002.

363. Rosen, K. H.: *Discrete Mathematics*, McGraw Hill, New York, 2013.

364. Rysdyk, R.: Course and heading changes in significant wind, *AIAA Journal of Guidance, Control and Dynamics*, vol. **30**, pp. 1168–1171, 2007.

365. Ruchti, J.; Senkbeil, R.; Carroll, J.; Dickinson, J.; Holt, J.; Biaz, S.: Unmanned aerial system collision avoidance using artificial potential fields, *AIAA Journal of Aerospace Information Systems*, vol. **11**, pp. 140–144, 2014.

366. Rupniewski, M. W.; Respondek, W.: A classification of generic families of control affine systems and their bifurcations, *Mathematics of Control, Signals, and Systems*, vol. **21**, pp. 303–336, 2010.

367. Ruz, J. J.; Arevalo, O.; Pajares, G.; Cruz, J. M.: Decision making along alternative routes for UAV in dynamic environments, *IEEE International Conference on Emerging Technologies and Factory Automation, ETFA*, Patras, Greece, 2007.

368. Sabo, C.; Cohen, K.; Kumar, M.; Abdallah, S.: Path planning of a fire-fighting aircraft using fuzzy logic, *AIAA Aerospace Sciences Meeting*, Orlando, FL, paper AIAA 2009-1353, 2009.

369. Sadovsky, A. V.: Application of the shortest path problem to routing terminal airspace air traffic, *AIAA Journal of Aerospace Information System*, vol. **11**, pp. 118–130, 2014.

370. Samad, T.; Gorinevsky, D.; Stoffelen, F.: Dynamic multi-resolution route optimization for autonomous aircraft, *IEEE International Symposium on Intelligent Control*, Mexico, pp. 13–18, 2001.

371. Santamaria, E.; Pastor, E.; Barrado, C.; Prats, X.; Royo, P.; Perez, M.: Flight plan specification and management for unmanned aircraft systems, *Journal of Intelligent and Robotic Systems*, vol. **67**, pp. 155–181, 2012.

372. Savla, K.; Frazzoli, E.; Bullo, F.: Traveling salesperson problems for the dubbins vehicle, *IEEE Transactions on Automatic Control*, vol. **53**, pp. 1378–1391, 2008.

373. Schmitt, L.; Fichter, W.: Collision avoidance framework for small fixed wing unmanned aerial vehicles, *AIAA Journal of Guidance, Control and Dynamics*, vol. 37, pp. 1323–1328, 2014.

374. Schouwenaars, T.; Valenti, M.; Feron, E.; How, J.; Roche, E.: Linear programming and language processing for human/unmanned-aerial-vehicle team missions, *AIAA Journal of Guidance, Control, and Dynamics*, vol. **29**, no. 2, pp. 303–313, 2006.

375. Seibel, C. W.; Farines, J. M.; Cury, J. E.: Towards hybrid automata for the mission planning of unmanned aerial vehicles, In: Antsaklis (ed.), *Hybrid Systems V*, Springer-Verlag, Berlin, pp. 324–340, 1999.

376. Sennott, L. I.: *Stochastic Dynamic Programming and the Control of Queuing Systems*, Wiley, Hoboken, NJ, 1999.

377. Shah, M. Z.; Samar, R.; Bhatti, A. I.: Guidance of air vehicles: A sliding mode approach, *IEEE Transaction on Control Systems Technology*, 2014. DOI: 10.1109/TCST.2014.2322773.

378. Sinha, A.; Tsourdos, A.; White, B.: Multi-UAV coordination for tracking the dispersion of a contaminant cloud in an urban region, *European Journal of Control*, vol. **34**, pp. 441–448, 2009.

379. Smith, J. F.; Nguyen, T. H.: Fuzzy logic based resource manager for a team of UAV, *Annual Meeting of the IEEE Fuzzy Information Processing Society*, Montreal, pp. 463–470, 2006.

380. Soler, M.; Olivares, A.; Staffetti, E.: Multiphase optimal control framework for commercial aircraft 4D flight planning problems, *AIAA Journal of Aircraft*, 2014. DOI: 10.2514/1C032677.

381. Song, C.; Liu, L.; Feng, G.; Xu S.: Optimal control for multi-agent persistent monitoring, *Automatica*, vol. **50**, pp. 1663–1668, 2014.

382. Sridharan, M.; Wyatt, J.; Dearden, R.: Planning to see: A hierarchical approach to planning visual action on a robot using POMDP, *Artificial Intelligence*, vol. **174**, pp. 704–725, 2010.

383. Stranders, R.; Munoz, E.; Rogers, A.; Jenning, N. R.: Near-optimal continuous patrolling with teams of mobile information gathering agents, *Artificial Intelligence*, vol. **195**, pp. 63–105, 2013.

384. Sujit, P. B.; Saripalli, S.; Sousa, J. B.: Unmanned aerial vehicle path following, *IEEE Control System Magazine*, vol. **34**, pp. 42–59, 2014.

385. Sun, X.; Gollnick, V.; Li, Y.; Stumpf, E.: Intelligent multi criteria decision support system for systems design, *AIAA Journal of Aircraft*, vol. **51**, pp. 216–225, 2014.

386. Sundar, K.; Rathinam, S. Algorithms for routing an unmanned aerial vehicle in the presence of refueling depots, *IEEE Transaction on Automation Science and Engineering*, vol. **11**, pp. 287–294, 2014.

387. Sussman, H. J.: The Markov Dubins problem with angular acceleration control, *IEEE 36th Conference on Decision and Control*, San Diego, CA, pp. 2639–2643, 1997.

388. Tang, J.; Alam, S.; Lokan, C.; Abbass, H. A.: A multi-objective approach for dynamic airspace sectorization using agent based and geometric models, *Transportation Research Part C*, vol. **21**, pp. 89–121, 2012.

389. Techy, L.: Optimal navigation in planar true varying flow: Zermelo's problem revisited, *Intelligent Service Robotics*, vol. **4**, pp. 271–283, 2011.

390. Temizer, S.: Planning under uncertainty for dynamic collision avoidance, PhD thesis, MIT, 2011.

391. Tewari, A.: *Advanced Control of Aircraft, Spacecrafts and Rockets*, Wiley Aerospace Series, Hoboken, NJ, 2011.

392. Toth, P.; Vigo, D.: *The Vehicle Routing Problem*, SIAM, Philadelphia, PA, 2002.

393. Trumbauer, E.; Villac, B.: Heuristic search based framework for on-board trajectory redesign, *AIAA Journal of Guidance, Control and Dynamics*, vol. **37**, pp. 164–175, 2014.

394. Turnbull, O.; Richards, A.; Lawry, J.; Lowenberg, M.: Fuzzy decision tree cloning of flight trajectory optimization for rapid path planning, *IEEE Conference on Decision and Control*, San Diego, CA, pp. 6361–6366, 2006.

395. VanDaalen, C. E.; Jones, T.: Fast conflict detection using probability flow, *Automatica*, vol. **45**, pp. 1903–1909, 2009.

396. Vanderberg, J. P.: Path planning in dynamic environments, PhD thesis, University of Utrecht, The Netherlands, 2007.

397. Vazirani, V.: *Approximation Algorithms*, Springer-Verlag, Berlin, 2003.

398. Wang, X.; Wei, G.; Sun, J.: Free Knot Recursive B Spline for compensation of nonlinear smart sensors, *Measurement*, **44**, pp. 888–894, 2011.

399. Wang, Y.; Wang, S.; Tan, M.; Zhou, C.; Wei, Q.: Real-time dynamic Dubins-helix method for 3D trajectory smoothing, *IEEE Transactiona on Control System Technology*, 2014. DOI: 10.1109/TCST.2014.2325904.

400. Weiss, A.; Petersen, C.; Baldwin, M.; Scott, R.; Kolmanovsky, I.: Safe positively invariant sets for spacecraft obstacle avoidance, *AIAA Journal of Guidance, Control and Dynamics*, 2014. DOI: 10.2514/1.G000115.

401. Wilkins, D. E.; Smith, S. F.; Kramer, L. A.; Lee, T.; Rauenbusch, T.: Airlift mission monitoring and dynamic rescheduling, *Engineering Application of Artificial Intelligence*, vol. **21**, pp. 141–155, 2008.

402. Williams, P.: Aircraft trajectory planning for terrain following incorporating actuator constraints, *AIAA Journal of Aircraft*, **42**, pp. 1358–1362, 2005.

403. Wu, P.: Multi-objective mission flight planning in civil unmanned aerial systems, PhD thesis, Queensland University of Technology (Australia), 2009.

404. Yakimenko, O. A.: Direct method for rapid prototyping of near optimal aircraft trajectory, *AIAA Journal of Guidance, Control and Dynamics*. vol. **23**, pp. 865–875, 2000.

405. Yang, I.; Zhao, Y. Trajectory planning for autonomous aerospace vehicles amid known obstacles and conflicts, *AIAA Journal of Guidance, Control and Dynamics*, vol. **27**, pp. 997–1008, 2004.

406. Yanmaz, E.; Costanzo, C.; Bettstetter, C.; Elmenreich W.: A discrete stochastic process for coverage analysis of autonomous UAV networks, *IEEE GLOBECOM Workshop*, Miami, FL, pp. 1777–1782, 2010.

407. Yanushevsky, R.: *Guidance of Unmanned Aerial Vehicles*, CRC Press, Boca Raton, FL, 2011.

408. Yokoyama, N.: Path generation algorithm for turbulence avoidance using real-time optimization, *AIAA Journal of Guidance, Control and Dynamics*, vol. **36**, pp. 250–262, 2012.

409. Gandhi, R.; Yang, L. G.: Examination of planning under uncertainty algorithms for cooperative UAV, *AIAA Infotech@Aerospace*, Rohnert Park, CA, paper AIAA-2007-2817, 2007.

410. Hantos, P.: Systems engineering perspectives on technology, readiness assessment in software intensive system development, *AIAA Journal of Aircraft*, vol. **48**, pp. 738–748, 2011.

411. Inigo-Blasco, P.; Diaz-del-Rio, F.; Romero, M.; Cargigas, D.; Vicente, S.: Robotics software frameworks for multiagent robotic systems developement, *Robotics and Autonomous Systems*, vol. **60**, pp. 803–821, 2012.

412. Jorgensen, U.; Skjetne, R.: Generating safe and equally long trajectories for multiple unmanned agents, *IEEE 28th Mediterranean Conference on Control and Automation*, Barcelona, Spain, pp. 1566–1572, 2012.

413. Karahan, I.; Koksalan, M.: A territory defining multiobjective evolutionary algorithms and preference incorporation, *IEEE Transaction on Evolutionary Computation*, vol. **14**, pp. 636–664, 2010.

414. Karaman, S.; Frazzoli, E.: Complex mission optimization for multiple UAV using linear temporal logic, *American Control Conference*, Seattle, WA, pp. 2003–2009, 2008.

415. Kamgarpour, M.; Dadok, V.; Tomlin, c.: Trajectory generation for aircraft subject to dynamic weather uncertainty, *IN 49th IEEE Conference on Decision and Control*, Atlanta, GA, pp. 2063–2068, 2010.

416. Karaman, S.; Frazzoli, E.: Linear Temporal logic vehicle routing with applications to multi-UAV mission planning, *International Journal of Robust and Nonlinear Control*, Vol. **21**(12) pp. 1–38, 2010.

417. Karimoddini, A.; Liu, H.; Chen, B.; Lee, T.: Hybrid 3D formation control for unmanned helicopter, Technical report, NUS-ACT-11-005, 2011.

418. Kloetzer, M.; Belta, C.: Temporal logic planning and control of robotic swarms by hierarchical abstraction, *IEEE Transaction on Robotics*, vol. **23**, pp. 320–330, 2007.

419. Kon Kang, B.; Kim, K. E.: Exploiting symmetries for single and multi-agent partially observable stochastic domains, *Artificial Intelligence*, vol. **182**, pp. 32–57, 2012.

420. Krebsbach, K.: Deliberative scheduling using GSMDP in stochastic asynchronous domains, *International Journal on Approximate Reasoning*, vol. **50**, pp. 1347–1359, 2009.

421. Kulkarani, A.; Tai, K.: Probability collectives: A multi-agent approach for solving combinatorial optimization problems, *Applied Soft Computing*, vol. **37**, pp. 759–771, 2010.

422. Lemaitre, C.; Reyes, C. A.; Gonzalez, J. A.: *Advances in Artificial Intelligence*, Springer, Berlin, 2004.

423. Liu, J.; Wu, J.: *Multi-Agent Robotic System*, CRC Press, Boca Raton, FL, 2001.

424. Liu, L.; Shell, D. A.: Assessing optimal assignment under uncertainty, In: Matsuoka, Y.; Durrant-White, H.; Neira, J. (eds) *Robotics, Science and Systems*, The MIT Press, Cambridge, MA, pp. 121–128, 2010.

425. Low, C. B.: A rapid incremental motion planner for flexible formation control of fixed wing UAV, *IEEE Conference on Decision and Control*, Maui, HI, pp. 2427–2432, 2012.

426. Lyons, D.; Calliess, J. P.; Hanebeck, U.: Chance constrained model predictive control for multi-agent systems, 2011. arXiv preprint arXiv:1104.5384.

427. Margellos, K.; Lygeros, J.: Hamilton-Jacobi formulation for reach-avoid differential games, *IEEE Transactions on Automatic Control*, vol. **56**, pp. 1849–1861, 2011.

428. Marier, J. S.; Besse, C.; Chaib-Draa, B.: A Markov model for multiagent patrolling in continous time. *ICONIP*, Bangkok, Thailand, vol. **2**, pp. 648–656, Springer, 2009.

429. Martin, P.; de la Croix, J.P.; Egerstedt, M.: A motion description language for networked systems, *In 47th IEEE Conference on Decision and Control*, Mexico, pp. 558–563, 2008.

430. Marvel, J.: Performance metrics of speed and separation monitoring in shared workspaces, *IEEE Transactions on Automation Science and Engineering*, vol. **10**, pp. 405–414, 2013.

431. Mesbahi, M.: On state-dependent dynamic graphs and their controllability properties, *IEEE Conference on Decision and Control*, Bahamas, pp. 2473–2478, 2004.

432. Mesbahi, M.; Egerstedt, M.: *Graph Theoretic Methods in Multiagent Networks*, Princeton Series in Applied Mathematics, Princeton University Press, Princeton, NJ, 2010.

433. Moon, J.; Oh, E; Shin, D. H.: An integral framework of task assignment and path planning for multiple UAV in dynamic environments, *Journal of Intelligent Robots*, vol. **70**, pp. 303–313, 2013.

434. Moses Sathyaraj, B.; Jain, L. C.; Fuin, A.; Drake, S.: Multiple UAV path planning algorithms: A comparative study, *Fuzzy Optimal Decision Making*, vol. **7**, pp. 257–267, 2008.

435. No, T. S.; Kim, Y.; Takh, M. J.; Jeon, G. E.: Cascade type guidance law design for multiple UAV formation keeping, *Aerospace Science and Tecnology*, vol. **15**, pp. 431–439, 2011.

436. Oberlin, P.; Rathinam, S.; Darbha, S.: A transformation for a multiple depot, multiple traveling salesman problem, *American Control Conference*, St. Louis , MO, pp. 2636–2641, 2009.

437. Ono, M.; Williams, B. C.: Decentralized chance constrained finite horizon optimal control for multi-agent systems, *In 49th IEEE Control and Decision Conference*, Atlanta, GA, pp. 138–145, 2010.

438. Parlangeli, G.; Notarstefano, G.: On the reachability and observability of path and cycle graphs, *IEEE Transactions on Automatic Control*, vol. **57**, pp. 743–748, 2012.

439. Ponda, S.; Johnson, L.; How, J.: Distributed chance constrained task allocation for autonomous multi-agent teams, *Proceedings of the 2012 American Control Conference*, Montreal, Canada, pp. 4528–4533, 2012.

440. Rabbath, C. A.; Lechevin, N.: *Safety and Reliability in Cooperating Unmanned Aerial Systems*, Springer, Berlin, 2011.

441. Rathinam, S.; Sengupta, R.; Darbha, S.: A resource allocation algorithm for multivehicle system with nonholonomic constraints, *IEEE Transactions on Automation Science and Engineering*, vol. **4**, pp. 4027–4032, 2007.

442. Rosaci, D.; Sarne, M.; Garruzzo, S.: Integrating trust measuring in multiagent sytems, *Internbational Journal of Intelligent Systems*, vol. **27**, pp. 1–15, 2012.

443. Saget, S.; Legras, F.; Coppin, G.: Cooperative interface for a swarm of UAV, 2008). arXiv preprint arXiv:0811.0335 - arxiv.org.

444. Shanmugavel, M.; Tsourdos, A.; Zbikowski, R.; White, B. A.; Rabbath, C. A.; Lechevin, N.: A solution to simultaneous arrival of multiple UAV using Pythagorean hodograph curves, *American Control Conference*, Minneapolis, MN, pp. 2813–2818, 2006.

445. Semsar-Kazerooni, E.; Khorasani, K.: Multi-agent team cooperation: A game theory approach, *Automatica*, vol. **45**, pp. 2205–2213, 2009.

446. Seuken, S.; Zilberstein, S.: Formal models and algorithms for decentralized decision making under uncertainty, *Autonomous agent multi-agent systems*, 2008. DOI: 10.1007/s10458-007-9026-5.

447. Shamma, J. S.: *Cooperative Control of Distributed Multi-Agent System*, Wiley, Hoboken, NJ, 2007.

448. Shi, G.; Hong, Y.; Johansson, K.: Connectivity and set tracking of multi-agent systems guided by multiple moving leaders, *IEEE Transactions on Automatic Control*, vol. **57**, pp. 663–676, 2012.

449. Shima, T.; Rasmussen, S.: *UAV Cooperative Decision and control*, SIAM, Philadelphia, PA, 2009.

450. Sirigineedi, G.; Tsourdos, A.; Zbikowski, R.; White, B.: Modeling and verification of multiple UAV mission using SVM, Workshop on formal methods for aerospace, pp. 22–33, 2010.

451. Stachura, M.; Frew, G. W.: Cooperative target localization with a communication aware- unmanned aircraft system, *AIAA Journal of Guidance, Control and Dynamics*, vol. **34**, pp. 1352–1362, 2011.

452. Surynek, P.: An optimization variant of multi-robot path planning is intractable, *In 24th AAAI Conference on Artificial Intelligence, Atlanta, GA,* 2010.

453. Turra, D.; Pollini, L.; Innocenti, M.: Fast unmanned vehicles task allocation with moving targets, *In 43rd IEEE Confernece on Decision and Control*, Nassau, Bahamas, pp. 4280–4285, 2004.

454. Turpin, M.; Michael, N.; Kumar, V.: Decentralized formation control with variable shapes for aerial robots, *IEEE International Conference on Robotics and Automation*, Saint Paul, pp. 23–30, 2012.

455. Twu, P. Y.; Martin, P.; Egerstedtd, M.: Graph process specifications for hybrid networked systems, *Discrete Event Dynamical Systems*, vol. **22**, pp. 541–577, 2012.

456. Ulusoy, A.; Smith, S. L.; Ding, X. C.; Belta, C.; Rus, D.: Optimal multi-robot path planning with temporal logic constraints, *IROS IEEE/RSJ International Conference on Intelligent Robots and Systems*, San Francisco, CA, 2011.

457. Ulusoy, A.; Smith, S. L.; Ding, X. C.; Belta, C.: Robust multi-robot optimal path planning with temporal logic constraints, *IEEE International Conference on Robotics and Automation* Saint Paul, MN, 2012.

458. Virtanen, K.; Hamalainen, R. P.; Mattika, V.: Team optimal signaling strategies in air combat, *IEEE Transactions on Systems, Man and Cybernetics*, vol. **36**, pp. 643–660, 2006.

459. Wen, G.; Duan, Z.; Yu, W.; Chen, G.: Consensus of multi-agent systems with nonlinear dynamics and sampled data information: A delayed input approach, *International Journal of Robust and Nonlinear Control*, vol. **23**, pp. 602–619, 2012.

460. Wu, F.; Zilberstein, S.; Chen, X.: On line planning for multi-agent systems with bounded communication, *Artificial Intelligence*, vol. **175**, pp. 487–511, 2011.

461. Yadlapelli, S.; Malik, W.; Darbha, M.: A lagrangian based algorithm for a multiple depot, multiple traveling salesmen, *American Control Conference*, Philadelphia, PA, pp. 4027–4032, 2007.

462. Zhang, H.; Zhai, C.; Chen, Z.: A general alignment repulsion algorithm for flocking of multiagent systems, *IEEE Transactions on Automatic Control*, vol. **56**, pp. 430–435, 2011.

463. Zhang, J.; Zhan, Z.; Liu, Y.; Gong, Y. et al.: Evolutionary computation meets machine learning: A survey, *IEEE Computational Intelligence Magazine*, vol. **6**, pp. 68–75, 2011.

464. Zhao, Y.; Tsiotras P.: Time optimal path following for fixed wing aircraft, *AIAA Journal of Guidance, Control and Dynamics*, vol. **36**, pp. 83–95, 2013.

465. Zhi-Wei, H.; Jia-hong, L.; Ling, C.; Bing, W.: A hierarchical architecture for formation control of multi-UAV, *Procedia Engineering*, vol. **29**, pp. 3846–3851, 2012.

466. Zou, Y.; Pagilla, P. R.; Ratliff, R. T.: Distributed formation flight control using constraint forces, *AIAA Journal of Guidance, Control and Dynamics*, vol. **32**, pp. 112–120, 2009.

Index

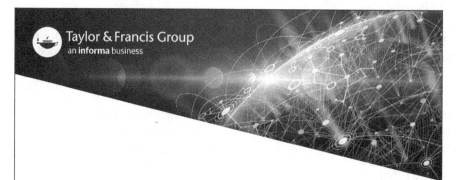